Values and Choices
In the Development of
The Colorado River Basin

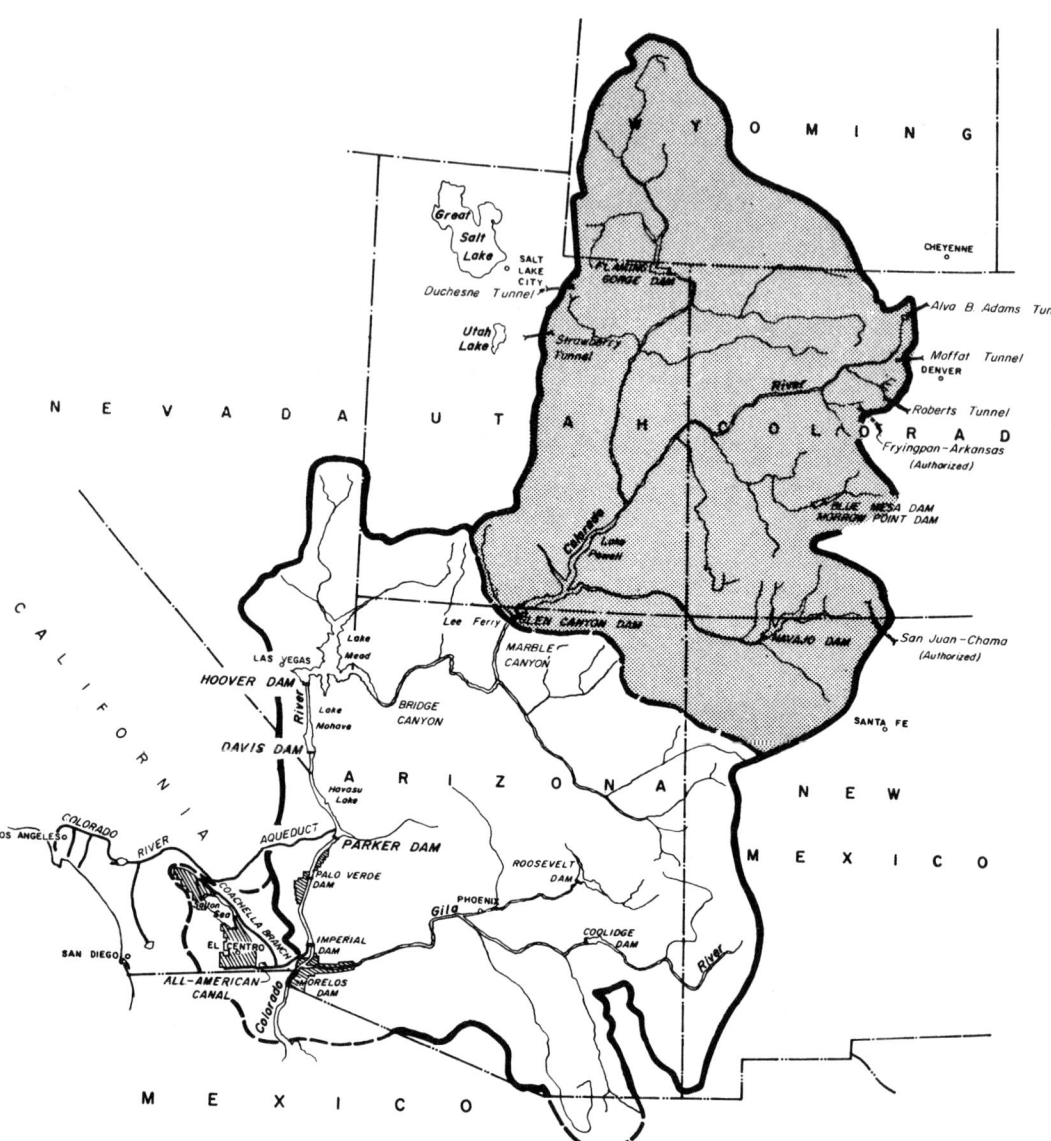

COLORADO RIVER BASIN
Upper Basin shaded.
Reservoirs and reservoir sites in the Colorado Basin are shown.

Courtesy Upper Colorado River Commission

Values and Choices In the Development of The Colorado River Basin

Collaborating Authors

Bahe Billy
A. Bruce Bishop
Jack M. Campbell
Jared Carter
Henry P. Caulfield, Jr.
A. Berry Crawford
Harold E. Dregne
B. Delworth Gardner
David W. Goodall
Ival V. Goslin
Russell Gum
Helen Ingram

Roger Kraynick
Myron B. Holburt
John D. Hunt
Dean E. Mann
W. Don Maughan
Dean F. Peterson
Theodore G. Roefs
Lawrence E. Royer
Thayer Scudder
Clyde E. Stewart
Gerald W. Thomas
D. Wynne Thorne

Bernard Udis

Dean F. Peterson
A. Berry Crawford
Editors

THE UNIVERSITY OF ARIZONA PRESS
Tucson, Arizona

About the Editors . . .

Dean F. Peterson in 1976 became director of the Soil, Water and Management Division of the Office of Agriculture's Development Support Bureau of the U.S. Agency for International Development. **A. Berry Crawford,** on leave from Utah State University as director of the Policy Research Program, became, in 1977, director of the Institute for Policy Research of the Western Governors Policy Office. He resigned this post in early 1978 to become consultant to the Office of Intergovernmental Relations in the White House.

The University of Arizona Press
Copyright © 1978
The Arizona Board of Regents
All Rights Reserved
Manufactured in the U.S.A.

Library of Congress Cataloging in Publication Data

Main entry under title:

Values and choices.

Papers and discussions selected from a symposium held in San Francisco, Feb. 28 and March 1, 1974, and sponsored by the Committee on Arid Lands of the American Association for the Advancement of Science.
    Bibliography: p.
    Includes index.
    1. Water resources development—Colorado River watershed —Congresses. I. Peterson, Dean F. II. Crawford, A. Berry. III. American Association for the Advancement of Science. Committee on Arid Lands.
HD1695.C7V34 1978   333.9′162′097913   77-13716
ISBN 0-8165-0643-4
ISBN 0-8165-0480-6 pbk.

Contents

PART II. FUTURE DIRECTIONS

Illustrations

TABLES

Preface

In the realm of the development of arid lands river basins, this volume examines the dynamics of a particular arid place. In broad perspective, it utilizes a specific case to examine the forces and considerations that have led to present conditions, and to discuss the consequences of alternative development futures in the light of present and projected societal needs and values. In pinpointing a particular watershed as a basis for regionality, the choice had to be arbitrary—since certainly the Colorado River Basin is not a discrete economic, social or ethnic entity. Moreover, the high mountain watersheds from which the river rises are neither arid nor sub-arid. But there are many rivers that flow through arid lands and these rivers usually dominate the arid lands scene.

As viewed in the mid 1970s, the situation in the Colorado River Basin is the result of a complex mixture of circumstances, determined partly by chance, partly by the particular nature of the resources of the area, and partly by societal choice. This "mix" has led to some lively conflicts. Two of the obvious and most important areas of conflict stem from circumstances of political geography (e.g., the allocation of the waters of the river), and from the value domain (e.g., the preservation of parks and monuments). One needs to know by what process these conflicts have been resolved, and what are the gains and losses. With increasing pressure upon the basin's resources, conflicts become increasingly serious. Most of these can be resolved only by human choice.

The framework for the presentations in this book came from the symposium in San Francisco, February 28 and March 1, 1974, sponsored by the Committee on Arid Lands of the American Association for the Advancement of Science (AAAS), on ''Values and Choices in the Development of an Arid Land River Basin (Colorado River Basin),'' from which invited papers and discussions were selected for inclusion in this volume.

Special thanks are given to the University of Arizona Press for the patient and expert assistance which has been provided in effecting this publication.

DEAN F. PETERSON
A. BERRY CRAWFORD

Contributors

BAHE BILLY, Planning Coordinator, Navajo Agricultural Products Industry, Farmington, New Mexico.

A. BRUCE BISHOP, Associate Professor of Civil Engineering, Utah Water Research Laboratory, Utah State University, Logan, Utah.

JACK M. CAMPBELL, President, Federation of Rocky Mountain States, Denver, Colorado.

JARED CARTER, Deputy Undersecretary, U. S. Department of the Interior, Washington, D. C.

HENRY P. CAULFIELD, JR., Professor of Political Science, Colorado State University, Fort Collins, Colorado.

A. BERRY CRAWFORD, Associate Professor of Philosophy and Associate Director of Research, Environment and Man Program, Utah State University, Logan, Utah.

HAROLD E. DREGNE, Professor and Chairman, Department of Agronomy, Texas Technical University, Lubbock, Texas.

B. DELWORTH GARDNER, Professor and Head, Department of Economics, Utah State University, Logan, Utah.

DAVID W. GOODALL, Professor, Range Science and Ecology, Ecology Center, Utah State University, Logan, Utah.

IVAL V. GOSLIN, Executive Secretary, Upper Colorado River Commission, Salt Lake City, Utah.

RUSSELL GUM, Associate Professor, Department of Agricultural Economics, University of Arizona, Tucson, Arizona; Agricultural Economist, Natural Resource Economics Division, Economic Research Service, U.S.D.A.

MYRON B. HOLBURT, Chief Engineer, Colorado River Board of California, Los Angeles, California.

JOHN D. HUNT, Assistant Dean and Chairman of the Institute for the Study of Outdoor Recreation and Tourism, College of Natural Resources, Utah State University, Logan, Utah.

HELEN INGRAM, Director, Institute of Government Research, University of Arizona, Tucson, Arizona.

ROGER KRAYNICK, Bureau of Economic Research, University of Colorado, Boulder, Colorado.

DEAN E. MANN, Professor of Political Science, University of California, Santa Barbara, California.

W. DON MAUGHAN, Member, State Water Resources Control Board, Sacramento, California.

DEAN F. PETERSON, Professor of Civil and Environmental Engineering and Vice President for Research, Utah State University, Logan, Utah.

THEODORE G. ROEFS, Associate Professor of Hydrology and Water Resource Administration, University of Arizona, Tucson, Arizona.

LAWRENCE E. ROYER, Assistant Professor of Outdoor Recreation, Utah State University, Logan, Utah.

THAYER SCUDDER, California Institute of Technology, Pasadena, California.

CLYDE E. STEWART, Agricultural Economist, U. S. Department of Agriculture, Logan, Utah.

GERALD W. THOMAS, President, New Mexico State University, Las Cruces, New Mexico.

D. WYNNE THORNE, Professor of Soils and Director, Agricultural Experiment Station, Utah State University, Logan, Utah.

BERNARD UDIS, Professor of Economics and Director, Bureau of Economic Research, University of Colorado, Boulder, Colorado.

Values and Choices
In the Development of
The Colorado River Basin

the equation may require that a full spectrum of life-sustaining and amenities resources be transported to the extraction site. King Solomon extracted the cuprous minerals of the Negev; in the late twentieth century most of the world's supplies of petroleum are being extracted from the arid lands of Arabia.

As noted by Kelso (1970), development of highly complex economies and urban centers from internal resources in arid and sub-arid regions is occurring throughout the world. Success seems to depend upon a relatively wide diversity of economic resources, a resource base not already over-exploited, and competitive advantages within the commercial linkages to a greater economic heartland. The mid-twentieth century has witnessed a new exploitation of arid and sub-arid lands based on utilization of the attractive features of arid environments to realize a desired lifestyle. As evidenced by the United States' Southwest, large cities and their environs are developing. Here undesirable features of the environment are overcome through importation: energy for air conditioning, water to provide lawns and trees, and so on. The modern survival requirement of personal income is provided by industrialization and associated commerce, services and tourism. The transport linkages of the arid zone to the contextual economy now become complex indeed; not only money and material goods, but also political power and air and water pollution flow through them.

EXTRACTION AND CONSERVATION

Arid and sub-arid regions often yield important mineral wealth. Since these resources are essentially non-renewable, economies based on them tend to flourish and then disappear. The ghost towns of the West testify to this process. Exploitation of resources that take a long time to renew, like soil and vegetative cover, may have consequences that are important to more than the local inhabitants. An example is the dust bowl of the 1930s. Recreational use of arid land by off-road vehicles could result in destruction of a resource having a regeneration time extending into centuries. Although the case is not clear, there is evidence that destruction of vegetative cover by overgrazing may result in the spreading of deserts on a continental scale and that increased dust in the atmosphere may be causing climatic shifts of global consequence (Cloudsley-Thompson, 1971; Bryson, 1973).

Wilderness, scenic landscapes, and other "natural" amenities associated with arid lands are increasingly valued by people and there is strong, and certainly defensible, feeling that these should not be destroyed. Exploitation cannot occur without some external environmental changes. An important question is how much and what degree of disturbance shall be tolerated. As long as the offenses were rather small in scale and remote from the heartland, there was little public concern. In America this is no longer true.

Considerations of exploitation and conservation introduce decision link-ages into the transport factor of the equation that extend far beyond the boundaries of the American arid West.

BACKGROUND FOR A CASE STUDY

The foreword of *Aridity and Man* (Hodge and Duisberg, 1963) docu-ments the beginnings of interest by the American Association for the Ad-vancement of Science in arid zone potentials and problems. This interest was manifest first in a symposium on the "Potentials of Desert and Arid Lands" held by the Southwestern and Rocky Mountain Divisions at El Paso, Texas, in 1951. This symposium followed UNESCO's formation of an Advisory Committee on Arid Zone Research by less than one year. From the El Paso meeting, the Division's Committee on Desert and Arid Zone Research (CODAZR) emerged. Subsequently a Committee on Arid Lands was formed at the national level. Over the years a number of conferences and symposia have been held at regional, national and international levels and most aspects of arid land science and development have been studied. Besides *Aridity and Man,* three other volumes have resulted: *Arid Lands in Perspective* (Mc-Ginnies and Goldman, 1969), *Arid Lands in Transition* (Dregne, 1970) and *Polar Deserts and Modern Man* (Smiley and Zumberge, 1974). These studies have dealt with a wide area of topics. While site specific material has been drawn upon in each instance, none has considered a specific regional case in a comprehensive way.

The Committee on Arid Lands decided to try such a synthesizing ap-proach with a symposium early in 1974; and, after considerable study, chose the Colorado Basin. Except for the higher mountains, which provide the source waters of the Colorado River, all of this basin is classed as "extremely arid" or "arid." The Committee felt that the proposed study should be placed in a broad historical perspective and that future possibilities should receive emphasis. What happens in the future might depend upon value-weighted public decisions, and certainly this basin will be in an arena in which the struggle over the values represented by the general concepts of development, progressivism, distributive justice, conservation and preservation will be highly evident. These considerations led to the theme, "Values and Choices in an Arid Land River Basin: The Colorado." The chapters that follow are the result. The title suggests the organizing idea not only of the symposium but also of this volume: that of considering how choices and the value com-mitments and trade-offs they embody have shaped the history of development in the Colorado River Basin and define options for the future.

A historical overview is provided in the first five chapters of the book. Don Maughan and Ival Goslin discuss, respectively, the physical setting in which development decisions have been made and how the implementation of water development projects has altered physical conditions and influenced subsequent development. The chapter by Helen Ingram views this development in political perspective. Specifically, she considers how opposing political forces have interacted in various decision making processes to shape development and water allocation policies. She contends that national and corporate values and perspectives have dominated in these decision-making processes and are the source of many inequities and problems borne by basin inhabitants. Some of the economic consequences of development decisions are analyzed by Bernard Udis and Roger Kraynick. Using a Shift-Share method of analysis to trace employment changes in the basin's industries and citing Wilbur Thompson's dictum ''Let me know your industry mix and I will tell your fortune,'' these authors discuss features of the basin's recent economic history in relation to employment changes. Finally, Henry Caulfield views development in the basin from a policy perspective. He characterizes the policies that have prevailed and shaped development, and discusses how changes in value emphasis have resulted in policy confrontations and accommodations.

In many ways, physical development and institutional arrangements have greatly reduced the number of options for use of the waters and lands of the Colorado River Basin. One commentator has observed that, for its size, the Colorado is probably the most utilized, controlled, and fought over river in the world. The chapters in Part II explore ''future directions'' for management of the Colorado. Together, they constitute an assessment of how management options have been foreclosed or still exist, and how factional struggles and attempts to resolve present-day problems may play major roles in influencing the future development of the basin.

Certainly one of the major problem areas in the region is the increasing salinity of the river. B. D. Gardner and Clyde Stewart address this problem, primarily in relation to irrigated agriculture, and discuss various policy options for coping with the problem. The chapter by Jared Carter outlines the main features of the Department of the Interior's policy concerning development of the basin's great stores of energy resources. Although written before the establishment of the Federal Energy Agency, Carter's chapter contains an insightful discussion of many of the problems associated with the development of these resources. T. G. Roefs, in his discussion of Carter's paper, urges a systems approach to allocation of costs and benefits of energy development and describes desirable characteristics of such a system.

Problems and prospects concerning agricultural, recreational and urban development are treated by Gerald Thomas, Lawrence Royer and John Hunt, and Dean Mann, respectively. These chapters are discussed by Wynne Thorne, David Goodall, and Thayer Scudder. Mann's paper deals primarily with the problem of achieving or maintaining high "quality of life" among the region's communities and concludes:

> Choices based on available information must be made by individuals in making their private decisions about where to live and work and what kind of lifestyle to adopt, and by communities in their corporate capacity in choosing the kind of environment they wish to fashion. If the past is any indicator, the choice will be made for growth: more economic development and more people with relatively modest concern for the quality of the environment in which they live. These choices will be made by individuals and firms and acquiesced in by communities as political and social units because the ethic of growth remains strong, because the population will continue to grow, *because the region is so vulnerable to economic forces over which the local population has little control.* (Italics ours.)

The tensions among competing kinds of land use in the basin—tensions aggravated by the scarcity of water and, as Gerald Thomas stated early in 1974, "Today and yesterday's environmental crisis, today and tomorrow's energy crisis, and tomorrow and the day-after-tomorrow's food and fiber crisis"—are analyzed in these chapters and viewed from a variety of perspectives. The Colorado is an international river providing water for irrigation and municipal use in the Mexicali and San Luis valleys of northwestern Mexico. Being last on the river, depletions upstream have reduced the quality and quantity of the water flowing across the border. The situation is described by Myron Holburt in his analysis of the history of negotiations leading to the "Brownell Agreement" and the implications of the terms of this agreement to upstream users in the United States.

The chapter by Bahe Billy contains an informative discussion of Navajo culture and land use traditions, and discusses many of the problems with which the Indian must deal in protecting basic values and rights and achieving a measure of self-determination.

Two chapters contain primarily methodological discussions. In one, by Russell Gum, an effort is made "to relate current planning methodologies to the problems of including aesthetic values in resource planning to speculate about possible improvements in the planning process." In the other, Berry Crawford and Bruce Bishop explore the "systems" concept of "human-oriented carrying capacity," in which continued or sustained "quality of life" is balanced against the use of resources, both locally and extra-regionally. They describe a possible framework for use of the concept as a regional planning tool and discuss some of the considerations involved in applying a carrying capacity-based planning process to the Colorado River Basin and its subregions.

Values and Choices
In the Development of
The Colorado River Basin

Introduction

Whither Goeth the River?

DEAN F. PETERSON and A. BERRY CRAWFORD

The past century has been a period of rapid change for the arid and semiarid lands of the American West. Almost certainly the pace of this change will continue, and perhaps even accelerate. The southwestern states have the highest rates of population growth in the country. Throughout much of the region, vast hydrocarbon fuel resources lie buried beneath the surface awaiting imminent, large-scale development. The basin of the Colorado River lies generally in the arid zone and this land is caught up in the quickening pace of change.

The Colorado River Basin is richly endowed with natural resources. Besides its extensive mineral resources and agricultural and grazing resources, these include recreational opportunities, wilderness, open space, magnificent landscapes, and archaeological sites. Still, however, the dominating resource is the *river*. Extraction of mineral hydrocarbons one day may dominate the region's economy, but even this vast potential will be limited by the availability of water. Since the river is already almost completely utilized, hard political decisions and shifting values appear to be the order of the day.

While the Colorado Basin, like any arid-zone basin, has its own special uniqueness, the transitional forces that are impinging upon it are not atypical. This element of commonality suggested to the Committee on Arid Lands that it might explore the transition processes through which arid lands are passing by focusing on a case study. The basin selected for this study, the Colorado River Basin, certainly qualifies as an interesting one.

[1]

SOME GENERALITIES— THE SURVIVAL EQUATION

A life-system survival equation having both time and space dimensions may be stated as follows: the minimum base production + *in situ* reserves (storage) + space transfers (transport) must equal or exceed sustenance requirements over any given time period. Minimum base production depends on the availability of sufficient resources, one of which is water or moisture and is defined as the amount that can be produced during a critical period of drought. Besides being hot and dry, the climate in arid regions is highly variable and it is variability that makes the second two terms in the equation particularly important in these regions. Few, if any, life systems evolve at only the productivity enforced during the stress of the drought, but rather they rely on storage and often transport to carry them through at a substantially higher level of production capability. Thus, in arid lands, biological and cultural adaptation emphasizes storage and transfer to offset variability. The equation applies to life systems ranging from the simplest plant species to the most complex technological civilizations.

Many of the adaptations are most remarkable, For example, seeds of annual plant species sprout only when soil moisture conditions are such that completion of a life cycle is likely. Similarly, perennials have physiological characteristics that preserve the vital spark through long periods of dormancy when transpiration is nil. In the animal kingdom, parallel physiological defenses have evolved. Storage of energy sources may be part of the defense system against the stress of drought, since the storage of water itself is much more difficult. Mobile creatures move about over the arid surface to take advantage of spatial shifts in productivity resulting from seasonal or even long-term variations.

For humans, space transfers are particularly important, and nomadism is the classical response with storage in the form of livestock on the hoof. Oasis culture, the polar opposite, is found only when there is a naturally stored and uniformly dispensed source of water. There are more primitive food-gathering adaptations, too: the Bushmen of the Kalahari and Shoshonean inhabitants of the Great Basin are examples. In the latter case, the survival range of a family unit of five persons required a territorial radius of about 35 kilometers (Woodbury, 1963). Technology has greatly enhanced the storage and transport portions of the equation. Drilling of wells and irrigation systems utilizing surface canals and ghanats predate history by several thousands of years. Long before the Christian era, the powerful nations of the Middle East and Egypt were nourished by extensive irrigation systems requiring highly sophisticated technology and administrative organization for their construction and maintenance. Then, as now, rivers whose sources were in non-arid mountainous areas flowed through arid regions or were diverted into them. When arid lands are exploited for their mineral resources, the transport part of

The concluding chapter, written by Jack Campbell, assesses the potential for energy resources development in the basin and calls for a "forum," a group representing the seven basin states and having strong linkages with the governors, to manage the development of the basin's resources on a comprehensive basis. He describes this proposal as encouraging a process of "accelerated evolution" and warns against the counter-positions of "no change" and "too much change." Given the realities of the public decision process, Campbell concludes:

> To those who prefer to insist rigidly upon the status quo, this approach may conjure up fear of future consequences; to those who want immediate radical overhaul of existing structures, it will seem far too timid. However, the alternatives...are not pleasant to contemplate. To stand fast, resisting change, would be to invite—almost to insure—increasing federal intrusion into matters historically reserved to the states. To catapult into a totally new management arrangement would accentuate the adversary approach and lead to confusion and turmoil when sound management is imperative.

Literature Cited

Bryson, Reid. 1973. *Climatic Modification by Air Pollution, II: The Sahelian Effect.* Report No. 9, Institute for Environmental Studies. Madison: University of Wisconsin.

Cloudsley-Thompson, J.J. 1971. Recent expansion of the Sahara. *Intern. J. Environmental Studies* 2:35-39.

Dregne, Harold E., ed. 1970. *Arid Lands in Transition.* Publication No. 90, American Association for the Advancement of Science. Washington, D. C.

Hodge, Carle and Duisberg, Peter C., eds. 1963. *Aridity and Man.* Publication No. 74, American Association for the Advancement of Science. Washington, D.C.

Kelso, M. M. 1970. Economic growth opportunities and constraints. In *Arid Lands in Transition,* ed., Harold E. Dregne. pp. 87-88. Publication No. 90, American Association for the Advancement of Science. Washingtion, D. C.

McGinnies, William G. and Goldman, Bram J. 1969. *Arid Lands in Perspective.* American Association for the Advancement of Science. Tucson: The University of Arizona Press.

Smiley, Terah L. and Zumberge, James M., eds. 1974. *Polar Deserts and Modern Man.* Tucson: The University of Arizona Press.

Woodbury, R. B. 1963. Indian adaptations to arid environments. In *Aridity and Man,* eds., C. Hodge and P. C. Duisberg. pp. 71-73. American Association for the Advancement of Science, Washington, D.C.

Part I

Historical Perspectives

1

Physical Setting

W. DON MAUGHAN

PHYSICAL CHARACTERISTICS

The Colorado River is one of the great rivers of the West. It extends 1,400 miles from its headwaters in north central Colorado to its mouth in the Gulf of California. Together with hundreds of tributaries and subtributaries—among them the Gunnison, the Green, the Dirty Devil, the San Juan, the Little Colorado, the Virgin and the Gila—it is the dominant river system in the entire southwestern corner of America, draining an area one-twelfth the size of the continental United States, including parts of Wyoming, Colorado, Utah, Nevada, California, New Mexico and almost all of Arizona. Transfers of its water outside of the basin have influenced the growth and development of additional large segments of most of these states.

The land through which it travels and which it helped shape contains three distinct geographic provinces. Throughout its devious course are found some of the most varied and spectacular natural formations on the face of the earth. Beginning in the high ranges and parklands of the Rocky Mountains, the river courses southwest through the sheer canyons that it has cut into the enormous plateaus of Utah and Arizona (the Kaiparowits, the Paria, the Kaibab, the Kanab, the Uinkaret, the Coconino, and the Shivwits). It winds finally through the Mojave, Colorado, and Sonoran deserts in its journey to the sea.

The Colorado River drains a vast area of 244,000 square miles—242,000 square miles are in the United States and 2,000 square miles in northern Mexico. The drainage basin from Wyoming to below the Mexican border is some 900 miles long and varies in width from about 300 miles in the upper basin to 500 miles in the lower basin. It is bounded on the north and east by the Continental Divide in the Rocky Mountains and on the west by the Wasatch Range and a series of lesser plateaus, ranges and divides.

The Salton Sea Basin, an additional area of 7,800 square miles which includes the Coachella and Imperial valleys in southeastern California, hydrologically is no longer a part of the basin, although as historians will recall, water will flow uncontrolled into this area if it has the opportunity. Once having entered, the water will not return to the river system.

The upper or northern portion of the Colorado River Basin in Wyoming and Colorado is a mountainous plateau, 5,000 to 8,000 feet in altitude, marked by broad rolling valleys, deep canyons, and intersecting mountain ranges. Hundreds of peaks in these mountain ranges rise to more than 13,000 feet above sea level and several exceed 14,000 feet. There are many picturesque mountain lakes in these headwater sections. The southern portion of the basin is studded with rugged mountain peaks, interspersed with broad, level, alluvial valleys and rolling plateaus. The main stream of the Colorado River and its principal tributaries flow for the most part in deep canyons. The Green River, the primary tributary of the Colorado River, flows in similar canyons in Wyoming, Colorado, and Utah, and its chief tributaries, the Yampa and White rivers from the east, and the Duchesne, Price, and San Raphael rivers from the west, flow through rolling hills and canyons to reach the Green. The San Juan River, a large tributary of the Colorado River from the east, drains mountain slopes and plateaus in southwestern Colorado, northwestern New Mexico, and northern Arizona, and flows through a formidable canyon in southwestern Utah before joining the Colorado River in Glen Canyon. The Glen Canyon section of the main stream and tributaries thereto are in deep canyons draining a series of plateaus and mesas. Below Glen Canyon is the awesome Grand Canyon where the Colorado has carved an unparalleled chasm. This canyon yawns above an inner gorge rising in gigantic cliff steps to the Colorado Plateau, a mile above the streambed. This great central plateau is a rolling expanse of brightly hued crags and cliffs, huge canyons, painted deserts, and extensive, almost inaccessible barren areas. Elevations on the mesas of the plateau section generally range from 4,000 to 6,000 feet. The principal tributaries in this section are the Little Colorado on the east and the Virgin River on the west. Emerging from the canyon country at the southeast corner of Nevada, the Colorado River courses through broad valleys bordered by mesas. The Gila River, main tributary in this section, rises in the mountainous regions of southwestern New Mexico and drains most of southern Arizona. Southwest of the Gila Basin, the Colorado River continues through its great delta area to the Gulf of California.

CLIMATE

The climate of the Colorado River Basin varies widely due to the large differences in elevation, a considerable range in latitude and the distribution of mountain ranges and highlands. The climate is semiarid to arid. In the northern portion and at the higher elevations elsewhere, a four-season climate prevails while at the lower elevations in the southern portion, seasonal differences are much less distinct.

Annual precipitation varies from less than 4 inches in the southern desert to more than 50 inches in the higher, northern mountains. In the upper basin most valleys and agricultural areas receive 10 to 20 inches of precipitation, much of it as snow. In the irrigated areas of the lower basin, including the Imperial and Coachella valleys, physically outside the basin, precipitation varies generally from less than five inches to eight inches, although some irrigated lands at high elevations receive more.

The region is subject to wide extremes of temperature ranging from less than -50 degrees to more than 115 degrees Fahrenheit. The frost-free period varies from 20 days or less at the high, northern elevations to about 180 days in northern irrigated and urban areas and extends from 270 to essentially 365 days in the southern parts of the basin.

GEOLOGIC HISTORY

Geologic history not only has shaped the physical characteristics of the Colorado River Basin, but also is responsible for many of the problems and important opportunities of the basin.

The Colorado River drainage system extends into five physiographic provinces. Its headwaters are in the southern and middle Rocky Mountains and it crosses the Wyoming Basin, the Colorado Plateau, and the Basin and Range provinces. To reach the Gulf of California, the river and most of its tributaries must flow through many mountains and high plateaus which are structural barriers of resistant rocks. No other river in the western hemisphere crosses so many. One of these is the Colorado Plateau, a tremendous structural block covering more than half the drainage basin, that has been uplifted and tilted northeast against the river. The Grand Canyon, where the Colorado River leaves the plateau, slices through the highest part of the rim—highest structurally and one of the highest topographically (Geological Survey, 1969).

Most valleys in the drainage system are narrow and steep-sided; broad, alluvial floodplains are uncommon. There are a few small floodplains in the Rocky Mountains and on the Colorado Plateau upstream from some of the structural barriers. But the extensive floodplains are in the Basin and Range

Province near Needles, in California; below the Bill Williams River and along parts of the Gila River, in Arizona. Throughout its 300-mile course across the Colorado Plateau, much of it in rock-vault canyons, the river meanders extensively, much more so than most rivers not located on alluvial floodplains. The meanderings apparently reflect the river's difficulties in maintaining its flow across rising folds and against the northeast tilt of the Colorado Plateau. The folding and tilting have continued, as minor movements, while the river has been cutting its canyons. Probably these minor movements are still going on.

As one crosses the Basin and Range Province from the Colorado Plateau to the Gulf of California, the whole aspect of the Colorado River valley changes. Alternately the river flows in unconsolidated gravels and sand that fill broad, open structural valleys and in deep, narrow rock-walled canyons through block-faulted mountains separating the valleys.

Rocks of a wide range of age are found in the Colorado River Basin. The high Rocky Mountains which dominate the region are composed of granite, schists, gneisses, lava, and sharply folded sedimentary rocks. Many periods of deposition and erosion have played a part in causing the present structure of these mountains. Ancient seas settled in the basin countless times, depositing beds of limestone, sandstone, and shale. Each time, crustal forces of the earth elevated the region above sea level and the process of cutting it down began again. The Rocky Mountains of Colorado, the Wind River Mountains of Wyoming, and the Uinta and Wasatch mountains in Utah all have been affected topographically by glacial action.

In contrast to the folded rocks of the mountains which border the basin, the plateau country of southwestern Wyoming, eastern Utah, and northern Arizona is composed principally of horizontal strata of sedimentary rocks. Many formations of hard sandstone and limestone, separated by softer shale, often highly colored, have resulted in topographic and geologic formations found in no other locality in the United States.

The present Gulf of California once extended much farther north and filled what is now the Imperial Valley. Over the years following the last glacial period, sediments from the heavily laden Colorado River built up a great delta which finally cut off an arm of the gulf to create a vast inland sea of brackish water. The Salton Sea as it was before irrigation began was the remnant of that sea.

Because of the relatively steep gradients and the erosive nature of many of the rocks, the sediment load of the Colorado is one of the highest in the United States. At Grand Canyon this load would be about twice the average of other rivers in the United States if it were not trapped in Lake Powell, and represents an average lowering of about 6.5 inches per 1,000 years of the basin upstream from the Grand Canyon. Most of the erosion is in the shale formations.

MINERAL RESOURCES

Just as the geological history of this region is unique in the United States, so is the combination of mineral resources resulting from that history. The basin has been an important source of mineral resources since before the arrival of the Spanish explorers. The general region felt the impact of the gold and silver booms of the nineteenth century, but over the years since then, these minerals have become relatively less important. They now are recovered mostly as a byproduct of the processing of other minerals. The Colorado River Basin, however, must be viewed as one of the really important mineral resource areas of the United States.

The basin has important deposits of crude oil, coal, copper, oil shale, uranium, molybdenum, trona (source of natural sodium carbonate), and limestone (for cement production). There also are less significant deposits of a number of other minerals. It is estimated that the basin contains about 7 billion barrels of crude oil, about 160 billion tons of coal of which about 80 billion tons are recoverable, and over one trillion recoverable barrels of shale oil (Upper Colorado Region State-Federal Agency Group, 1971a). Over 50 million tons of copper are in the known reserves not including submarginal rock, which provides an increasingly significant part of production in the 1970s (Lower Colorado Region State-Federal Interagency Group, 1971a). Moreover, conditions are favorable for the discovery of additional copper deposits. Significant quantities of other minerals are in the known or predicted reserves; these include molybdenum, uranium, potash, sand and gravel.

Undoubtedly, the fact that energy sources cannot satisfy all of the desires of our society at 1970 costs is going to increase very substantially the rate at which the coal and oil shale resources in this region are put to use. Bid prices for publicly owned oil shale mineral rights indicate the seriousness with which oil shale is being viewed in the mid-1970s. The construction of a number of new powerplants in the basin and plans for other very large plants lend emphasis to the mineral and energy producing potential of the basin.

SALINITY SOURCES

Just as the geological history of this region was largely responsible for the formation of mineral resources, so it has been responsible for the location in the basin of tremendous deposits of salts, such as potash, which can be mined, as well as large beds of highly salted clays and shales. In many parts of the basin, these salt-laden beds or salt deposits are exposed to erosion or to contact with percolating waters from precipitation or irrigation, and thus contribute to the salt load of the river. In addition, in both the upper and lower portions of the basin highly mineralized springs introduce substantial quantities of soluble solids into the river and its tributaries.

LAND RESOURCES

All of the more than 160 million acres in the basin are being used for one or more purposes. Some 120 million acres are used for grazing, the production varying greatly according to the amount of precipitation. Some 75 million acres are important to wildlife production, and about 15 million acres are commercially significant for timber production. Irrigated cropland as of 1965 amounted to about 3 million acres; non-irrigated cropland totals about 600,000 acres, mostly in the upper basin. Over 45 million acres have soil and topographic characteristics which are adaptable to irrigation agriculture. Much of the basin is suitable for designation as wilderness areas, primitive areas, national parks, historical monuments, scenic views or similar types of land use (Upper Colorado Basin State-Federal Agency Group, 1971b; Lower Colorado State-Federal Agency Group, 1971b).

About 65 percent of the area is in public ownership, principally federal.

WATER SUPPLY

By far the greatest portion of the water supply of the Colorado River Basin arises in the upper portion (Fig. 1.1). The virgin or undepleted water supply of the upper Colorado basin during the period 1914-1965 has been estimated at about 14.9 million acre-feet (Upper Colorado Region State-Federal Interagency Group, 1971a), and for the period 1931-1969 it was about 13 million acre-feet (Colorado River Board of California, 1970). Analyses for other historical periods would produce somewhat different results. An additional 3,100,000 acre-feet historically has been produced in the lower basin and is, of course, available for use only in the lower basin or Mexico. For all practical purposes, the water supply of the Colorado River is completely depleted since only minor quantities of essentially unusable water reach the Gulf of California. Some areas outside the basin, such as the eastern slope of Colorado, are supplied with water diverted from the Colorado River, but the entire diversion to such areas is a depletion of the total Colorado River supply.

POPULATION

Population trends are quite different in the two parts of the basin. The lower basin is part of the fastest growing area in the United States. In the 15-year period from 1950 to 1965, the population more than doubled. In 1965, about two-thirds of the 1,877,000 population resided in the three metropolitan areas of Las Vegas, Nevada, and Phoenix and Tucson in Arizona (Lower Colorado Region State-Federal Interagency Group, 1971c).

Fig. 1.1. Rocky Mountain snowpack—a source of Colorado River water.

In the upper basin, the population increased from 281,000 in 1950 to 338,000 in 1960, a change of 20 percent; but from 1960 to 1965, it decreased about one percent (Upper Colorado Region State-Federal Interagency Group, 1971d).

HISTORY OF WATER QUALITY PROBLEM

This great river, which produced the Grand Canyon and the Colorado River delta, was recently described by a critic as an "effluent trickle of brine." That is an overstatement, but it indicates the extent to which the water resource is being used. There is no doubt that the Colorado River has been effectively converted into an immense continental resource for man's use. Now, however, increased attention must be paid to the consequences.

In a state of nature, the water of the Colorado at Lee's Ferry was of quite good quality. Studies have indicated that the salinity at that location before irrigation commenced was probably about one-half its concentration today. In recent years, based on actual measurements, the average concentration at Lee's Ferry is about 600 mg/l and for Parker and Imperial Dams it is about 730 and 880 mg/l, respectively (Colorado River Board for California, 1970). Those numbers are worrisome in themselves, but the really alarming thing is that projections indicate the water above Imperial Dam in the year 2000 will have a salinity well above 1000 mg/1, and possibly as high as 1350 mg/1 unless something is done to reverse this trend.

About 23 percent of the salt in the water at Hoover Dam originates from point sources of natural origin, such as thermal springs. Some of the springs and other sources could be controlled, and specific project plans for such control have been drawn. However, the financial costs are high and environmental factors associated with controlling the salinity may be difficult to overcome.

The activities of man which have caused progressively increased levels of salinity are far more difficult to control. Large acreages of crops have caused concentrations of salt by evapotranspiration of applied water, and irrigation return flow has washed soluble salts from the soil profile or substrata into the river. Mining and other soil disturbances have also caused soluble salts to enter the river.

Unless great care is taken, the salinity problem will be amplified by actions associated with untapping the great energy resources of the upper basin, especially if mass development of oil shale and coal reserves occurs. Difficulties will be associated not only with avoiding saline concentrations in the river, but also with the availability of water for these purposes under the limitations of interstate agreements and compacts and widespread public concern about the land use effects of mining and processing these minerals.

The immediacy and significance of salinity in the Colorado River Basin are indicated by Mexico's experience. Mexico has claimed that the high salinity in the Colorado River at Morelos Dam, the diversion point for Mexicali Valley, threatened not only the economic welfare of a large section of Mexico, but also the well-being of a large Mexican population. This, of course, must not occur and steps are being taken to assure better quality water for that region. Since this topic is to be discussed further in this volume (see, e.g., chapters 2, 6, 11 and 15), I will not review the measures being proposed as a solution, but this is a matter demanding immediate attention.

The long-range solution to the salinity problem in the Colorado River Basin lies outside the control of any one state. Either the states reach an agreement or the Environmental Protection Agency may assume responsibility for the problem by default. Recognizing this, water quality and water resource representatives of the seven Colorado River Basin states formed a cooperative group known as the Colorado River Basin Salinity Control Forum. In 1972, the basin states and the Environmental Protection Agency set as their goal the development of a basin-wide plan to maintain salinity concentrations at or below levels in the lower main stem while the upper basin continues to develop its compact-apportioned water. The action preceded provisions in the Federal Water Pollution Control Act Amendments of 1972 requiring the states to set proposed water quality standards and a plan of implementation in more specific terms. The Environmental Protection Agency required that procedures be established not later than June 30, 1974, that will lead to adoption by the states on or before October 18,1975, of criteria for water quality and an implementable plan for quality control.

Literature Cited

Colorado River Board of California. 1970. Need for controlling salinity of the Colorado River. Los Angeles.

Geological Survey, U. S. Department of the Interior. 1969.The Colorado region and John Wesley Powell. Geological Survey Professional Paper 669, pp. 63, 65. Washington, D.C.: U. S. Government Printing Office.

Lower Colorado Region State-Federal Interagency Group for Pacific Southwest Interagency Committee and Water Resources Council. 1971a. Lower Colorado region comprehensive framework study, Appendix VII, Mineral resources, p. 27.

——. 1971b. Lower Colorado region comprehensive framework study, Appendix VI, Land resources and use, pp. 58–121.

——. 1971c. Lower Colorado region comprehensive framework study, main report, p. 18.

Upper Colorado Region State-Federal Interagency Group for Pacific Southwest Interagency Committee and Water Resources Council. 1971a. Upper Colorado region comprehensive framework study, Appendix VII, Mineral resources, p. 32.

——. 1971b. Upper Colorado region comprehensive framework study, Appendix VI, Land resource and use, pp. 35–64.

——. 1971c. Upper Colorado region comprehensive framework study, Appendix V, Water resources, p. 47

——. 1971d. Upper Colorado region comprehensive framework study, Main report, p. 32.

2

Colorado River Development

IVAL V. GOSLIN

The history of the development of the Colorado River should read like a novel. Certainly such a history has all the elements of a heart-rending story —tragedy, pathos, comedy, romance, and human interest. An outline, such as this chapter must be, will lack the qualities of a novel, yet it is hoped that part of the events enumerated will serve as a background for understanding the influences that have led to present conditions and for analyzing the consequences and relative values of alternative choices with respect to future development.

A word of caution is offered. This chapter reflects the viewpoint of an author who has spent over 21 years on Colorado River problems. The events selected to outline the historical development of the river naturally reflect my biases and judgment. Many other events of historical significance could have been included, and another writer might have placed different emphasis or other interpretations on certain parts of the material discussed.

EXPLORATION—NAVIGATION

Navigation was the first use made of the lower main stem of the Colorado River by white men. As early as 1539, Francisco de Ulloa, exploring what was believed to be a strait, sailed to the head of the Gulf of California. He noted the turbid condition of the water, and guessed that a great river entered the gulf near its head. Ulloa did not see this stream, the Colorado, but indicated its supposed position on a sketch map. The actual discovery of the river occurred the next year, 1540, when three explorers, one by sea, and two by land, reached it. Captain Hernando de Alarcon, the first of the three to be on the scene, sailed up the Gulf of California to its head. Alarcon proceeded up the Colorado in small boats to a point about 100 miles above the mouth of the Gila River (La Rue, 1916).

Melchior Diaz, from Coronado's main expedition, journeyed overland to the mouth of the Colorado, proceeded up the river to a point several leagues above the Gila River, crossed and explored some of the country to the west.

Garcia Lopez de Cárdenas, another of Coronado's lieutenants, traveled through what is now northern Arizona and "arrived at a river the banks of which seemed to be more than three or four leagues apart in air line." This is the first written description of the Grand Canyon of the Colorado River.

As noted by Lieutenant J. C. Ives (1861), "In less than 50 years after the landing of Columbus, Spanish missionaries and soldiers were traveling upon the Colorado, following its course for a long way from its mouth, and even attaining one of the most distant and inaccessible points of its upper waters. More information was gained concerning it at this time than was acquired during the three subsequent centuries."

Juan de Oñate, during his expedition in 1604-05, from the Rio Grande to the mouth of the Colorado, arrived on the banks of a stream flowing north-westerly, which he named Colorado. This stream is now known as the Little Colorado River. It appears that Oñate was the first person to use the name "Colorado," but this name for the main river did not come into permanent use until 1775 when Padre Francisco Tomas Garces used it.

Steamboating began on the lower Colorado River in 1851, mostly for the purpose of carrying freight from the head of the Gulf of California to Yuma.

From 1846 to the start of the Civil War, the lower Colorado was explored by surveying and exploring parties under the auspices of the U. S. Department of War. The most detailed examination of the river made during this period was by Lieutenant Joseph C. Ives in 1857-58. He ascended the river in a steel, stern-wheel steamboat, 50 feet in length, which had been constructed in Philadelphia and transported in sections via ship and the Panama Railroad Company to San Francisco, and thence to the mouth of the Colorado River where it was assembled. A detailed examination was made of the river with the objective of determining how far it was navigable for steamboats. Ives turned back at the mouth of Las Vegas Wash, which he called the head of navigation. Within a few years, steamboats carried cargos farther upstream to the Mormon settlement of Callville, which was founded in 1864.

Although practically nothing was done to develop the river for naviga-tion, except periodically to remove sandbars, blast rock obstructions, and construct a few docking facilities, it is of historical significance that early proposals were under consideration to make parts of the stream commercially navigable as a transportation artery. The coming of the railroad sounded the death knell of water navigation on the Colorado.

After the two expeditions of John Wesley Powell in 1869 and 1871, from Green River, Wyoming, through the canyons of the Green and Colorado rivers, others began thinking of the Colorado River in terms of another form of transportation. Frank M. Brown conceived the idea that the deep canyons might provide a practicable route for a railroad. He and Robert Brewster Stanton organized an expedition to survey the river for this purpose from Grand Junction, Colorado, to the Pacific Coast. Brown believed that such a railroad at approximately river grade all the way would carry enough coal to the southwest to justify its construction. The railroad line was incorporated as the Denver Colorado Canyon and Pacific Railway. During the surveying expedition by boat in 1889-90, Brown lost his life in a whirlpool. Stanton's party carried out the remainder of the survey to the head of the Gulf of California. Due to lack of financing and economic justification, construction of this railroad was never initiated, although Stanton's colleagues at the time agreed that his survey had established the engineering feasibility of a railroad down the Colorado.

In addition to explorations mentioned, many others could be related to subsequent development of the Colorado River system. These played their part in spreading knowledge of the region and undoubtedly stimulated many of the hundreds of thousands of immigrants to the Pacific Southwest in the latter half of the nineteenth century.

EARLY IRRIGATED AGRICULTURE

Ancestors of the Cocopah and Yuma Indians should probably be credited with being the first to use the waters of the river for increasing their food supply over the amount provided by nature alone. If nature is regarded as reproducing plants by chance because only those seeds that "accidentally" encounter satisfactory conditions of moisture, soil, light and heat, succeed in growing, the ancient Indians can be regarded as having improved upon nature by removing part of their dependency upon "accidents." Similar to the early Egyptians in the Nile Valley of Africa, these Indians undoubtedly scattered seeds in the mud when the river receded after seasonal overflow. Nature was then permitted to take its course. Nothing was done by the first Indian "agriculturists" to alter the flow of the river itself.

Progenitors of the cliffdwellers appear to have been the first to artificially convey water to the land in order to raise crops in the Colorado River Basin. Lands below many of the abandoned cliff houses show unmistakable signs of having been traversed by ditches and irrigated. In the valleys of the Gila and

Little Colorado rivers, traces of old irrigation systems may still be discerned. Ancient canals of Arizona's Salt River Valley would have been capable of watering a quarter of a million acres of land, although probably not all at the same time, but in blocks, one block of land being abandoned when it became non-productive and another farmed in its place.

Catholic priests of the Jesuit Order were the first white men to irrigate. Missions were established at Guevavi and San Xavier, in southern Arizona, by 1732. During the period from 1768 to 1822, considerable irrigation was practiced along the Santa Cruz River near the missions and the Spanish presidios of Tubac and Tucson. Orchards were planted and annual crops of wheat, barley, corn, beans, melons, squash, peppers and tobacco were harvested. The diversion of water to mission grounds transformed the surroundings from arid deserts to gardens. Headworks and canals of this period were small and of the simplest construction, but the native workers were skillful irrigators. They also adopted certain ideas of equity and custom relating to the distribution and utilization of the water resource, some of which have persisted in irrigation practice to the present time. One of their rules of equity was that water is appurtenant to the land.

Father Silvestre Escalante in his diary of the Domínguez expedition from Santa Fe, New Mexico, into western Colorado and Utah, to Utah Lake, and south to the Virgin River, east and across the Colorado River and back to Santa Fe in 1776 recorded his speculations upon the possibility of diverting water for irrigation at various points during the journey. He did not mention any use of water for irrigation from the deep canyon reaches of the river. There is no evidence that he contemplated any large impoundments of water behind huge dams, like lakes Mead and Powell, which we take for granted today.

MODERN IRRIGATION

Modern irrigated agriculture utilizing Colorado River water may be regarded as dating from 1853 when the United States consummated the Gadsden Purchase from Mexico. Increasing numbers of Americans made up of military men, immigrants en route to California, and other hardy pioneers suddenly began to establish permanent homes in the Southwest. The first relatively modern irrigation works were started in the 1850s in the Colorado River Basin in Colorado, Utah and Wyoming. At first only stream bottom lands were reclaimed, and the facilities were simple.

About the middle of the nineteenth century, the Mormons, often regarded as the fathers of American reclamation, began irrigating lands from the upper reaches of numerous tributaries of the Colorado River as rapidly as their migrations from the basin of the Great Salt Lake would permit. The first irrigation facilities in the fertile tributary valleys consisted of simple diversion dams and ditches to water the most convenient river bottoms. The towns of Santa Clara in southern Utah and Fort Supply in Wyoming were established in 1854. Settlements spread into Arizona, Nevada, and many parts of Utah in the 1860s and '70s. Later these Mormon pioneers constructed reservoirs to store water for use on more extensive areas at higher elevations and during periods of low streamflow.

In the early 1890s, it was recognized that in many parts of the basin storage reservoirs were needed in order to provide a reliable supply of water during the latter part of the crop-growing season and to retain water from the years of plenty for use in following low-water years.

Prior to 1902, irrigation in the Colorado River Basin was first by the individual farmer and later by communities of farmers who joined to construct one diversion channel to water lands that otherwise would have required a number of ditches. Still later, local, private water companies were organized. These were able to construct large canals and lateral distribution systems.

In the early 1900s, irrigation in the upper Colorado River Basin was principally in scattered small developments on the main stream and many tributaries. General farm crops were cultivated. On a large portion of the irrigated areas, particularly in the Green River Basin, hay was raised for livestock feed. Irrigation was continuing to expand, limited by short growing seasons, extremes of temperature range, topography, low-value crops, and high cost of building projects.

After Congress passed the Reclamation Act of 1902, the Reclamation Service (predecessor of the U. S. Bureau of Reclamation) began investigations to determine the feasibility of constructing large irrigation projects in western states. Some of the earliest projects investigated and constructed were: Salt River Project on the Salt River in Arizona, started in 1903; Uncompaghre on the Gunnison and Uncompaghre rivers in Colorado, 1904; Yuma, on the Colorado River in Arizona, 1905; Strawberry on the Spanish Fork and Strawberry rivers in Utah, 1906; and the Grand Valley Project on the Colorado River in Colorado, in 1912. Two of these, one in the lower and one in the upper Colorado River Basin, are mentioned below because they represent the initiation of concepts later incorporated into other federal reclamation projects.

Salt River Project, Arizona

Historically, at least as early as 200 B.C., the ancient Hohokam people, by means of miles of hand-excavated canals, were irrigating their corn and cotton fields in the Phoenix area of the Salt River Valley, Arizona. Probably due to some combination of climatic change, erosion, lack of water storage reservoirs, or other factors, these people who created an early agrarian empire were forced to leave the valley.

The first white men promoted irrigation in central Arizona prior to 1869. Their facilities were primitive by modern standards, consisting of temporary brush and rock diversion weirs which were scoured from the stream beds annually during high water. They had no storage reservoirs in which to conserve water for delivery to their land when streamflows were low. They badly needed capital and organization in order to provide storage and a system of regulated water delivery.

Within a month after the Federal Reclamation Law was enacted on June 17, 1902, the federal government was examining the feasibility of the Salt River Project. Surveys and estimates were promptly made for construction of Roosevelt Dam, which was completed in 1911.

Early in 1903, the water users of the valley formed a Salt River Valley Water Users Association to contract with the government. An agreement covering construction of the dam was executed in mid-1904. Surveys were also made for an estimate of the cost of hydroelectric power generation for pumping. In 1910, a contract was executed between the association and the United States for construction of power canals and electric generating facilities, with the cost to be borne by the association because there was not enough money in the Federal Reclamation Fund. The two original purposes of the power generating facilities were to produce power necessary for the construction of the dam, and to provide power for pumping underground water.

As construction progressed, it became apparent that the possibilities for power generation were greater than originally contemplated. Electric energy could be generated in excess of requirements of the Salt River Project itself. The application of the revenues received from the sale of power in excess of that needed by the project was a bone of contention between the federal government and the water users. In later years, this issue was settled by public law in favor of the Salt River Project, which then became one of the forerunners of the concept of using revenues derived from the sale of power to retire that part of the construction costs of an irrigation project beyond the financial ability of water users to repay.

Strawberry Valley Project, Utah

The irrigation of the lower part of the Strawberry Valley on the south side of Spanish Fork River and of the area adjacent to Utah Lake on the north side of the river in the Great Salt Lake Basin was commenced by early settlers prior to 1860. Before 1900 the need for supplemental late season water became evident and placed a rigid limit on further development of irrigable lands. The federal government started the first reconnaissance surveys and assessment of irrigable lands in 1903, making Strawberry one of the earliest federal projects investigated under the Reclamation Act. The Strawberry Valley Water Users' Association was organized in 1903 to contract for repayment of construction costs.

As part of the construction process, a diversion dam, power canal, and powerplant were built, and the electric energy generated was used in constructing other project features.

Constructed facilities included a dam and storage reservoir with an active capacity of 270,000 acre-feet, a tunnel 3.7 miles long, and a main canal and distribution system of 77 miles of which 62 miles were concrete lined.

The Strawberry Valley Project is distinctive in two respects. It was the first large-scale diversion of water from the Colorado River drainage basin into the Great Basin. It was also one of the first of the federal projects to generate hydroelectric energy.

With the beginning of construction on federal reclamation projects, interest in the development of irrigation by private capital was renewed. During the period from 1903 to 1909, many irrigation systems were planned, most of them under another federal law known as the Carey Act, and under various state irrigation district laws. One of the earliest of the private developments to become firmly established was in the Palo Verde Valley of California. Probably the most important was that of the California Development Company in the basin of the Salton Sea, the Imperial Valley of California.

Palo Verde Valley

About 1856, Thomas H. Blythe acquired about 40,000 acres of land for his Blythe Rancho in the Palo Verde Valley, west of the Colorado River in California. He made water filings, constructed a gravity intake from the river known as the Blythe Intake, a main canal and laterals, and irrigated a considerable area by gravity. By 1877 permanent irrigation development from the lower main stem of the river had become a reality.

Imperial Valley of California

The title of "Father" of the Imperial Valley could rightfully be bestowed upon Oliver M. Wozencraft, a medical doctor from San Francisco, who assumed the position of Indian Agent for the federal government. In 1849 Wozencraft, with several men, mules, and a pack train, carefully investigated that section of the then almost unknown Colorado Desert, now called Imperial Valley. While enduring many hardships due to the heat and blowing sand, Wozencraft conceived the idea of reclamation of the desert. Ten years later, in 1859, the doctor induced the California State Legislature to pass a bill giving him all state rights to 1,600 square miles of the Salton Sink.

Wozencraft had to gain permission from the federal government before he could proceed further. A bill was presented to Congress in the fall of 1859 and referred to the proper committee which reported favorably upon it. In its report on the bill, the committee said:

> This tract embraces (according to Lieutenant Brigland) about 1,600 square miles in the basin of what now is and must remain, until an energetic and extensive system of reclamation is inaugurated and brought to successful completion, a valueless and horrible desert. The labor of reclamation must be commenced within two years and be completed within ten years. As fast as water shall be introduced, upon a report to that effect being made by a duly appointed commission, patents shall issue for the parts reclaimed, and when all the conditions are fulfilled, then, and not until then, shall the title rest in said grantee.

The Civil War caused Congress to sidetrack the legislation. After the war, Wozencraft appealed to one Congress after another, but that body was too engrossed in affairs of reconstruction of the nation to listen. After expending all of his personal assets, including his family home, the doctor died in 1887, still trying to fulfill his dream to bring water to the parched Salton Basin.

Charles R. Rockwood rediscovered Imperial Valley in the Salton Basin in 1892. He immediately became obsessed with the idea of bringing water to it as a reclamation process. During the 1880s another gentleman, George Chaffey, had shared with Wozencraft the vision of irrigating the Salton Sink. Chaffey was an organizer, practical engineer, financier, and the contractor hired by the California Development Company in April, 1900, to construct canals capable of bringing 400,000 acre-feet of water from the Colorado to the desert at a cost not to exceed $150,000. In spite of the financial and organizational impediments, difficulties with Mexico over canal routes, several changes from the original plans of the parent company, and friction among company officers, water was turned through the headgates on May 14, 1901. Canal construction continued without interruption until February, 1902. In 22 months the California Development Company had been transformed from bankruptcy to a concern worth millions of dollars, with 400 miles of canals and laterals, 100,000 acres of land ready for water, 2,000 enthusiastic homeseekers ready to start, and the towns of Imperial and Calexico in embryo stage.

The original canal from the Colorado River started at Hanlon's Crossing in California about 500 feet north of the Mexican boundary. The canal crossed the boundary and extended thence about four miles south to the dry channel of the Alamo River, which was cleared and enlarged. The canal traveled for about 55 miles in Mexico before it was turned back into the United States so that the water could be used in Imperial Valley.

In some respects, the construction of this Alamo Canal was ill-conceived and foolhardy. It was built before the dangerous, erratic, and unpredictable flows of the river were controlled. It is difficult to forget that its construction was accomplished in the face of a stern warning by an eminent authority, F. H. Newell, later director of the Reclamation Service, who wrote in the Smithsonian Institution Report that:

> If we go into this depression below sea level and interfere with natural conditions, or—as we say—"develop the country," we are brought face to face with the great forces of the river and the uncertainty as to whether it will desire to continue in the channel in which we happen to have found it.

The Alamo Canal was abandoned in 1904 in favor of an alternate diversion several miles downstream in Mexico because difficulty had been encountered in establishing a firm water right from the United States, and there was a tendency for the upper reaches of the canal to become clogged with silt deposits. In granting a license for the alternate diversion site, the Mexican Government demanded several severe conditions, one of which was that up to one-half of all water diverted would be used for irrigation of Mexican lands. As a result of this condition, there occurred periods of water shortage when American water users were forced to decrease their irrigation in order that Mexican lands could be watered.

A large river-regulating reservoir was needed on the main Colorado upstream to store water in seasons of plenty for deferred use during low natural water flow periods.

Unprecedented floods destroyed the Mexican heading in the spring of 1905. Due to poorly organized and incompetent operation and maintenance, control of the river was lost. The entire flow poured through two enormous canal breaks and eroded cavernous channels on its way to the Salton Sink, creating the Salton Sea on the site of ancient Lake Cahuilla. The flow of the Colorado River into the Salton Sea continued for almost two years, raising its surface from about 250 feet below sea level to 195 feet below sea level, and creating a lake surface of 330,000 acres. The break was finally closed in 1907 with the financial backing and efforts of the Southern Pacific Railroad.

Closing the break did not terminate the struggle for the Imperial Valley people. Danger from floods that might occur at any time, increasing demands for water, constant difficulties associated with diverting the river and maintaining the Alamo Canal, created what appeared insurmountable problems. The river did break westward again in 1909, requiring a levee system to be constructed in Mexico at a cost of about $6 million, paid by Imperial Valley landowners and the United States. Furthermore, the Mexicans were not paying a fair proportion of costs of operating the canal. Flood control work was impeded by Mexican requirements for paying duty on equipment and supplies sent across the border, and the requirement that a large amount of inefficient foreign labor had to be utilized. These factors, in combination with the fact that Americans had to decrease their use of water due to shortage of streamflow, led to agitation and support for a canal to be entirely located on American soil.

The Imperial Valley development was a project constructed by private enterprise that encountered severe problems due to the vagaries of the river. At the same time, other developments were taking place in both the upper and lower Colorado River basins.

NEED FOR WATER STORAGE RESERVOIRS

Irrigation Insurance

E. C. LaRue of the U. S. Geological Survey estimated that in 1913 378,000 acre-feet of water were being taken, or proposed to be removed annually, by transmountain diversions from the Colorado River Basin (LaRue, 1916). Four ditch systems in Utah accounted for 120,000 acre-feet per year from the headwaters of the Duchesne, Price, and Virgin rivers. Four ditches were taking 21,000 acre-feet from the headwaters of the Colorado River to the South Platte and Arkansas River basins. Four other systems that were in the planning stage would divert 237,000 acre-feet per year to the eastern slope of Colorado for a total of 258,000 acre-feet. According to a report by F. E. Weymouth (1924), Chief Engineer of the Reclamation Service, in 1922 the approximate irrigation development in the Colorado River Basin was as shown in Table 2.1. Additional large diversions, amounting to

TABLE 2.1

Irrigation Development in the Colorado River Basin (1922)

	Area irrigated (acres)	Area irrigable (acres)	Total (acres)
Upper Basin	1,450,000	2,750,000	4,200,000
Lower Basin	950,000	1,350,000	2,300,000*
Total in United States	2,400,000	4,100,000	6,500,000
Mexico	200,000	800,000	1,000,000
Grand Total	2,600,000	4,900,000	7,500,000*

(Weymouth, 1924)

*Includes 430,000 acres irrigated and 400,000 irrigable in the Gila River Basin.

almost half a million acre-feet, were under consideration for development in Utah and Colorado, including transmountain diversion for Denver's municipal water system.

Although climatic conditions were most favorable for growing high-value crops with superior yields, such as melons, lettuce, cotton, alfalfa, and semi-tropical fruit in the lower basin, its development was limited by restricted late season low river flows. By 1922, irrigation in the Gila River Valley of Arizona was well advanced. The Imperial Valley in California had over 400,000 acres in cultivation, watered by direct diversion from the Colorado River. The Imperial lands suffered water shortages in every low-water year. In addition, Imperial had to supply water for 200,000 acres of land in Mexico. At this point, when the water was available, about 3 million acre-feet per year were being channeled from the Colorado River Basin for use in the United States and Mexico.

To insure that an adequate supply of water for irrigation would be available seasonally and annually a large equalizing reservoir was needed upstream from the agricultural lands in Arizona, California, and Mexico.

Flood Protection

The lower reaches of the river were constantly in danger of prolonged flooding from the melting snows of Colorado, Utah, and Wyoming mountains. Floods originating on the lower tributaries, while of shorter duration, could also be extremely damaging. As described above, the tragic menace from floods was fully realized in 1905-06 when 30,000 acres of valuable land were inundated, homes were destroyed, farms were ruined, highways and the railroad were washed away. Millions of dollars worth of damage resulted.

The construction of levees in both the United States and Mexico was required to protect lands in the Imperial Valley and others being farmed on the river delta. Each year new floods attacked the levees, and they had to be built higher and stronger. International problems with Mexico complicated levee maintenance and caused an excessive financial burden. Levees to protect the Yuma Project gave way several times with disastrous consequences. In 1922 levees along the Palo Verde Valley were breached. To protect the lands along the lower Colorado, about 150 miles of levee system had to be maintained. Menace of the flooding river remained even after $10.25 million dollars had been spent in levee construction and maintenance between 1906 and 1924. At least 100,000 people in the area lived in constant fear that they might be inundated.

A mammoth river-regulating, flood-storage reservoir was needed on the main river upstream to protect the lives of thousands of persons and millions of dollars of capital investments in the lower basin.

Silt Retention

The Colorado River annually deposited over 100,000 acre-feet of silt in the delta region between the levees, raising its bed higher and higher, and making larger and continuous expenditures of funds necessary to maintain levees that protected Imperial Valley. By 1923-24, the Imperial Irrigation District was spending over half a million dollars per year to remove silt from its canal system. In addition, it was costing Imperial Valley farmers $1 million annually to repair damages caused by silt on their farms.

A large silt retention basin was needed upstream on the main stem of the Colorado River.

Reliable Water Supply for Municipalities and Industries

The population of the coastal plains adjacent to the Los Angeles metropolitan area experienced phenomenal growth, doubling between 1920 and 1930. This area needed a new source of water. Los Angeles had absorbed its entire supply from Owens Valley and had studied other sources for water. The only practicable and adequate source at that time was the Colorado River by conservation of its flood waters in a storage reservoir. The city of Los Angeles made a reconnaissance survey of a route to the Colorado River in 1923 and justified construction of an aqueduct.

The need for municipal and industrial water in southern California, therefore, was another link in the chain of necessity for construction of a large conservation reservoir in the canyon of the Colorado River.

Increasing Demands for Electric Energy

In the early 1920s hydroelectric power developments in the Colorado River Basin were mostly confined to tributaries of the river. There were 36 powerplants with the combined installed capacity of only 37,000 kilowatts. The largest of these were the Reclamation Service plant at Roosevelt Dam on the Salt River in Arizona (10,300 kilowatts) and the Shoshone Plant of the Central Colorado Power Company on the main stem of the Colorado River upstream from Glenwood Springs, Colorado (10,000 kilowatts). Power generated at Roosevelt Dam helped to fulfill the energy needs of the Phoenix area. The increasing population of the Los Angeles metropolitan area caused a rapidly growing power market, which, together with advances made in the technology of power transmission, created a demand for large blocks of electrical energy greatly in excess of the capability of the hydroelectric resources available. Southern California was badly in need of another source of power for its burgeoning industries.

A large dam with hydroelectric power generators was a logical installation to be superimposed upon the wild Colorado River when harnessing it.

Pyramiding Problems

Water resource development in the upper basin lagged considerably behind that in the lower Colorado River Basin in the 1920s. Rapid progress had been made in the Gila River Basin. As we have seen, potential developments on the main stem of the lower river were impeded by lack of storage facilities. Existing developments suffered frequent water shortages and were threatened by floods aggravated by enormous silt deposits and there were increasing demands for more irrigation in California and Mexico and for large quantities of municipal and industrial water and electrical energy in California. The situation was rapidly becoming tense.

FALL–DAVIS REPORT

The U. S. Reclamation Service transmitted to Congress its comprehensive report entitled ''Problems of Imperial Valley and Vicinity'' (Fall-Davis Report) in 1922. Three of the six recommendations of this report that were of great significance in the subsequent development of the river are:

1. Through suitable legislation the United States would undertake construction, with government funds, of a high line canal from Laguna Dam to the Imperial Valley and be reimbursed by the lands benefited.
2. Through suitable legislation the United States would undertake construction with government funds of a reservoir at or near Boulder Canyon on the lower Colorado River to be reimbursed by the revenues from leasing the power privileges incident thereto.
3. Every development hereafter authorized to be undertaken on the Colorado River by federal government or otherwise be required in both construction and operation would give priority of right and use: first, to river regulation and flood control; second, to use of storage water for irrigation; third, to development of power.

It was rapidly becoming apparent that the natural flow of the Colorado River could not supply all of the uses contemplated by the seven Colorado River Basin states. In addition, it was obvious that the lower basin states, particularly California and Arizona, were growing much more rapidly in population and water use than were the upper basin states. The latter were becoming apprehensive that if the lower basin continued to be developed at such a rapid rate, there soon would be no water left for consumption in the upstream states under the western doctrine of fixing the right to use water by prior appropriation. The lower basin pressed for river development through the aid of the federal government. The upper basin objected. In 1919 and

1920, bills were introduced in the Congress to authorize building this canal and a storage reservoir on the main river somewhere downstream from the junction of the Green and Colorado Rivers.

EVOLUTION OF THE "LAW OF THE RIVER"

Colorado River Compact (45 Stat. 1057)

Proposals for storage in the lower basin without protective guarantees for the upper basin states were regarded by upper basin water authorities as threats to establish priorities that would effectively prevent future utilization of the water in the upper basin.

In the 1920s, laws with respect to rights to use water from interstate streams were not well fixed. Each state claimed the exclusive authority to regulate the appropriation of water within its borders. The federal government claimed jurisdiction over water of interstate streams. The lower end of the Colorado River was considered navigable and subject to federal laws. At the same time there was widespread desire for federal aid for financing a large multiple purpose development believed to be necessary for optimum use of the waters of the lower Colorado.

If a stalemate of long duration was to be avoided, some type of agreement allocating the use of the river's waters among the seven basin states was necessary before a comprehensive plan could proceed. The lower basin states wanted an interstate agreement because they needed the political support of the upper basin states to secure passage of authorizing legislation by Congress. The upper basin states favored a compact in order to protect their deferred use of water against prior appropriations in the lower basin.

On November 24, 1922, the compact commissioners of the seven basin states and Herbert Hoover, Secretary of Commerce, representative of the United States, signed the Colorado River Compact at Santa Fe, New Mexico. This Compact among the states did several fundamental things necessary before further river development could proceed.

1. The Colorado River Basin was divided into the upper basin and the lower basin, with the line of demarcation located at Lee's Ferry, Arizona, which was defined as a point one mile below the mouth of the Paria River and a few miles south of the Utah-Arizona boundary. Here the waters of the entire upper basin system, including the Paria River and return flows from the upper basin projects converge into one stream.

2. The annual beneficial consumptive use of 7.5 million acre-feet of water was apportioned to each sub-basin with the lower basin granted the right to use another million acre-feet annually if it is available.

3. States of the basin were aligned into two divisions. The upper division states include Colorado, New Mexico, Utah, and Wyoming. The lower division states are Arizona, California, and Nevada.

4. Rights of Mexico to use water were recognized in that each basin was to provide water for *one-half of any deficiency* that might occur in any amount granted to Mexico by a future international treaty.

5. The upper division states were not to cause the flow of the Colorado River at Lee's Ferry to be less than 75 million acre-feet in any period of ten consecutive years. Because state boundaries do not coincide with the drainage basin boundaries of the upper and lower basins, two upper division states—New Mexico and Utah—have part of their territory in the lower basin, while Arizona, a lower division state, has a small portion in the upper basin.

6. Under a definitive term of the Compact, the Colorado River Basin included "all of the drainage area of the Colorado River system and all other territory within the United States of America to which the waters of the Colorado River system shall be beneficially applied."

7. The Compact negotiators, believing they were dividing the use of only a part of the river's flow, provided that at any time after October 1, 1963, if and when either basin had reached its total consumptive use as apportioned, the use of the remaining waters could be further apportioned between the two basins.

8. The Colorado was recognized as a navigable river, but, "The use of its waters for purposes of navigation shall be subservient to the uses of such waters for domestic, agricultural, and power purposes."

9. Consumption of water for agricultural and domestic purposes was made dominant over impoundment and use of water for generation of electric energy.

10. Each state was permitted to regulate and control the appropriation, use, and distribution of water within its boundaries, subject to other provisions of the Compact.

The Colorado River Compact was supposed to remove causes for disagreement and rivalry between the two basins in the development of the river's resources, because prior development in the lower basin could create no prior right to the use of water there as against future uses in the upper basin, supposedly leaving the latter basin free to develop at its own slower pace. The Compact certainly did clear away roadblocks that had previously prevented Congressional legislation to authorize major projects in the lower basin.

Boulder Canyon Project Act (45 Stat. 1057)

Many outstanding people were involved in long strenuous efforts before Congressional authority could be obtained for constructing major water developments on the Colorado River. As early as 1914, Congress had provided funds to the Reclamation Service for reconnaissance studies of reservoir sites, irrigation projects, and water rights. John T. Whistler, a reclamation engineer, made reports in 1918-19 in which he advocated the necessity for 10 to 12 million acre-feet of storage to supply all future irrigation requirements in the basin and some flood control and power development. Whistler's principal reservoir sites were located in the upper basin.

In 1918, a joint venture by the Imperial Irrigation District and the United States under an All-American Canal Board, planned the construction of an All-American Canal to head at Laguna Dam. Two bills to authorize its construction failed to pass Congress.

The Kinkaid Act of 1920 directed the Secretary of the Interior to report on the condition and possibilities for irrigation in Imperial Valley, with half the cost of the investigations charged to local interests. The Reclamation Service made a detailed survey of dam sites in the lower basin in Black and Boulder canyons because problems in the Imperial Valley could more readily be solved here with dams in the upper basin.

The Fall-Davis report demonstrated the feasibility of a dam in Boulder Canyon from an engineering standpoint. This report also presented data on flood control, water supply, upstream water uses, and hydroelectric power. It showed conclusively that construction of a dam at Boulder or Black canyon was the key to future downstream river development. This report was the first to recommend construction of a dam to such an unprecedented height as 600 feet.

Immediately after the Fall-Davis Report reached the Senate, Congressman Phil D. Swing and Senator Hiram Johnson of California introduced bills to authorize the construction of a Colorado River Development Project. Swing-Johnson bills were introduced in the 67th, 68th, 69th, and 70th Congresses. The last of these became the Boulder Canyon Project Act.

Influential citizens of the lower basin continued urging the construction by the federal government of Boulder Dam and an All-American Canal. These proposals were strongly opposed by those who were against public power development.

In 1924 a report by F. E. Weymouth, Chief Engineer of the Reclamation Service, stressed the need for flood protection and for storage of water to prevent shortages and losses of crops in Imperial Valley. The Weymouth Report urged construction of a dam in Black Canyon to create a reservoir with

Fig. 2.1. Hoover Dam and Lake Mead. This first large multiple-purpose project provides irrigation, power and flood control.

capacity of 34 million acre-feet. Also in 1924, the flow of the Colorado River dropped to such a low stage that for several weeks Imperial Valley received barely sufficient water for domestic and stock watering purposes and suffered severe crop losses. The immediate construction of a dam was demanded.

During committee hearings on the third Swing-Johnson Bill in 1926, testimony was offered for the first time on the proposal to construct an aqueduct from the Colorado River to Los Angeles and use the river for domestic water supply for southern California. Such a scheme required a large amount of electric energy and thus provided a potential market for power to be generated. This proposal received substantial support in Congress. The Boulder Canyon Project Act became part of the "law of the river" when it was signed by President Calvin Coolidge on December 21, 1928, after a special Colorado River Board had reported that a dam on the Colorado at either Black or Boulder Canyon was feasible, but that the Black Canyon site was preferable (Fig. 2.1).

The Boulder Canyon Project Act also created the Colorado River Dam Fund to accomplish provisions of the act. An appropriation not to exceed $165 million was authorized to be repaid with 4% interest, except for $38.5 million to be used for construction of the All-American Canal. The law also provided that the act could not take effect, nor would any work be done, unless and until all seven states had ratified the Colorado River Compact or, as an alternative, unless and until California and five of the other States had ratified the Compact, and the State of California had agreed to limit its annual consumptive use of Colorado River water not to exceed 4. 4 million acre-feet of the water apportioned to the lower basin by the Colorado River Compact, plus not more than one-half of any excess or surplus waters unapportioned by the Compact.

Arizona refused to ratify the compact. On June 25, 1929, President Herbert Hoover proclaimed the Act in effect since the other six States had ratified it, including California, whose legislature had also adopted the limitation provision of a consumptive use not to exceed 4.4 million acre-feet per year. Contracts having been executed to guarantee disposition of the firm energy to be generated at the dam, construction was started after President Hoover signed an appropriation act in 1930 carrying $10,660,000 for the Boulder Canyon Project. The first water was stored in Lake Mead behind Boulder Dam in 1935.

The United States leased power privileges to the Department of Water and Power of the City of Los Angeles and the Southern California Edison Company. These two entities operate the power facilities at the dam.

Under a 1930 contract, the United States agreed to deliver 1.1 million acre-feet of water per year from storage in Lake Mead to the aqueduct of the Metropolitan Water District of Southern California. The delivery is made in accordance with priorities fixed in a 1931 seven-party agreement, the seven parties being Metropolitan Water District of Southern California, Palo Verde Irrigation District, Imperial Irrigation District, Coachella Valley County Water District, City of Los Angeles, and City and County of San Diego. A charge of 25¢ per acre-foot is made for water delivered to the Metropolitan Water District of Southern California and to the City and County of San Diego.

Under another contract executed in 1933, the United States constructed Parker Dam on the Colorado River with money supplied by the Metropolitan Water District of Southern California. Parker Dam is owned and operated by the United States and provides a regulated forebay for diversion of water into the Colorado River Aqueduct which was constructed by the Metropolitan Water District to convey water to the Southern California coastal plain.

All contracts for water and power are subject to their availability and to the terms of the Colorado River Compact and the Boulder Canyon Project Act.

Boulder Canyon Project Adjustment Act (54 Stat. 774)

The Boulder Canyon Project Adjustment Act of 1940 reduced the interest rate from 4% to 3%, removed competititon as the basis for establishing power rates, specified that power income must be sufficient to operate and maintain the Boulder Canyon Project, and provided for payment of certain annual amounts from power revenues to Arizona and Nevada and into the Colorado River Development Fund. Monies advanced by the United States to the Colorado River Dam Fund were to be repaid during a 50-year period, except for $25 million allocated to flood control and deferred until after the end of the repayment period.

The Colorado River Development Fund of $550,000 per year authorized by the Adjustment Act was established to pay for "studies and investigations by the Bureau of Reclamation for the formulation of a comprehensive plan for the utilization of the waters of the Colorado River system for irrigation, electrical power, and other purposes in the States of the Upper Division and the States of the Lower Division."

Mexican Water Treaty (Treaty Series 994; 59 Stat. 1219)

As late as the time of the Gadsden Purchase in 1853, the Colorado River was regarded as being valuable only for navigation. As the West was settled, thriving communities were established in the United States and in Mexico. Many of these were primarily dependent upon diversion of water from the river for irrigation.

Irrigation developments had been made without any agreement between the United States and Mexico providing for consumption of the waters of the international river. Neither country insisted upon maintenance of navigability in the border reaches of the Colorado River as had been contemplated in early treaties. Both countries appeared to acquiesce to the concept that irrigation benefits were superior to benefits that might be derived from river transportation.

As previously noted, the Colorado River Compact among the seven Colorado River Basin states recognized that if an international treaty were executed between the United States and Mexico, the upper and lower divisions would each be called upon to make up *one-half of any deficiency* that might exist in the amount of water agreed to be delivered annually to the Republic of Mexico.

A treaty between the United States of America and the United Mexican States pertaining to the division of waters of the Colorado and Tijuana rivers and of the Rio Grande became effective in 1945. This treaty established an International Boundary and Water Commission. The treaty allotted to Mexico from the waters of the Colorado River a guaranteed annual quantity of 1.5 million acre-feet to be delivered in accordance with certain conditions and specifications as to place and rate. If it is determined by the United States section of the International Boundary and Water Commission that a surplus exists over the amount necessary to supply users in the United States, the delivery to Mexico can provide a total quantity not to exceed 1.7 million acre-feet per year. Mexico can acquire no rights by use of the waters of the Colorado River system for any purpose whatsoever in excess of 1.5 million acre-feet annually. In the event of extraordinary drought or serious accident to the irrigation facilities in the United States, the water allotted to Mexico will be reduced in the same proportion as the consumptive use in the United States is reduced. The treaty also provides that the water of the Colorado River to be delivered to Mexico "shall be made up of the waters of the said river, whatever their origin" and shall be delivered by the United States in the limitrophe portion of the river. Under other terms of the treaty, the two nations agreed to construct certain diversion works, measuring devices, and flood control works in the two countries under specified divisions of responsibilities and cost sharing.

The Mexican Water Treaty was supposed to be a step forward in international cooperation and to settle for all time the almost century-old dispute over allocation of the waters of the Colorado River. One cannot help but question why Mexico was guaranteed an annual delivery of twice as much water as had ever been used (750,000 acre-feet) in that country prior to the construction of Hoover Dam from a river well known to be water deficient. Increased consumptive uses of Colorado River water by Mexico between 1935 and the Mexican Water Treaty in 1945 were made possible because Lake Mead behind Hoover Dam could store the erratic flows of the river.

Basin-wide Comprehensive Planning

During the 1940s, engineers, economists, and political leaders in the upper basin states, particularly in Colorado and Utah, began to realize that development of the water resources of the Colorado River was needed to relieve economic distress in local areas, stabilize highly developed agriculture, and create opportunities for agricultural and industrial growth and economic expansion. They advocated comprehensive basin-wide planning designed to lead to the ultimate development of all water resources of the basin. A similar movement was on foot in the lower basin, especially in California and Arizona. Before detailed planning could take place, it was necessary to have a basin-wide inventory of potential irrigation projects, power generating projects, and possible municipal and industrial uses of water. Emphasis was still on the development and utilization of water for irrigated agriculture. The realization that the cost of such developments would be beyond the capability of water users to repay, even under the liberal reclamation laws, crystallized the support behind the concept of using excess revenues from the sale of hydroelectric power for paying portions of the costs of irrigation.

The advantages of having an abundant supply of low-cost electric energy were not overlooked by proponents of water projects. Such energy would stimulate industry in the entire power market area, create new taxable values, new opportunities, and increased purchasing power.

Under constant urging by the states, the Bureau of Reclamation in 1946 published a comprehensive departmental report by the regional directors of the two regions in the Colorado River Basin on the development of the water resources of the Colorado River Basin (U.S. Department of the Interior, 1947). Called ''The Colorado River'' and often referred to as the ''blue book,'' the report's theme was ''a natural menace becomes a national resource.'' It was sent to the interested states and federal agencies for comments, and in 1950, as a report of the Secretary of the Interior, was transmitted to Congress.

One of the far-reaching conclusions in the report was that there was not sufficient water available in the Colorado River system for full expansion of existing projects and those authorized for construction and for all potential projects studied. The report presented an array of 134 potential water use projects or units of projects, mostly multiple purpose in nature, within the Colorado River drainage basin. Possible diversions of water out of the Colorado River Basin to adjacent basins were also considered in the report. The potential estimated annual average depletion of those projects within the natural drainage basin, plus exports to other basins, was estimated to be about 20.2 million acre-feet, considerably more than the available supply.

Probably the most pertinent recommendation in the report was "that the states of the Colorado River Basin determine their respective rights to deplete the flow of the Colorado River consistent with the Colorado River Compact." The Secretary of the Interior and the President refused to recommend authorization of any projects until such a determination had been made.

Early Litigation Attempts by the State of Arizona

Although the Colorado River Compact had apportioned the use of water between the upper and lower basins, agreement among the states relative to how much Colorado River water each could consume, consistent with the Compact, appeared unattainable. In the lower basin, Arizona and California were constantly at odds over how much water each should have. Starting in the early 1930s, California obtained contracts for delivery of an aggregate quantity of 5,362,000 acre-feet annually. Arizona's exclusive use of the Gila River, without including it as a part of the Compact allotment, was a point in dispute. In 1930, Arizona brought suit against the Secretary of the Interior and the other six states to enjoin enforcing or carrying out of the Boulder Canyon Project Act which was supposed to effectuate the Compact. In 1934 and 1936 Arizona instituted two more suits against California and the other five states seeking to perpetuate testimony of negotiators of the Colorado River Compact and seeking judicial apportionment of the unappropriated water. These two suits were initiated by Arizona after all of the California water storage and delivery contracts had been executed. None of the three lawsuits reached the hearing stage.

During the 79th, 80th, 81st and 82nd Congresses, Arizona sought Congressional authorization for construction of her vast Central Arizona Project under which 1.2 million acre-feet of water would be diverted from the river. Some of the bills were passed by the Senate. None were passed by the House of Representatives. In 1951 the House Committee on Interior and Insular Affairs adopted a resolution that consideration of bills relating to the Central Arizona Project "be postponed until such time as use of the water in the lower Colorado River Basin is either adjudicated or binding or mutual agreement as to the use of the water is reached by the States of the lower Colorado River Basin."

Upper Colorado River Basin Compact (63 Stat. 31)

Instead of using the courts to apportion the consumptive use of water allocated to them by the Colorado River Compact, the five states (Arizona, Colorado, New Mexico, Utah, and Wyoming) having interests in the upper basin, negotiated and signed the Upper Colorado River Basin Compact in 1948. After each state's legislature had ratified this Compact, Congress gave its consent to it in 1949.

The Upper Colorado River Basin Compact apportioned a fixed quantity of 50,000 acre-feet of consumptive use of water per year to Arizona for use in the small portion of the state in the Upper Basin. Of the remainder, 51.75% was apportioned to Colorado, 11.25% to New Mexico, 23% to Utah, and 14% to Wyoming. This Compact created the Upper Colorado River Commission as an administrative agency for the four upper division states, Colorado, New Mexico, Utah, and Wyoming. Arizona is not a member of the Commission. A federal representative appointed by the President serves as chairman. The Upper Basin Compact contains rules and regulations for determining curtailment of water uses during any year in which such curtailment is deemed necessary by the Commission to meet delivery requirements under the Colorado River Compact to the lower basin. It also recognizes certain agreements as to the use of water of interstate streams between its member states within the upper basin, and specifies that consumptive use of water in the upper basin and in each state thereof shall be measured by the inflow-outflow method in terms of man-made depletions of the virgin flow at Lee's Ferry, as contrasted with the method of diversion of water less return flows used in the lower basin.

Colorado River Storage Project Act (70 Stat. 105)

The Bureau of Reclamation's report, "The Colorado River," that inventoried the potential opportunities for river regulation, irrigated agriculture, and power generation in the basin was the catalyst that stimulated the technical and political leaders in the upper Colorado River Basin to accept and aggressively promote the concept of comprehensive upper basin-wide development. The Upper Colorado River Basin Compact, signed in 1948, provided the vehicle for formulating the plan of development known as the Colorado River Storage Project and participating projects. As early as 1948 and 1949, members of Congress from the State of Utah introduced bills to authorize the federal government to construct the Central Utah Project which would use waters from the Colorado River system. These bills could not be enacted into law because they lacked the support of the other upper basin states, and because they were introduced prior to approval of the Upper Colorado River Basin Compact which did not occur until April 6, 1949. In 1950 a bill was introduced to authorize the Colorado River Storage Project—about six months prior to the publication of an interim report on such a project by the regional directors of the Bureau of Reclamation.

It was not until December, 1952, that the Department of the Interior finally submitted its report to Congress proposing a basin-wide plan of development for the upper Colorado River Basin to be known as the Colorado River Storage Project and participating projects. Numerous bills were introduced in the House of Representatives and the Senate of the United States

Congress. It was not until January of 1955 that any of these bills received solid support of the Executive Branch of the federal government. This first occurred when President Dwight Eisenhower urged passage of a Colorado River Storage Project Bill in his State of the Union Message in January, 1955, followed by a request for $5 million in his Budget Message to initiate construction, contingent upon favorable actions by Congress. In spite of the bitter opposition of certain water and power interests in California, a group of vociferous conservation organizations, and anti-reclamation members of Congress from the eastern part of the United States, the Colorado River Storage Project Act became part of reclamation law in April, 1956. This Act was the result of many compromises among the member states of the Upper Colorado River Commission, and with conservation organizations with respect to a proposed dam at Echo Park on the Green River, which was withdrawn from the legislation, and with respect to the Rainbow Bridge National Monument in southern Utah on an arm of Lake Powell, behind Glen Canyon Dam.

The Colorado River Storage Project Act authorized the construction of four large storage units capable of holding 33,583,000 acre-feet of water for river regulation, power generation, and consumptive use by exchange with downstream water users. These storage units are Glen Canyon Dam and Lake Powell on the Colorado River in Arizona and Utah; Navajo Dam and Reservoir on the San Juan River in New Mexico and Colorado; Flaming Gorge Dam and Reservoir on the Green River in Utah and Wyoming, and the Curecanti Storage Unit on the Gunnison River in Colorado consisting of three dams and reservoirs—Blue Mesa, Morrow Point and Crystal (Figs. 2.2 and 2.3). The authorizing act also provided for the construction of eleven participating irrigation projects. Ten participating projects have been added by enactment of subsequent amendatory laws in 1962, 1964, and 1968: San Juan-Chama and Navajo Indian Irrigation Projects Act (76 Stat. 96); Savery-Pot Hook, Bostwick Park, and Fruitland Mesa Projects Act (78 Stat. 852); Colorado River Basin Project Act (82 Stat. 885).

Several unique features in the Colorado River Storage Project Act are related to repayment, accounting, and funding requirements not found in previous reclamation law. Some of these innovations are having great influence on natural resource development in the Colorado River Basin. This law provided:

1. For the creation of an Upper Colorado River Basin Fund to which all appropriations from the general fund of the U.S. Treasury shall be credited as advances, except those for recreational purposes which are nonreimbursable;
2. That all revenues (power, municipal water, irrigation, or other) derived from storage units or water-using participating projects shall be

Fig. 2.2. Glen Canyon Dam and Lake Powell provide reservoir storage to equalize the highly variable flow of the Colorado River.

Fig. 2.3. Reservoir above Morrow Point Dam fills the narrow gorge of
the Gunnison River, Curecanti Unit Colorado River Storage Project.

[43]

credited to the Basin Fund and shall be available for paying operation, maintenance, and repair and emergency costs, and costs of power and municipal water features within fifty years with interest;

3. That each participating consumptive-use project must pay its own operation, maintenance, and emergency charges from its own revenues;

4. That costs of storage units allocated to irrigation shall be returned from revenues in the Basin Fund within fifty years;

5. That revenues in the Basin Fund in excess of amounts needed to defray costs under items 2, 3, and 4 above shall be apportioned within the Basin Fund to the credit of the states as follows:

Colorado	46.0%
New Mexico	17.0%
Utah	21.5%
Wyoming	15.5%

for the purpose of returning the cost of irrigation allocations of participating projects within fifty years that are beyond the capacity of repayment by water users;

6. That if a participating project has power and/or municipal water facilities in addition to irrigation facilities, it must pay from its own revenues the operation, repair, maintenance and emergency charges for all of its own facilities, repayment of its power costs and its municipal water costs, and interest on the power and municipal water investment;

7. That after all the costs under item 6 are paid from the revenues of a given participating project, if excess revenues remain in the Basin Fund that were derived from its own power and/or municipal water facilities, the excess is credited within the Basin Fund for use within that state wherein the project is located before the percentage apportionments are made to the four states;

8. That excess power revenues credited within the Basin Fund to each State may be used for repaying costs of irrigation projects only within the state and may not be used within another state unless appropriate consent is obtained; and

9. That business-type budgets must be submitted each year to Congress.

Among other innovative features of this law was the manner of treating costs of a project that benefits the Navajo Indians. In recognition of the fact that assistance to these Indians is the responsibility of the entire nation and not of any one state or group of states, the law specified that when the Navajo

Indian Irrigation Project was authorized, the cost of irrigation allocation beyond the capability of the land to repay should be nonreimbursable. Another part of the law provides that payment of construction costs within the capability of the land to repay will be deferred for as long as the land remains in Indian ownership. In addition, the Navajo Dam and Reservoir, required for the irrigation of Indian lands, without hydropower generators and having limited value in river system regulation, was classified as a storage unit. By virtue of this classification, and because its cost has been allocated mostly to irrigation, the cost of this dam and reservoir would be almost entirely paid by power revenues from other storage units.

This Act also placed emphasis upon the potential recreational values of the upper Colorado River Basin. For the first time the development of recreational facilities as a valuable social asset for the general welfare of the citizens of the nation was specifically recognized by making the costs of such facilities constructed in conjunction with the water resource development nonreimbursable. In this sense, this part of the law evinced a recreational consciousness on the part of its proponents and helped to pave the way for passage by Congress of subsequent laws relating to federal responsibility with respect to outdoor recreation.

Fryingpan-Arkansas Project Act (76 Stat. 389)

For the past fifty years there has been a heated intrastate controversy in Colorado between water interests on the two sides of the Continental Divide. The root of the argument is over the diversion of water from the Colorado River and its tributaries from the west slope of the Rocky Mountains where there are few people but plenty of water to east slope municipalities and irrigation projects where there is a very limited natural water supply and most of the population of the state is located. A number of important cities, including Aurora, Colorado Springs, Pueblo, and the sprawling Denver metropolitan area are dependent upon importation of large quantities of water from western Colorado. The Colorado-Big Thompson Reclamation Project in northern Colorado, near the cities of Greeley and Loveland, obtains a large supply of water from the upper Colorado River system through a tunnel extending from the western slope.

The Fryingpan-Arkansas Project is mentioned here because it represents the result of agreement after prolonged and painful negotiations between representatives of water interests on the two slopes of Colorado. It incorporated into the authorizing legislation a historic agreement between water authorities of eastern and western Colorado, reached many years before, known as Senate Document No. 80 of the 75th Congress. It also incorporated the project operating principles adopted by the State of Colorado and published in House Document No. 130 of the 87th Congress. The Fryingpan-

Arkansas Project was badly needed to supply supplemental water to the Pueblo area. The Arkansas River Basin is also the last transmountain diversion project to be authorized by Congress in the upper Colorado River Basin.

Last Arizona vs. California Lawsuit

In 1952 the State of Arizona initiated the last Arizona vs. California lawsuit in the U.S. Supreme Court in compliance with the mandate of the House Committee in 1951 to obtain an adjudication of a water supply for the Central Arizona Project. The decision in this famous case did not come until 1964. Arizona won the case. Her right was confirmed for the delivery of water by the Secretary of the Interior from the lower main stem of the Colorado River.

This suit was not decided under the Colorado River Compact, but under the Boulder Canyon Project Act. The four upper division states, as such, were not parties in the case. The states of New Mexico and Utah were parties, but as lower basin states only. The highlights of the Supreme Court's decree (376 U.S. 340) included:

1. Control of the river below Hoover Dam was given to the Secretary of the Interior who was authorized to deliver from the main stem of the river 4.4 million acre-feet of water per year to California, 2.8 million acre-feet to Arizona, and 300,000 acre-feet per year to Nevada, when 7.5 million acre-feet were available, with any surplus to be divided between Arizona and California.
2. The federal reservation theory of reserved water rights for Indian lands and other federal reservations, such as national forests, was fully recognized.
3. A determination was made of water rights on the Gila River between New Mexico and Arizona.

The decision in this lawsuit was destined to have far-reaching effects, not only in the Colorado River Basin but also in all river basins where there are federal reserved or Indian-owned lands.

Colorado River Basin Project Act (82 Stat. 885)

In 1942 and 1944, contracts were executed between the Secretary of the Interior and the State of Nevada for the delivery of water to Nevada from Lake Mead, subject to its availability for use in Nevada under the Compact and the Boulder Canyon Project Act.

Not until 1944 did the State of Arizona enact a statute which purported unconditionally to approve, ratify, and confirm the Colorado River Compact. By contract in that same year the United States agreed to deliver certain quantities of water from storage in Lake Mead for use in Arizona, subject to its availability for use in Arizona under the Compact and Project Act. As seen above, Arizona earlier made several attempts to persuade Congress to authorize the construction of a Central Arizona Project but without success because of strong opposition from California under whose interpretation of the ''law of the river'' there was not enough water available for such a development in Arizona.

In 1963, preceding the United States Supreme Court's 1964 decision in Arizona vs. California, members of Congress from Arizona introduced bills which, if enacted, would have authorized the Secretary of the Interior to construct the Central Arizona Project.

California, having lost the lawsuit, immediately served notice that unless a priority were to be given to her consumptive use of 4.4 million acre-feet of water per year over the uses by Arizona, she would again oppose authorizing legislation.

The upper division states—Colorado, New Mexico, Utah, and Wyoming—also had a vital interest in the Central Arizona Project.

For the longest period of streamflow records (1896-1973) for the Colorado River at Lee's Ferry, the annual average virgin flow is only 14.8 million acre-feet. In only one decade (1941–1950) following signing of the Colorado River Compact did the ten-year average virgin flow exceed this figure. This phenomenon is of particular importance to the upper basin because of the Compact requirement to deliver to the lower division states 75 million acre-feet in every period of ten consecutive years. The trend in river flow for more than forty years at the time the last Central Arizona Project legislation was pending in Congress had been downward. For the total period since 1922 the annual average had been only 13.8 million acre-feet, and in two unrelated ten-year periods, 1931-1940 and 1954-1963, the annual average virgin flow for each ten-year period amounted to only 11.8 million acre-feet. The annual average for one 12-year period, 1953-1964, amounted to only 11.6 million acre-feet. It was obvious to the four upper division states that if the Compact-required delivery to the lower basin of 75 million acre-feet in each ten years was to be met, the annual average amount remaining for consumptive use in the upper basin was about 20% less than the 7.5 million acre-feet apportioned by the Colorado River Compact.

The upper division states had received more than their share of opposition and harassment from both Arizona and California during the initial filling period of the storage units of the Colorado River Storage Project, especially during the filling of the dead storage space in Lake Powell behind Glen Canyon Dam.

The upper basin water users could also plainly see that if a full supply of water were to be given to the Central Arizona Project, a large proportion of that water, especially in low water years, would have to come from the upper basin, which had not yet put its allocation to beneficial consumptive use. Protection of the rights of the upper division states against excessive consumptive uses in the lower basin that would encroach further into their Colorado River Compact allotment was of vital concern. After Arizona agreed to give a priority to California's 4.4 million acre-feet per year the danger to the upper basin became even more critical. The upper division states feared that if their water were used by Arizona and California, they might forever be precluded from using their legal entitlement by superior political pressures from the lower basin states, in spite of the terms of the Compact. They also had learned from experience that the Secretary of the Interior could not be trusted to operate the river in their best interest if left to manipulate according to his own whims and desires or under the influence of political forces stronger than their own from the outside. The upper division states wanted the secretary to be controlled in his river operations by definite guidelines.

The pending Central Arizona Project legislation was viewed by the upper division as an opportunity for effective interstate cooperation in resolving several outstanding operational problems involving all of the Colorado River Basin states.

In order to protect their future interests in the use of water and the generation of power, the upper division states insisted upon having a number of their projects authorized for construction at the same time the Central Arizona Project was authorized, and demanded that the legislation contain specific parameters for operating criteria for governing the actions of the Secretary of the Interior in order to provide equitable operations of Lake Powell and Lake Mead.

After several years of negotiations between the states of the upper and lower divisions, the Colorado River Basin Project Act became law on September 30, 1968. This Act accomplished the primary purpose of authorizing construction of the Central Arizona Project. In addition:

1. It gave a priority to California for the consumptive use of 4.4 million acre-feet per year of lower main stream water over uses by the State of Arizona, thus requiring the Central Arizona Project to assume shortages when the annual supply of water for use downstream from Hoover Dam falls below 7.5 million acre-feet. In effect, California gained by legislation what had been lost under the decision in Arizona vs. California in 1964.

2. A Lower Colorado River Basin Development Fund was created which was patterned after the basin fund authorized in 1956 for the upper basin's Colorado River Storage Project.

3. Five participating water-use projects were authorized to be con-

structed in Colorado. One potential project in Utah received conditional authorization and another Utah project in the lower basin was authorized to participate in the use of revenues of the Lower Colorado River Basin Development Fund for repayment purposes. The Secretary of the Interior was directed to expedite feasibility studies of several other projects in the upper basin.

4. The new law described explicit guidelines for the formulation of long-range operating criteria for Lake Powell and Lake Mead.

5. The Secretary of the Interior was directed to "conduct full and complete reconnaissance investigations for the purpose of developing a general plan to meet the future water needs of the western United States." Proponents of the legislation, due to opposition of the northwest states, failed to have included any kind of scheme to import water into the Colorado River Basin. In fact, their opponents managed to have a provision included in the Act that prohibits the Secretary from even studying an importation of water for a period of ten years, or until after September 30, 1978.

6. Congress declared that satisfaction of the requirements of the Mexican Water Treaty from the Colorado River constitutes a national obligation that will become the first obligation of any effective water augmentation project authorized by Congress in the future.

7. The purpose of the Colorado River Development Fund that originally, under the Boulder Canyon Project Adjustment Act, was designated to be used for investigating water resource projects in the basin was changed to repaying the Upper Colorado River Basin Fund ($500,000 per year) for deficiencies in power generation charged against it during the filling period of reservoirs of the Upper Basin's Colorado River Storage Project.

The Colorado River Basin Project Act, more than any other law on the river, including the earlier Colorado River Storage Project Act, is a truly basin-wide comprehensive law. It could never have come into existence except as a result of intensive negotiations by capable, dedicated water statesmen representing each of the seven Colorado River Basin states. It is hoped that it will prove to be the forerunner of future seven-state, basin-wide endeavors to solve mutual problems of the entire Colorado River Basin.

The Colorado River Basin Project Act was the last of the major reclamation laws aimed directly at the Colorado River Basin and having great impact upon the historical development of its water resource. For all practical purposes, if all of the water utilization contemplated under this Act were to occur, there would be very little unused water left for future development. The bulk of what would be left would be in Wyoming and Utah. Figure 2.4

shows the location of many of the reservoirs now located in the Colorado River including those authorized by the Colorado River Basin Project Act.

If one were to ask, "Where are we?" and "How did we get here?" with respect to water development in the Colorado River Basin, one would certainly have to acknowledge that the events described above provide historical background; but they do not reflect all of the reasons for the present situation. Today water development is in a state of chaos, confusion, and frustration.

The disarray into which all types of resource development have been thrown is due to a great extent to the extremely rapid social changes in the 1960s and '70s compared with the more predictable expanding progress of the previous fifty years. There has been an abrupt wide swing of the pendulum by society from the earlier emphasis on an economic ethic to a social ethic with emphasis on such things as environment and ecology. Whether one regards this pendulum swing as good, bad or indifferent is beside the point. The point is that the change in thinking has affected the actions of members of Congress, state legislatures, and even the courts, which are being accused of making laws as well as interpreting them. As in all new social movements in history, the proponents of a new ethic become zealous and sometimes push the pendulum until the clock either overturns or a more reasonable equilibrium is assumed. This point may have been reached already with respect to water development in the Colorado River Basin.

At least four new nationwide laws passed by Congress in the 1960s and '70s and a number of other events had impacts upon history related to water development in the basin. Limited space will not permit even a short discussion of each of these laws and events, or of their influences, but a brief introduction may provoke the reader to further contemplation.

Water Resources Planning Act (79 Stat. 244)

The Water Resources Planning Act of 1965 created the Water Resources Council at presidential cabinet level to provide for the optimum development of the nation's natural resources through the coordinated planning of water and related land resources. This Act also provided for the establishment of the Federal State River Basin Commissions and endowed the Water Resources Council with the authority to establish principles, standards, and procedures for federal participants in the preparation of comprehensive regional or river basin plans and for the formulation and evaluation of federal water and related resource projects. For the Colorado River Basin, no river basin commission has been authorized under the Act which provides that in the event the upper Colorado River Basin is involved at least three of the four states of Colorado, New Mexico, Utah and Wyoming must concur. The principles, standards, and procedures enunciated by the Council have been so rigid and restrictive

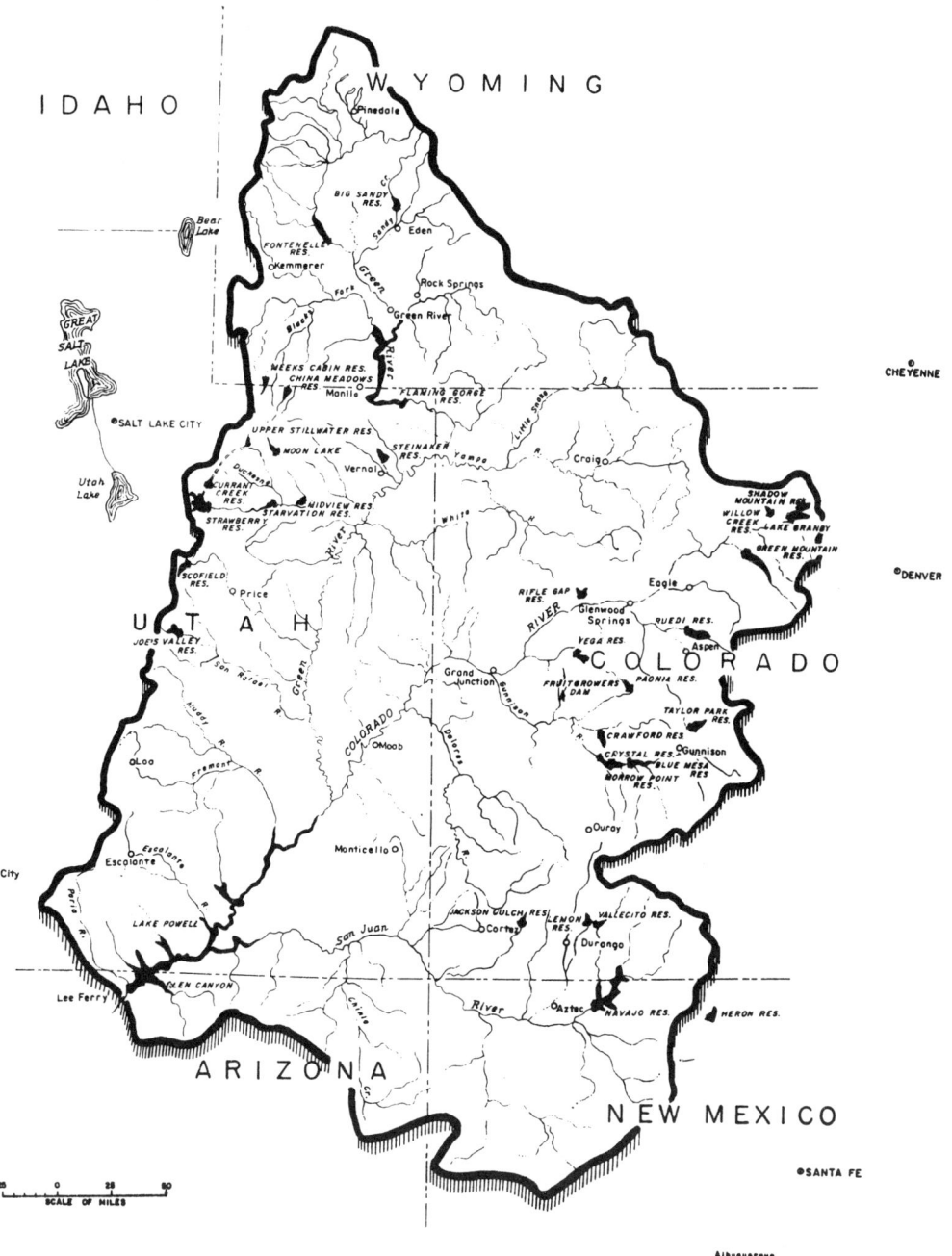

Fig. 2.4. Principal reservoirs and reservoir sites in the Colorado River Basin.

[51]

that projects already authorized by Congress for construction are being re-studied and reformulated. It may prove to be impossible to continue federal water development in the basin. There is reason to believe that the Office of Management and Budget of the Executive Branch of the federal government dominates the thinking and actions of the Water Resources Council.

National Water Commission Act (82 Stat. 869)

The National Water Commission Act of 1968 established the National Water Commission to study present and anticipated water resource problems and to make recommendations related thereto. The Commission's report had 290 conclusions and recommendations that for the most part were ill-conceived and not conducive to the promotion of water development. The Commission concluded that all increased demands for United States' agricultural products can and should be met by intensified management on non-irrigated farms. It suggested that non-agricultural water needs (domestic, municipal, industrial, etc.) in the arid west can be provided from a substantial reduction in irrigated agriculture. The Commission recommended that, in the main, future water resource development be the responsibility of state and local entities. Future federal water programs, including irrigation, for the most part would be completely reimbursed with interest. In short, the National Water Commission recommended that our nation repudiate basic principles that have guided water resource programs since they were initiated in this country.

National Environmental Policy Act (P.L. 91-190)

The National Environmental Policy Act of 1969, intended by Congress to protect and enhance man's environment, is being used as a vehicle in court actions to delay and prevent water development. The Colorado River Basin has been a favorite target. The Central Utah Project in the upper basin and the Central Arizona Project in the lower basin have received the most high velocity bullets of the so-called environmentalists.

Federal Water Pollution Control Act (P.L. 92-500)

The Federal Water Pollution Control Act Amendments of 1972 were enacted by Congress to make possible the badly needed protection and improvement of the quality of water in the nation's rivers and streams through federal-state cooperation administered under the jurisdiction and authority of an Environmental Protection Agency. The intent of the Act certainly has merit, but some of the specific goals and procedures may be questionable. As far as the Colorado River is concerned, the principal application of this law will be to enforce the maintenance of reasonable levels of total dissolved solids in the lower basin.

Increasing concentrations of salts from headwaters to the mouth of the river is not a problem that is unique to the Colorado River. It is found on every river where man has put water to use—and even on rivers where man has not used the resource. The loading of salts into the river from natural salt springs and seeps and from precipitation flowing over saline soils, or by discharges into the river of waste water from irrigation or other activities of man also loads salts into the river. Preliminary estimates indicate that for the Colorado River system about half the salt loading may be from natural sources and half from man's influence.

Fortunately, on the Colorado River, pollution of its water by natural or man-made salt discharges into it or by the removal of water from it has caused no health problems. The salinity problem is an economic one.*

The fact that increases in salt concentration would cause economic problems was recognized early by water users in California. Certainly, well prior to the Mexican Treaty negotiations of the 1940s, Californians were aware of the implications of using the last drop of water from the river system and increasing the salt concentration (Colorado River Commission of the State of California, 1930).

At its headwaters, the average salinity in the Colorado River is less than 50 mg/l and it increases progressively downstream until, at Imperial Dam below Lake Mead in the Lower Basin the present modified condition averages about 865 mg/l.† Projections of future salinity levels with a control program indicate that values of 1200 mg/l or more may occur at Imperial Dam by the year 2000. Should these increases in salinity levels happen, agriculture in the Imperial, Coachella, Gila, and Yuma valleys would be further threatened. A poorer quality water would also be diverted to the coastal cities in California by the Metropolitan Water District of Southern California and by the Las Vegas Valley Water District in Nevada causing further economic losses. Upon completion of the Central Arizona Project, water users in the Tucson and Phoenix areas would be similarly affected.

Several agencies, including the U. S. Geological Survey, Bureau of Reclamation, Colorado River Board of California, and Environmental Protection Agency have analyzed this problem. Obviously, one way to prevent further increases in salinity would be to use no more water from the river. It is also obvious that this solution is both physically and politically impossible.

---

*Salinity as used herein refers to the concentration of total dissolved solids in the water and is measured in milligrams per liter (mg/l), which is approximately equivalent to parts per million (ppm) to concentration of 7,000 mg/l.

†"Present modified" refers to historic conditions for the period 1914–1968 modified by the effects of all upstream projects present in 1974 as if they had been in operation for the entire period.

As the result of the "Conclusions and Recommendations" of the Reconvened Seventh Session of the Conference in the Matter of Pollution of the Interstate Waters of the Colorado River and Its Tributaries held on April 26-27, 1972 under the authority of the Federal Water Pollution Control Act and approved by the Administrator of the Environmental Protection Agency (EPA), a salinity policy objective was unanimously adopted by the seven Colorado River Basin states and the federal government for the Colorado River system. Under this policy, "The salinity problem must be treated as a basin-wide problem that needs to be solved to maintain lower basin water salinity at or below present levels while the upper basin continues to develop its compact-apportioned waters." A salinity control scheme conceived by the Department of the Interior, entitled, "Colorado River Water Quality Improvement Program," has been designated to implement basin-wide salinity control.

Under the Federal Water Pollution Control Act Amendments of 1972 water quality criteria were to be established for the Colorado River system at various points on the river and its tributaries. Obviously, if the criteria are too severe, further use of water from the river would be precluded without an effective program of salinity control. Criteria can be formulated by the states and approved by the Environmental Protection Agency. If the states fail to do the job, the EPA is required by law to promulgate water quality standards. Through an ad hoc Colorado River Basin Salinity Control Forum the seven States are working with officials of the EPA to write standards that will effectuate the "Conclusions and Recommendations" of the April, 1972 agreement.

Although the Mexican Water Treaty became effective on November 8, 1945, operations under the Treaty started in November, 1950. The annual deliveries of water to Mexico from 1950 to 1962 ranged from about 1,780,000 acre-feet to about 10,186,000 acre-feet.

Prior to 1961, Mexico, without protest , accepted the water as scheduled and delivered. In August, 1961, the drainage conveyance channel for the Wellton-Mohawk Project, a federal reclamation enterprise in Arizona near the Mexican border, was placed into full operation, and the storage of river flows in Lake Mead was increased in anticipation of filling Lake Powell upstream. During the 1961-62 winter, the water delivered to Mexico showed a marked increase in dissolved solids, including the highly concentrated return flows from the Wellton-Mohawk area.

Mexico protested that the water was not "usable." For a time, pending a solution by other measures, the Secretary of the Interior released 50,000 acre-feet of stored Colorado River waters in excess of the Treaty allotment in order to reduce the concentration of salts. As the result of negotiations, the two governments in 1965 approved Minute No. 218 of the International

Boundary and Water Commission, United States and Mexico. Under the agreement of this Minute, the United States spent approximately $11.25 million (U.S. Department of the Interior, 1963) for main conveyance channels, collector lateral drains, drainage wells, tile drainage, electric power facilities, and a by-pass channel to convey drainage water around the point of diversion to Mexico.

In 1972 under another agreement embodied in Minute No. 241, the delivery of water by the United States in excess of the treaty allotment was increased from 50,000 to 118,000 acre-feet per year.

By 1972-73, Mexico noted that the Environmental Protection Agency and the seven states of the Colorado River Basin had entered into an agreement under which efforts would be made to maintain the salinity of the Colorado River at Imperial Dam at water quality levels as of that date, while the upper division states would proceed with developing their compact apportionments of water use. Mexico was then not satisfied with the agreement and operations under Minutes No. 218 and 241, and again objected, demanding that the quality of water delivered to Mexico should be as good as that delivered to other lower basin water users.

A State Department negotiator, Herbert Brownell, appointed by President Richard Nixon with diplomatic status, negotiated with his Mexican counterpart another agreement embodied in Minute No. 242. Under Minute No. 242, which is supposed to constitute "a permanent and definitive solution of the international problem of the salinity of the Colorado River," the United States would be required:

1. To adopt measures no later than July 1, 1974, to deliver to Mexico upstream from Morelos Dam 1,360,000 acre-feet of water of the 1.5 million acre-feet treaty allotment with the salinity of no more than 115 ppm ± 30 ppm over the annual average salinity of water at Imperial Dam

2. From September 1973 until the United States adopts such measures, to discharge 118,000 acre-feet of Wellton-Mohawk drainage waters to the river downstream from Morelos Dam and substitute therefor an equal volume of other waters from reservoirs in the United States

3. To pay the cost of extending the concrete-lined, Wellton-Mohawk by-pass drain from Morelos Dam to the Santa Clara Slough near the Gulf of California in Mexico, and to operate and maintain that portion of it in the United States

4. To limit pumping of groundwater within five miles of the Arizona-Mexico border to 160,000 acre-feet of water per year pending an agreement on groundwater (Mexico is already pumping 160,000 acre-feet per year)

5. To support the efforts of Mexico to obtain financing for improvement and rehabilitation of Mexicali Valley
6. To provide non-reimbursable assistance to Mexico for those aspects of rehabilitation of Mexicali Valley related to the salinity problem, including tile drainage. (The Imperial Irrigation District, across the border in the United States, has installed tile drains at a cost of $40.5 million and lateral lining at a cost of $10.3 million, while the land-owners themselves have spent $15.9 million on lining farm ditches (Valantine, 1974).

Among the measures that the Nixon Administration recommended to Congress to implement Minute No. 242 was the world's largest desalination complex to remove salt from Wellton-Mohawk Project drain waters, lining of 49 miles of the Coachella Canal, and purchase of approximately 25,600 acres of land on Yuma Mesa at a cost, including interest during construction, of about $152 million. Operation and maintenance costs of the desalination plant were estimated to be almost $10 million per year. For a 50-year period of 5⅝% interest the annual equivalent costs were estimated to be almost $18 million. Not included in legislation proposed by the administration was the pumping of 160,000 acre-feet of water per year on the United States' side of the border.

If Minute No. 242 does not constitute a new treaty, the best that can be said for it is that it is a serious amendment of an existing treaty. It incorporates groundwater which is not covered in the original treaty. It changes, at least for an indefinite period, the amount of water to be delivered to Mexico, and would cause the Secretary of the Interior to violate the reservoir operating criteria of the Colorado River Basin Project Act of 1968. It gives a water quality (salinity) guarantee to Mexico which was specifically omitted from the Treaty, and for all practical purposes precluded by the inclusion of the language saying that the water to be delivered to Mexico "shall be made up of the waters of said river, whatever their origin." In addition, Minute No. 242 provides for a unilateral unspecified amount of foreign aid commitment without prior approval of the Congress, which may be a violation of the treaty itself.

Certainly Minute No. 242 goes far beyond the definition of a Minute describing operations to implement a substantive agreement. As a new Trea-ty, or as an amendment to an old one, Minute No. 242 should go before the United States Senate for ratification before it can be implemented, if for no other reason than to provide a hearing for interested parties.

The administration on February 7, 1974 sent its proposed legislation (H.R. 12834 and S. 3094) to Congress to implement Minute No. 242. Its legislation did not include a salinity control program for United States citizens using Colorado River water. Whether the administration's bills can be supported by the Colorado River Basin states will probably depend to a large extent upon the pressures the federal government is willing to take to protect interests within those states.

Because the Nixon Administration failed to support legislation to authorize implementation of a Colorado River Basin salinity control effort for basin states, twelve members of the House of Representatives from those states introduced H. R. 12165 in the Second Session of the 93d Congress. This bill, if enacted, would implement Minute No. 242 of the International Boundary and Water Commission and provide for a salinity control program upstream from Imperial Dam in the United States. A counterpart bill (S. 2940) was introduced in the Senate. Hearings on the House bill were scheduled for March, 1974, by the Interior and Insular Affairs Committee. The seven Colorado River Basin states were prepared to carry the fight for protection of their interests to the U.S. Congress.

Legal Problems

In November, 1970, an organization known as Friends of the Earth, et al., filed a complaint in the U.S. District Court for the District of Columbia to restrict the Secretary of the Interior to filling Lake Powell behind Glen Canyon Dam to a level not to exceed elevation 3606 feet above sea level in order to keep water from entering Rainbow Bridge National Monument, later stipulating that there would be no damage to Rainbow Bridge. Such a restriction would destroy the use of approximately half the 27 million acre-feet of water storage capacity in the reservoir. On February 27, 1973, Judge Willis Ritter of the U.S. District Court for Utah ordered the removal of 1.9 million acre-feet of water above elevation 3606 feet from the reservoir and further ordered the Secretary of the Interior to prevent the water from exceeding that elevation.

On August 2, 1973, the Tenth Circuit Court of Appeals, after having stayed the order of the District Court on May 1, 1973, decided that the reservoir could be utilized to its full design capacity of 27 million acre-feet at elevation 3700 feet above sea level, but left the issue of damages to the Rainbow Bridge under the jurisdiction of the District Court for ten years.

On January 21, 1974, the U. S. Supreme Court refused to grant a *writ of certiorari* to the plaintiffs and allowed the decision of the circuit court to remain in effect.

This lawsuit and its effects cost the taxpayers millions of dollars and delayed filling Lake Powell to its maximum storage capacity and power generating efficiency.

There may be further action in the courts or in Congress with respect to the Rainbow Bridge issue.

On January 7, 1974, the Sierra Club, et al., filed a suit against the Commissioner of Reclamation seeking to prevent construction of certain facilities of the central Utah participating project of the Colorado River Storage Project. The plaintiffs alleged inadequacy of the Environmental Statement required under the National Environmental Policy Act. This lawsuit combined with the one and one-half years lost in construction while the environmental statement was in preparation has cost, and is continuing to cost, many millions of dollars of taxpayers' money.

Recognition having been given by the U.S. Supreme Court in Arizona vs. California to the reservation of water rights dating from the time Indian reservations or other federal reservations such as national forests were created has thrown a legal cloud over many existing water uses in the Colorado River Basin. Federal agencies are in the process of quantifying potential water uses for Indian and other federal reserved lands, presumably for submission to the courts for adjudication. What the ultimate effects of the Supreme Court's decision will be on future water developments in the Colorado River Basin are unknown.

NEW REGIONAL TRENDS

One cannot leave a discussion of water development in the Colorado River Basin without recognizing that in the 1950s and 1960s many water projects have been oriented towards the conservation and utilization of water for purposes other than irrigation. Navigation as a commercial transportation business was long ago abandoned. Recreational, municipal, and industrial purposes have come to the forefront, and in many instances have evolved into paramount purposes supporting water project development.

More and more municipal water has had to be developed for people congregated in growing cities and rural areas. In addition to the Los Angeles-San Diego metropolitan areas, large population centers such as Phoenix, Arizona; Denver, Colorado; Albuquerque, New Mexico; Salt Lake City, Utah, and others of smaller size require more water for consumption by more people. Many of the water projects authorized for construction by Congress in recent years have included works for the development of municipal water supplies. The Central Utah Project, the San Juan-Chama Project of New Mexico, and the Central Arizona Project are prime examples of this trend.

More people with more money to spend have discovered the beauty and space of the Colorado River Basin. Hundreds of millions of dollars are spent every year on recreational pursuits. Hunting, fishing, boating, and camping have become the backbones supporting multimillion dollar businesses. A large proportion of recreational activities revolve about water. Consequently, in response to public demand, many water developments have incorporated recreational facilities as part of the projects involved. This trend received its greatest impetus from the Colorado River Storage Project Act which provided for non-reimbursability of costs of certain recreation features of water projects.

Many new industries have either moved to the Colorado River Basin or have originated there. Water that in 1950 would have been developed only for irrigation is being conserved and utilized by industries that show higher economic and social returns. Water being used for agriculture in the 1970s is also being seriously considered for transfer to uses by other industries. This trend has been particularly noticeable where chemical or electric industries have needed water supplies for manufacturing plants or thermal-electrical powerplants and for people in the area to operate them.

There is a great imbalance in the amount of water available in the Colorado River Basin when compared with other natural resources needing water either directly or indirectly for their extraction or production. As has been mentioned before, in each of the seven basin states the water of the Colorado River has either been put to use or has been committed to future uses of which there are considerably more potentialities than there is water to supply them. This does not mean that there is no water unused or available. Some of the commitments have not materialized, and, in some instances projects that have been authorized for construction, principally for irrigation, have not been initiated or completely constructed. Water is available for other uses if the states and/or Congress should decide not to fulfill the original commitment.

With the nation in the grips of an energy crisis and the upper basin abundantly supplied with such energy resources as coal, oil shale, natural gas, oil, and radioactive elements, it is anticipated that some of the contemplated future uses of water will be changed in order that these resources can be converted to usable forms of energy to fulfill national requirements. It is also anticipated that some of the water being consumed in marginal production of irrigated crops will be transferred to such higher uses in the general welfare as thermal-electric energy generation, coal gasification, oil extraction from shales, production of radioactive elements, etc. and demands of associated people. In the national order of priorities there are many reasons to believe that the production of energy may override the priorities associated with irrigation, recreation, and many manufacturing industries. Sometime in the future perhaps only domestic and municipal consumption of water will demand a higher priority than energy—and some of the municipal uses of water may be forced to lower positions on the list of preferred uses.

Water will be on hand from the Colorado River system for meeting the energy requirements of U.S. citizens simply because of the national interest involved. Water will continue to be used in food and fiber production in the basin for as long as such production is essential and competitively efficient.

Because there is no certain method of augmenting the limited water resource of the river system, future water development in the basin will depend upon sophisticated, intensive mangement whose goal will be the best conservation and utilization of water in order to get the greatest economic and social values from it for the welfare of society.

Literature Cited

Colorado River Commission of the State of California. 1930: *The Boulder Canyon Project "To Convert a Natural Menace Into a National Resource."* San Diego, California.

Fall-Davis Report. 1922. *Problems of Imperial Valley and Vicinity.* A Letter From the Secretary of the Interior. Senate Doc. No. 142, 67th Congress, 2d Session, Washington, D.C.: Government Printing Office.

Ives, J. C. 1861. *Report upon the Colorado River of the West.* Senate Document 90, 36th Congress, 1st Session. War Department, Washington, D.C.

La Rue, E. C. 1916. *Colorado River and its Utilization.* United States Geological Survey, Water Supply Paper 395. Washington, D.C.: Government Printing Office.

U.S. Department of the Interior, Bureau of Reclamation. 1947. *The Colorado River.* Interim report on the status of the investigations authorized to be made by the Boulder Canyon Project Act and the Boulder Canyon Project Adjustment Act. House Doc. 419, 80th Congress, 1st Session. Washington, D.C.: Government Printing Office.

Valantine, Vernon E. 1974. Impacts of Colorado River salinity within the United States. Paper presented to the American Society of Civil Engineering national meeting on Water Resources Engineering 21-25 January, 1974. Los Angeles, California.

Weymouth, F. E. 1924. *Report on the Problems of the Colorado River Basin.* Bureau of Reclamation. Washington, D.C.

3

Politics of Water Allocation

HELEN INGRAM

The locus of decisions on the allocation of water in the Colorado River Basin is not within the basin or even in the capitols of basin states. It is in Washington. This chapter will argue that federal domination of decisions on water allocation has imposed costs upon the basin in terms of the quality of the decision making process and the rationality of water policy.

According to the Supreme Court, Congress initially asserted its power to divide the waters of the Colorado in the Boulder Canyon Project Act of 1928 (Arizona vs. California et al., 373 U. S. 550-52, 595-601, 1963). Whether the court's interpretation of legislative history is correct, and there are arguments that it is not (Hundley, 1972), the long-term effect of federal water development, begun in Hoover Dam, was federal control of allocation. The basin states have been motivated toward legal settlements of water rights by opportunities for federal water development. The Colorado River Compact was a necessity if there was to be a Boulder Canyon Project Act. Anticipation of the Colorado River Storage Project Act motivated upper basin states to allocate rights to percentages of available water among themselves under the Upper Basin Compact. Desire for federal funding of the Central Arizona Project prompted Arizona to request adjudication of her water rights in the Supreme Court. Further, as the latecomers to federal water development, the Indians, Arizona and the upper basin states of Utah and Wyoming have discovered, compact entitlements and court judgements have little significance without federal help in diverting water to use. The broad geographic jurisdiction, availability of funding and expertise of the federal government have eclipsed whatever capacity the states and localities might exercise in constructing water projects.

Federal water development has been characterized as classic distributive politics (Mann, 1973a). In distributive politics, participants have little perception of costs, only benefits. The basic support for specific water projects comes from local backers who anticipate various returns from project features or more broadly from the growth and development generated. Wider support for water development is achieved by stringing projects together in a package. Such combinations in omnibus or basin-wide bills facilitate logrolling behavior where senators and representatives vote favorably on projects outside their state or district in exchange for other legislators' support for constituency projects they may sponsor. Demands for benefits are dealt with by adding projects to the package or new features to projects. For instance, traditional fish and wildlife interests have been accommodated in distributive water projects by fish ladders, fish hatcheries and wildlife sanctuaries. Preservation and environmental groups have disrupted distributive politics because they perceive real cost and because it is more difficult to include items which benefit them.

Many critics, particularly economists, have long been critical of the failure to perceive the real costs to the nation in water development. They have characterized the game as "raid the treasury" where benefits are heaped on the interests of the project area while the more general, diffuse interest of the federal taxpayer suffers. Evaluation criteria have been criticized for exaggerating project benefits, and financial arrangements have been scored for extended pay-out periods and low interest rates. Even when stringent benefit-cost criteria and realistic interest rates are employed, it is argued that the national interest is sacrificed. There are opportunity costs involved in investing federal money in one program rather than another, and some allocation other than water projects might well produce greater public benefits. Further, the large federal bureaucracies with skilled personnel who plan, construct, and manage local projects represent a significant federal resource which might be employed to pursue different national goals.

COSTS TO BASIN OF FEDERAL WATER DEVELOPMENT

The perception of no real costs to the basin in federal water development may be disputed. The argument here is that the promised benefits of federal water politics have lured the basin into following the old adage of not looking a gift horse in the mouth. In accepting federal largesse the basin has incurred substantial costs. There have been costs involved in foregoing choice. The basin has made sacrifices in information, and has had to act without sufficient knowledge of the implication of its actions. The states in the basin have forfeited the development of their own independent planning and decision making capability. Finally, there have been real costs involved for the basin of the fragmented, inconsistent water development pursued under federal sponsorship.

As a general proposition, the quality of the best possible policymaking increases in direct proportion to available policy knowledge (Dror, 1968). Similarly for any individual or set of decision makers, the possibility for action consistent with self-interest is a function of choice and information. The wider the range of alternatives available and the more complete the information about the implications of various alternatives, the greater the likelihood of rational action. Conversely, the quality of decision making is likely to decline where choice and information are not available, the process of information gathering is curtailed, and decision makers face high levels of uncertainty.

Existing histories and analysis of two landmark pieces of legislation—the Colorado River Storage Project Act of 1956 (CRSPA) and the Colorado River Basin Act of 1968 (CRB Act)—will provide examples to illustrate the constraints under which the Colorado River Basin has been placed in federal water politics. The earlier act mainly affected the upper basin. Four major storage projects, Glen Canyon, Navajo, Flaming Gorge and a conditional Curecanti, were authorized. The more recent CRB Act included the Central Arizona Project (CAP) in the lower basin, but projects affecting Utah, Colorado, New Mexico and Nevada were also packaged into the legislation. Together these two pieces of legislation made large numbers of decisions about allocation of water in the Colorado Basin which will be terribly difficult, if not impossible, to reconsider (National Academy of Science, 1968).

Limitations of Choice Imposed on Regional Equity

Western interests perceive water projects as the vehicle through which they get their fair slice of a much larger national subsidies pie that is distributed across the country:

> That the West has historically relied on water projects to balance the equities in the political system that has seen tariffs provided for industry, price supports for midwestern and southern agriculture and public projects and welfare programs for urban interests seems beyond question (Mann, 1973a).

Where reclamation projects are perceived as the share of federal largesse, there is reluctance among recipients to be choosey about the wrappings of the gift.

In the eyes of western congressmen who negotiate bargains on packages of reclamation projects and shepherd them through committee and floor action, federal aid to western water development is a vehicle to obtain equity for eastern flood control. Dean Mann quotes Senator Arthur Watkins of Utah in a heated exchange with Pennsylvania Senator John Saylor during the hearings on the CRSPA:

> For many years, a long time before reclamation was started by the federal Government . . . the people of this United States, the taxpayers everywhere, have been contributing to flood control. As I remember, they did contribute to flood control in the very district the Congressman comes from up in Johnstown.... All we want to do is to get out the water, which is beyond our capacity to do. We do repay the costs, but the people in the East do not repay the cost of flood control (Mann, 1973b, p. 404).

The Interior and Insular Affairs committees (with jurisdiction over reclamation) are stacked with legislators who see their constituency interests served in these strategic posts. Data compiled from Congressional Directories, 80th Congress, 1947–48, through the 90th Congress, 1967–68, show that more than one-half of the members of the House Committee came from the seventeen western states during the period between 1947 and 1968. In the Senate, where sparsely populated western states have equal representation, the proportion of westerners on the Interior Committee is even larger.

If regional residents had a choice, they might prefer that federal contributions to regional development and welfare come in the form of military installations, highways, hospitals or welfare programs. Such a choice is not, in fact, presented. Once regional delegates to the national legislature have committed their careers to a specialization in water, the influence they might have developed in other policy arenas is necessarily foregone. If states and localities turn down a project, they cannot expect a compensatory federal investment to be made under another program. As Bromley, Schmid and Lord (1971, p. 4) have said about water development, "the political system within which piecemeal decisions are made, and especially the cost-sharing rules now in force, militate against a local realization that all federally-financed projects and programs are mutually competitive in the budgetary sense; we indeed have a tyranny of small decisions."

In order for water development projects to fulfill the function of equity to the West, any perception which localities might have of economic costs to themselves must be limited. Cost sharing and repayment provisions have traditionally minimized or disguised the obligation of the project area. From the time of the original reclamation act in 1902 up until the passage of the Colorado River Storage Project Act, extension of the repayment periods and liberalization of interest charges provided attractive lures to local and state sponsorships of water projects. Since localities have felt they were getting more than they were spending, they have not seen water projects as a trade-off against other local investment. For irrigation interests, the national reclamation program has provided means to transfer costs to others. Basin-wide accounting systems have allowed the application of power revenues to the repayment of irrigation costs rather than the reduction of power rates. Because the impact is indirect and diffuse, electric energy consumers have scarcely noted the hidden subsidy.

Other vehicles for regional equity, besides federal development of water resources, might have extended and clarified choice. Revenue sharing or bloc grants provide a clearer accounting of the way in which various regions benefit from federal revenue. States or localities could choose among various priorities in the expenditure of federal funds. Alternatively, if federal financial involvement were lessened or withdrawn, increased state and local cost sharing would encourage investment choices on the basis of state and local priorities and budgetary restraints.

Limitations on Choice Imposed by Federal Initiatives

The key choice for residents of the basin to make is among alternative futures. Is the basin to experience continued population growth and industrial expansion, or is it to maintain wide open spaces with low population densities and a limited number of industries? Is the basin to exploit its rich mineral resources for use of the nation as a whole, or is it going to forego development to preserve its relatively unspoiled environment? Is the existing distribution of wealth to be maintained or will resource development be used to redistribute wealth to less privileged groups including Indians and marginal farmers? Specific water development projects make sense when they provide means of achieving the desired future. In the politics of federal water development, the focus from the very beginning is upon specific local projects, shortcutting a general discussion of where the basin should or could be going. The issue is whether or not a state or locality will achieve the reward of a federal water project.

Federal agencies typically play the crucial role in selecting the projects to which locally oriented interests attach themselves. Various water agencies have, since the turn of the century, been identifying dam sites and studying various water works which might be built. One of the initiatives for the CRSPA was a Bureau of Reclamation reconnaissance planning report. The so-called "Krug Report" listed a possible 134 potential water development projects (Mann, 1973b). Similarly, the Pacific Southwest Water Plan set the stage for political bargaining on the Colorado River Basin Bill (Englebert, 1965). Without question, potential local support is an important consideration in agency study and recommendation of specific projects. At the same time, quite separate considerations of agency doctrine and mission affects what the Bureau of Reclamation aspires to construct. The Bureau has preferred large dam projects worthy of the skills of its engineers. The agency has shied away from projects where the negotiation of contracts and the coercion of beneficiaries to pay their share of costs promise to be difficult.

There is ample evidence that the state of Colorado has sometimes seen more of Bureau of Reclamation interest than state interest in the Bureau's studies and reports. Colorado water officials, according to Mann, commented on the draft "Krug Report" as follows:

> Colorado says that the so-called potential projects listed in Colorado are as yet largely uninvestigated by the Bureau, that the data contained in the report concerning such possible projects is (sic) inadequate for the purpose of determining their feasibility and desirability (Mann, 1973b, p. 31).

During Congressional Hearings in 1954, water officials continued to complain bitterly about the Bureau's penchant for working on big downstream reservoirs and power plants:

> Frankly, we have been and are of the opinion that the reclamation officials have been so engrossed with their large downstream reservoirs and power plants that they seriously neglected plans to make the water involved available for consumptive use (Mann, 1973b, p. 163).

Bias of Choice Introduced by Federal Evaluation

Federal authorization and appropriations have meant compliance with federal evaluation procedures. Although numerous attempts have been made to provide multiple objective evaluations including broad social and environmental considerations, in actual practice decision makers have tended to return to national economic efficiency as the primary criterion (Peterson, et al., 1971). While benefit-cost analysis has sometimes been manipulated to approve politically desirable projects, the need to somehow meet the test has structured the consideration of water projects (Wildavsky, 1968).

Cost-benefit analysis has been used so persistently because it serves a variety of federal interests. Distributive politics is made a good deal more reputable when it can be argued by participants that all projects are good projects with returns which exceed their costs. The benefit-cost ratio gives federal agencies a way of eliminating projects with local support but with promise to cause long-range contract, pay out and management problems. The Office of Management and Budget is partisan to benefit-cost analysis because it provides a clear-cut rationale for turning back water projects. Congress in general is given a means of eliminating those projects with such poor features that if endorsed would discredit its decision making. The individual congressman sponsoring a constituency project which fails the benefit-cost has an acceptable excuse for not getting his project authorized (Wildavsky, 1968).

The benefit-cost analysis has not served the interests of the basin in the same way it has aided actors on the federal level. The impact of the go/no-go determination has been a bias toward certain allocations attractive to the basin mainly because a better justification can be made for national economic efficiency. Social objectives have sometimes been forgone to satisfy national standards of evaluation. The Animas-La Plata project, a part of the Colorado River Basin Bill package, provides a not uncommon illustration. The local support for the project came from farmers who hoped to improve the quality of their lives.

In answering his own question, "Why do people stay with the soil, and suffer hardships caused by poor crops which [are] due to lack of enough moisture," Paulek of the La Plata Conservancy District said,

> They see the water flowing down the rivers from the snow melting high in the mountains and have hopes of getting a project, such as the Animas-La Plata project, constructed which would make it possible to apply the water to the excellent loam soil and produce sufficient crops to make an adequate living . . . . Our Fathers before us have sought for projects to stabilize the flow of the river. We are still living in faith that sufficient water can come to this area so that our children, and our children's children can have some of the advantages we have missed (Ingram, 1969, p. 36).

The Southern Ute Indians and the Ute Mountain Ute Indian Tribe also expected to better their existence through the Animas-La Plata project. Chief Jack Horse told a House Interior Subcommittee:

> My tribe has had a feasibility study done of the Fort Lewis Mesa area of the Animas-La Plata project and it showed that lack of water is the only thing that is keeping my tribe from developing a tribal herd of cattle on these lands. Such a herd of cattle would certainly be of benefit to my people. Also, we are very strongly interested in the recreation and other benefits that the tribe can have by use of the Meadows Reservoir of the Animas-La Plata project (Ingram, 1969, p. 41).

The feasibility report recommended a multipurpose Animas-La Plata project where irrigation benefits were to be primary, including some non-reimbursable benefits to Indian lands. Without taking into account the indirect social benefits, the benefit-cost ratio was below unity, 0.97 to 1. On the basis of insufficient economic justification, the Bureau of Budget (now the Office of Management and Budget) rejected the project. This action was particularly disappointing to Congressman Wayne Aspinall of Colorado, who saw the Animas-La Plata and four other projects as the means to put to use the remainder of Colorado's entitlement in the upper basin allotment. Aspinall feared that because of shortages in the basin, the Central Arizona Project could not be operated without borrowing the unused water flowing down the river to which the upper basin was entitled. Once Arizona put the water to use, Aspinall suspected it would be impossible to reclaim it. Consequently the Congressman, whose position as Chairman of the House Interior and Insular Affairs Committee gave him political clout, threatened to block the Central Arizona Project unless the five Colorado projects were authorized along with it. The Bureau of Reclamation was hurriedly set to rework the project.

In order to improve the benefit-cost ratio and skirt the opposition of the Budget Bureau, substantial additions were made to the allocation for municipal and industrial purposes. Marginal land, much of it belonging to the Indians, was dropped from the irrigation service area. The city of Farmington was approached by the Bureau of Reclamation and it agreed to purchase 15,000 acre-feet of municipal water in order to improve the feasibility of the Animas-La Plata and four other projects as the means to put to use the remain-industrial expansion and population growth. There was little time or opportunity to consider whether this growth would or should occur. The deprivation of Indian irrigation benefits was to be made up for by anticipated coal development and a portion of industrial allocation in the Animas-La Plata project was supposed to go to Peabody Coal development on the Ute reservation. The Indians' avenues for economic development were thus narrowed. Agricultural development was not to take place. There was very little time to debate, or information to consider the implications of this change.

Information Affected by Federal Standards

Not only have the criteria for national evaluation biased the allocation of benefits on water projects, but also the need to prove favorable benefit-cost ratios has tempered questions and suppressed debate on larger questions of basin development. From a policy standpoint, the most rational course for basin decision makers would have been to measure off all the potential projects in the upper and lower basin against a realistic projection of streamflows. Decision makers would then know how many projects could be built and could decide which had high priority. However, as the CRSPA illustrates, the need to pass muster under federal criteria makes it terribly difficult for the basin to consider future basin plans along with present proposals.

According to Stratton, Owen and Sirotkin (1959), it was impossible for the upper basin states to demonstrate the significance of evaporation savings during the debate on the Colorado River Storage Project Act. To do so would have taken the top off a hornet's nest of attacks on the economics of upper basin development. All the potential projects which various localities hoped might eventually get federal aid in the distributive process, but upon which economic justification had not yet been constructed, would have been discredited as not contributing to national economic efficiency. Basin-wide support for the CRSPA depended upon faith that interests not served in the particular bill would be served later. It would also have damaged basin unity essential to the building of congressional majorities. It can be argued that benefit-costs analysis is supposed to discourage the consideration of projects which do not contribute to national economic efficiency. The effect here, though, was simply to postpone consideration of projects and contribute to piecemeal decision-making on the basis of limited information.

Information Limited by Pressure Toward Unity

In American politics reliance is generally placed on contests to clarify alternatives and to generate information. Electoral campaigns and legislative debates are based on the notion that adversary proceedings are a good way to uncover the public interest. Water politics does not follow this prescription. The political challenge in distributive politics is to project support for a water development proposal, which is essentially local, into national consent. The usual strategy is to fashion agreement at each stage in the authorizations and appropriations process before moving the proposal onto the next stage. Strong disagreement is devastating to a project's fortunes.

Absence of controversy is especially necessary on the local level in order to succeed in federal water politics. A strong, united local pressure is necessary to move from early planning to actual construction. Construction agency officials estimate the strength and depth of local commitment to a project in determining how much work goes into planning and evaluation (Ingram, 1972). Evidence of local enthusiasm is very important at congressional hearings. Hearing records are filled with supporting testimony from local public office holders, businessmen, water district officials, chambers of commerce, and so forth. The pressure toward project area unity and agreement which is generated by federal water politics puts a strong damper on debate. The treatment of the Central Arizona Project within the State of Arizona during the period when the CRB Act was under active consideration by Congress is a case in point.

In the major newspapers of Tucson and Phoenix, the Central Arizona Project was a prominent feature of news analyses, editorial pages and letters to the editor from 1965 until the project was authorized in 1968. Substantially all of the coverage was in favor of the project. Whatever doubts were expressed by Arizonans about the need for or the justification of the project met with stiff rebukes. The *Arizona Review,* published at the University of Arizona, printed an article by two agricultural economists, Robert Young and William Martin, early in 1967. On the basis of research they asserted that Arizona had enough ground water to last 170 years and that redistribution of water from agricultural to municipal and industrial uses was preferable to the CAP. *The Arizona Daily Star* in Tucson accused the economists of advocating water cannibalism and asserted that the best thing for Arizona would be a new source of water and a great deal less noise from critics Martin and Young. When the representatives of a small, conservation oriented group took the Martin and Young article to congressional hearings, they were essentially disowned by Congressman Morris Udall of Arizona, the core activist behind the CAP. He stated, ''I think you people have done Arizona a disservice'' (Ingram, 1969, p. 21).

Basin-wide agreement is as important in getting a reclamation package through Congress as local agreement is in getting a single project into the package. Wayne Aspinall spelled out this rule of federal water politics to Arizona water leaders in 1964.

> Finally, I cannot emphasize enough the importance of unity with Arizona and agreement among basin states, and especially Arizona and California. My committee and the Congress have been following a policy of not deciding differences within a state and hesitate to consider a basin water development program when there is a serious controversy between or among the states involved. The problems of successfully moving a large reclamation program through the House of Representatives are so great under the best of conditions that the addition of a serious intrabasin controversy would present a very difficult task (Ingram, 1969, p. 26).

The need to preserve basin unity has discouraged the basin from addressing and making determinations on fundamental matters such as water supply. Within the basin, only California, which has used far more than its entitlement, has been persistently concerned whether there would be enough water for proposed development. They have feared sharing shortages. In both 1956 and 1968 most other basin actors chose to protect their allocations by getting projects authorized. At one time in debate on the CRB, Congressman Aspinall raised the spectre that the CAP would run short when the upper basin got its projects constructed. The Arizonas responded that instream storage would tide the basin over during the period when the CAP became operative but inter-basin transfers had not yet been constructed. Congressmen John Rhodes and Morris Udall responded personally in a letter to Aspinall expressing the hope that the basin states would not be caught up in a numbers game on Colorado River water. They noted that the assessment of available water depended very much upon the period of recorded history of river flows chosen for analysis. Arizonans preferred the period 1906-59 from which the largest supply could be projected (Ingram, 1969).

In order to achieve the necessary basin unity, the tendency has been toward mutual accommodation. Each state has accepted whatever projects in other states have support and possibility for authorization. Again, the effect has been to suppress full discussion of basin-wide development options.

Dean Mann cites a statement from Senator Arthur Watkins in the legislative history of the CRSPA which illustrates the predisposition of basin leaders toward mutual accommodation.

> Of course, if the State of Wyoming wants to put the water that is allocated to it by the Colorado River Compact to use on farms, that ought to be the business of Wyoming. Wyoming has recommended the Eden project. If you were just going to take the Eden project all by itself and stand it out here, you could say, "That does not seem feasible." But, when you consider the fact that the waters of Wyoming, in order to have title to them finally, must be put to a beneficial use, Wyoming would be wise and the people of that state would be all wise if they would say, "As far as we are concerned—and it should not be anyone else's business as long as we pay our way—we are willing to pay back the costs of this project, the power users and also the farmers" (Mann, 1973b, p. 29).

Information Limited By Pressure for Early Action

Senator Watkin's statement illustrates another imperative of federal water politics which works to the disadvantage of full debate and airing of information. States have a strong motivation to sew up entitlements, that is, to make sure the federal money does not result in a state's water being forever lost to allocations outside the state's boundaries. There are compelling political reasons for states to settle upon whatever project might pass muster under federal evaluation procedures to put in an omnibus basin bill which is up for consideration. The effect is to rush projects into authorization or conditional authorization pending further study, without sufficient time or information. The hasty revision of the Animas-La Plata project for inclusion in the CRB Act has already been recounted. In fact, all but one of the five Colorado projects inserted to use up the remainder of Colorado's allocation were either months or years away from being ready for authorization under normal timing. The Bureau of Reclamation went on overtime schedule to prepare these projects to accompany the CAP (Ingram, 1969).

The Hooker project in New Mexico provides another example of how pressure for action limits information. It was the means by which New Mexico was to use an additional 1,800 acre-feet of water bargained from Arizona. It or "a suitable alternative" was authorized in the CRB Act. Under such a process it was impossible for the people of New Mexico or the basin to know what they were getting. While the Hooker Dam reservoir threatened to back water into the Gila Wilderness Area, perhaps an alternative would not involve a wilderness invasion. Further, studies only a little better than reconnaissance level had been completed on Hooker Dam as a basis for decision-making.

Lack of Incentive for State Planning

Dependence upon federal water development and focus upon the complexities of federal water politics have discouraged states from developing their own planning capabilities. Except for California and perhaps Colorado, states in the basin have not had strong water planning agencies. Ival Goslin, Executive Secretary of the Upper Colorado River Basin, speaking generally about states, said in 1964:

> Many states have poor organizations for long-range planning and their water resource agencies lack financial support. Some states even appear to lack the proper agencies that can do their share in the overall planning job. In many instances, initiative in planning rests with federal agencies. State and local governments are often in the position of having to approve or disapprove plans without having made adequate studies for major decisions needed in the field of water resources (U. S. Senate, 1964, pp. 196-7).

Even as late as 1973, only two of the basin states reported to have completed state water plans. Although substantial gains were recognized in the last ten years, partly with the help of federal grants to state water planning, most states admitted that they had not yet developed the capability for an independent position vis-à-vis the federal government (Ingram, Bradley and Ingersoll, 1973).

Part of the basin states' lack of planning capability must be ascribed to general lack of financial resources. As important have been the imperatives of federal water politics. Federal water agencies have often related directly to local project areas and irrigation districts, shortcutting state involvement. More important, state water planners, understaffed and underfinanced, have felt outnumbered and outclassed in basin planning efforts. According to Felix Sparks (1969), Director of the Colorado Water Conservation Board, ''States have been content to let someone else do their planning for them and cry about the results.''

Even in the states that have managed to construct water agencies with staffs and funds, there have been difficulties in developing an independent state position. The orientation of many offices of the state engineer, interstate stream commissions or state water boards has been to cooperate with federal agencies and state congressional delegations in pushing projects through the authorization process. The dominant questions have been how to get the Bureau of Reclamation to pay attention to the planning of projects within the state; or how to marshal the governor and state congressional delegation at House of Representatives and Senate hearings. The focus of state attention has not been, at least during the period of the CRSPA and the CRB Act, upon identifying the goals and objectives of the state and determining how water development might serve them.

Inadequacies of Federal Water Policies Pursued

The kind of legislation which emerges from Congress at work at distributive politics reflects the process by which it was put together. The constraints on choice and the limitations on information have had an adverse effect upon the quality of legislation, especially from the viewpoint of basin interest. Neither the Colorado River Storage Project Act nor the Colorado River Basin Bill paid sufficient attention to the physical limits of the basin. The constraints of available water supply were not adequately recognized. There has been more difficulty in filling Lake Powell, the reservoir behind Glen Canyon Dam authorized by the CRSPA, than anyone envisioned. Fantastic dodges, such as making the Mexican Water Treaty a national obligation, were used to avoid facing the limits of supply in the CRB Act. The result has been continuing bitter complaints about the quality of water supplied to Mexico, which, if it comes in promised quantities, has high levels of salinity. It appears that the federal government may underwrite a desalinization plant on the Gulf of California in order to fulfill the U.S. obligation.

More general environmental impacts were not taken into account. Only where conservation groups made an issue of invasion of parks and monuments such as the Echo Park encroachment upon Dinosaur National Monument, or Marble and Bridge canyon dams in the Grand Canyon, were the effects of water development on the natural environment recognized and discussed. Information on the irreversible effects of some water resource investments which destroy or permanently alter natural environments ordinarily was not a part of the decision making process.

Neither of the two fundamental pieces of water legislation considered here was very logical or consistent. They encompassed no overall plan of development. Different projects and provisions were not complementary. The result is a river which is overcommitted and overmanaged. The crisis for the upper basin prompted by the short-lived district court decision to limit the filling of Lake Powell to protect Rainbow Bridge is a recent illustration of how tenuous and fragile arrangements are. The root of the problem was the contradiction built into the CRSPA which promised to have reimbursable expenditures repaid and Rainbow Bridge protected.

EVIDENCE OF CHANGE

There is convincing evidence that the grip which federal water politics has had upon the Colorado River Basin is loosening. For a variety of reasons, the water development bazaar has far fewer prizes to offer. The proportion of federal investment in western water development has shown a decline. The growth of grass roots environmental movements has caused state officials to have second thoughts about projects which threaten degradation of natural areas in the wake of economic development. State water planning agencies show increasing capability and independence. For instance, Wesley Steiner (1971), Director of the Arizona Water Commission, contended that the capability of his state had developed impressively. In the past, federal government planning was dominant and the Bureau of Reclamation established goals for the state. As of 1974, Arizona was claiming the right to allocate its own water resources.

An increasing number of water projects in the West are privately financed and developed. The report of the National Water Commission, even if only partially implemented, bodes for even further changes away from federal projects. The Commissioners recommended that beneficiaries be required to pay the full cost of federal development under most circumstances. The Commission asserted that a great many water development decisions should be turned back to the states (U.S. National Water Commission, 1973). In the view of some, the West has been reclaimed, and it is time for federal investment to be directed elsewhere. More realistically, the basin may be said to be taking over the mastery of its own house after very many remodelings, not always designed to meet the needs of the owners, had already taken place.

At a time when the basin has finally begun to assert a real independence from the Bureau of Reclamation and federal water development, new lures are being cast to attract basin decision-makers. The development of oil shale, which involves the use of large quantities of water, is one opportunity. The strip mining of coal for electic energy generation and coal gasification, also water consumptive, is another possibility for the vast coal reserves of the basin. Because of the relatively low density of the West, its mineral endowments, and its attachment to growth and economic development, the Colorado River Basin is likely to be presented with a number of schemes to put its natural resources to work. The danger is an incoherent, inconsistent development of other natural resources as well as water.

Unfortunately much of the future planning is taking place on the federal level or within corporations, often based outside the Colorado River Basin, which basin decision-makers find difficult to hold accountable. Hopefully, the basin has learned from the politics of federal water development that manna sometimes turns into mirage. At the very least, politicians in the region should know that the perspective of decision-makers outside the basin is likely to be different, if only because they do not have to live with the long-range physical consequences of projects constructed. Whatever proposals are placed before the basin should be subjected to full scrutiny of debate. The claim that localities will lose out and forego great opportunities if they do not immediately take the bait should be too old to attract. Even more rational behavior, on the basis of past experience, would be a basin-wide setting of goals and objectives so that proposed resource projects could be evaluated in light of the kind of future the Colorado River Basin envisions for itself.

Acknowledgment
Without intending to transfer responsibility for the argument made here to others, the author wishes to acknowledge helpful comments made by Paul R. Portney, Resources for the Future, and Robert McCain, Lake Powell Research Project.

Literature Cited

Bromley, Daniel W., Schmid, A. Allan, and Lord, William B. 1971. *Public Water Resources Project Planning and Evaluation: Impacts, Incidence, and Institutions.* Working Paper No. 1, Center for Resource Policy Studies and Programs, School of Natural Resources. Madison: University of Wisconsin.

Dror, Yehezkel. 1968. *Public Policy Reexamined.* San Francisco: Chandler Publishing Company.

Englebert, Ernest A. 1965. The origins of the Pacific Southwest water plan. In *New Horizons for Resources Research: Issues and Methodology.* Boulder: University of Colorado Press.

Hundley, Norris, Jr. 1972. Clio nods: Arizona vs. California and the Boulder Canyon Act—a reassessment. *Western Historical Quarterly* 3(1):19-51.

Ingram, Helen. 1969. *Patterns of Politics in Water Resource Development: A Case Study of New Mexico's Role in the Colorado River Basin Bill.* Division of Government Research. Albuquerque: University of New Mexico.

——. 1972. The changing decision rules in the politics of water development. *American Water Resources Bulletin* 8(6):1178.

Ingram, Helen, Bradley, Michael, and Ingersoll, David. 1973. *An Evaluation of Title III Water Resources Planning Grants to States.* United States Water Resources Council, Washington, D.C.

Mann, Dean E. 1973a. Political incentives in U. S. water policy: the changing emphasis on distributive and regulatory politics. Paper, August 1973, for International Political Science Association.

——. 1973b. Colorado River storage project. Unpublished manuscript prepared for the Lake Powell Research Project, National Science Foundation (RANN). Washington, D.C.: U.S. Government Printing Office.

National Academy of Sciences. 1968. *Water and Choice in the Colorado Basin.* Washington, D. C.

Peterson, Dean F., et al. 1971. *Water Resources Planning and Social Goals: Conceptualization Toward a New Methodology.* Logan: Utah Water Research Laboratory Publication PRWG-94-1. NTIS NO. PB 204228.

Sparks, Felix. 1969. Unpublished speech. Colorado State University, 10 September 1969, Fort Collins.

Steiner, Wesley. 1971. Interview conducted for the National Water Commission and notes retained in Helen Ingram files.

Stratton, Owen, and Sirotkin, Phillip. 1959. *The Echo Park Controversy.* University: University of Alabama Press.

U. S. National Water Commission. 1973. *Water Policies of the Future.* Washington, D. C. : U. S. Government Printing Office,

U. S. Senate, Committee on Interior and Insular Affairs. 1964. *Hearings on the Water Resources Planning Act.* 88th Congress, 2d Session. Washington, D. C.

Wildavsky, Aaron. 1968. The political economy of efficiency: cost-benefit analysis, systems analysis, and program budgeting. In *Political Science and Public Policy,* ed., Austin Ranney. p. 64. Chicago: Markham Publishing Co.

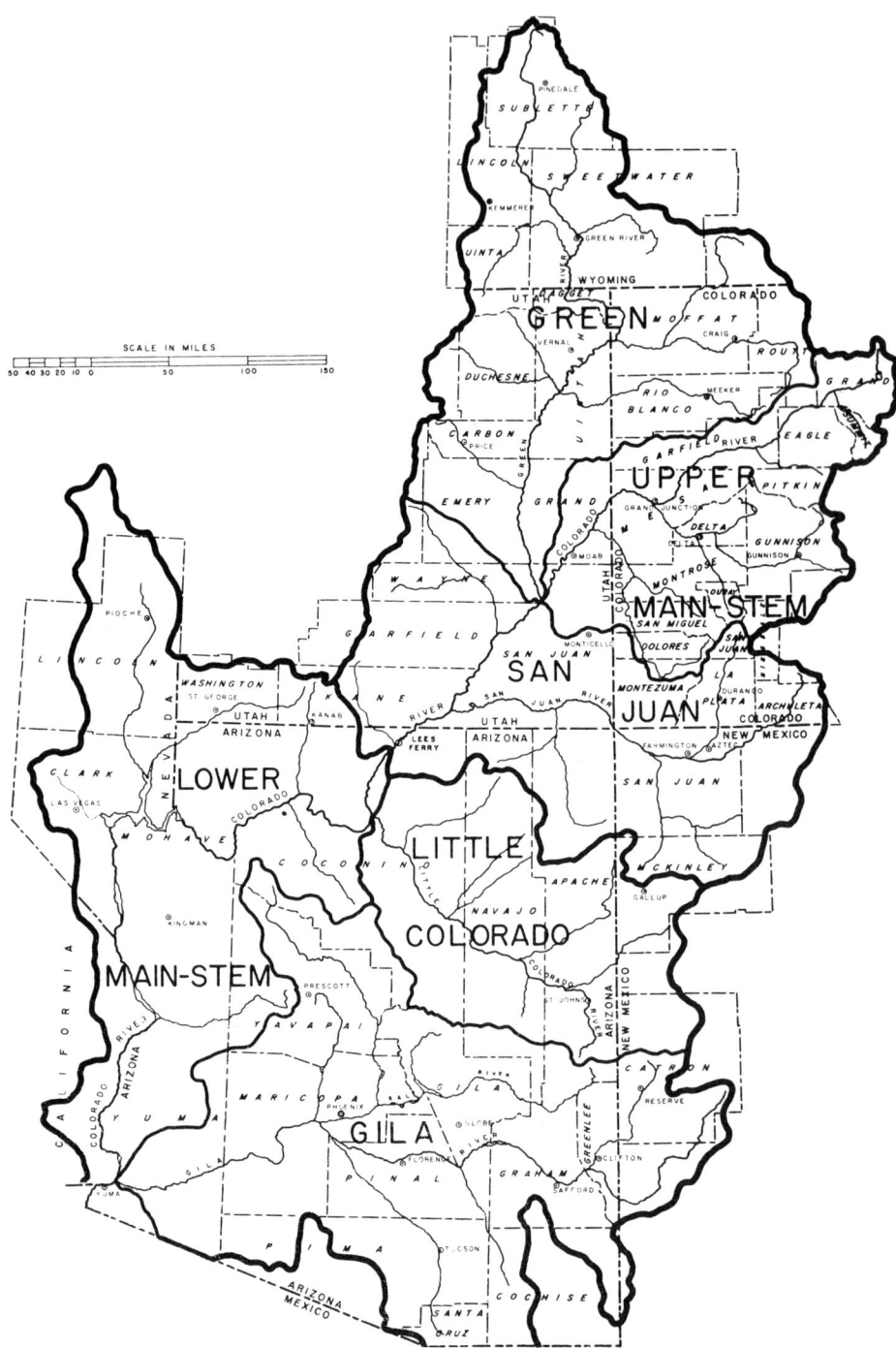

Fig. 4.1. Major sub-basins of the Colorado River Basin.

4

Comparative Analysis of Employment Change

BERNARD UDIS and ROGER KRAYNICK

There are almost too many options open to someone charged with the task of analyzing economic change in the Colorado River Basin (CRB). One could trace the early history of the region primarily as an exporter of minerals, agricultural raw materials and so forth, and an importer of most of its manufactured products, and then in the latter part of the twentieth century as a provider of scenic and recreation activities for tourists. We have chosen, however, to stress changes in the industrial structure of employment, following Wilbur R. Thompson (1973), whose belief in the usefulness of such data has led him to observe: "Let me know your industry mix and I will tell your fortune."

It is probably true that detailed information about the industrial structure of a region and how its people earn their living will reveal much about that region, its characteristics, and its problems. Our earlier work on the economy of the Colorado River Basin emphasized interindustry analysis of the industrial structure. In this effort, we focus on the changing industrial distribution of employment and the so-called shift/share method to analyze the components of such changes. However, it may be of interest before introducing this analysis to present some general characteristics of the changes which have occurred in regional employment. Table 4.1 indicates the percentage change in employment, decade by decade in nine major industrial categories and certain of their components over the 1940–1970 period for the Colorado River Basin and three other geographic regions (Fig. 4.1): the United States, the six states of the Colorado Basin, and the aggregate of the counties of the six states not included in the Colorado Basin (hereafter referred to as the Non-CRB

TABLE 4.1
Comparative Percentage Change in Employment By Decade, 1940–1970

Sector	COLORADO RIVER BASIN			Non-CRB AREA OF SIX STATES		
	1940–50	1950–60	1960–70	1940–50	1950–60	1960–70
Agriculture	− 0.54%	− 19.68%	− 31.36%	− 5.36%	− 32.37%	− 23.96%
Mining	− 3.78	30.97	2.37	− 11.04	36.99	6.39
Contract Construction	108.31	88.87	16.68	133.17	18.58	2.28
All Manufacturing*	51.58	142.55	60.17	64.54	75.26	11.67
Food	61.28	95.11	− 29.03	37.77	44.98	− 20.02
Textile & Apparel	− 38.86	10.53	82.63	48.34	40.24	86.10
Lumber, Wood, Furniture	45.91	12.34	3.94	45.19	6.55	3.37
Printing & Publishing	73.36	115.10	12.75	56.51	51.33	4.52
Chemical & Allied	193.07	78.02	65.44	152.22	35.29	− 0.43
All Machinery	213.54	754.86	236.95	150.49	157.66	113.41
Transportation Equip.	117.39	1961.13	− 24.91	125.00	540.56	55.03
Manufacturing, N.E.C.†	92.94	144.20	57.86	71.52	87.18	4.81
All Transportation*	57.97	7.96	7.21	43.95	0.92	− 3.36
Railroad	42.12	− 26.62	− 33.14	26.28	− 26.17	− 35.20
Trucking & Warehousing	63.36	82.33	11.47	60.49	55.61	9.09
Transportation, N.E.C.	193.73	67.88	81.03	118.95	16.31	44.04
Communication & Utilities*	148.91	73.76	41.65	88.19	37.75	27.61
Communications	155.63	104.54	35.09	87.90	39.05	21.82
Utilities	145.34	56.76	46.37	88.46	36.52	33.16
All Trade*	83.02	66.19	46.72	57.78	31.72	33.37
Wholesale Trade	84.73	71.23	49.32	75.17	28.12	44.83
Retail Food Stores	46.91	51.87	26.25	36.78	20.27	3.77
Eating & Drinking Places	105.47	51.67	49.55	75.52	33.07	45.04
Retail, N.E.C.	87.78	74.44	50.11	54.31	35.90	33.51
All Services*	69.52	109.09	71.13	46.65	64.11	58.93
Finance, Insurance & Real Estate	133.56	187.93	60.44	62.33	78.49	43.10
Lodging & Personal Services	63.51	93.78	63.96	27.33	25.73	24.48
Household Services	6.87	76.20	− 37.82	− 13.92	61.67	− 44.37
Business & Repair Services	101.41	66.69	88.96	65.55	52.86	42.02
Recreation Services	105.70	70.21	54.68	72.79	36.81	49.88
Professional Services	75.28	123.48	99.39	66.11	80.89	95.11
Public Administration	98.10	91.89	49.25	122.80	56.81	30.75
Total Employment Percentage Change	47.69	64.04	41.51	44.91	33.60	28.04

SIX STATES OF CRB   UNITED STATES

Sector	1940–50	1950–60	1960–70	1940–50	1950–60	1960–70
Agriculture	−3.82%	−28.18%	−26.69%	−17.55%	−36.60%	−36.74%
Mining	**−7.83**	**34.22**	**2.45**	**1.31**	**27.87**	**6.61**
Contract Construction	**125.99**	**37.29**	**7.55**	**67.07**	**13.30**	**16.54**
All Manufacturing*	**61.53**	**89.90**	**25.14**	**37.80**	**23.12**	**8.14**
Food	41.67	54.45	−22.17	27.81	32.30	26.95
Textile & Apparel	−8.11	27.44	84.81	18.05	5.86	0.81
Lumber, Wood, Furniture	45.49	1.31	3.63	26.64	7.95	−12.95
Printing & Publishing	59.90	65.24	6.86	35.21	36.97	0.63
Chemical & Allied	160.02	44.49	17.03	49.74	34.60	9.46
All Machinery	159.18	257.24	162.71	94.28	50.47	23.14
Transportation Equip.	132.11	771.06	24.34	53.23	38.06	15.09
Manufacturing, N.E.C.†	74.85	96.96	8.51	32.83	21.89	17.78
All Transportation*	**47.28**	**1.34**	**−0.50**	**35.10**	**−4.80**	**−.13**
Railroads	30.24	−26.29	−34.64	22.04	−30.40	−34.68
Trucking & Warehousing	61.15	61.83	9.72	38.77	33.15	14.87
Transportation, N.E.C.	131.66	27.42	54.54	59.04	5.47	21.91
Communications & Utilities*	**102.69**	**48.31**	**32.43**	**59.22**	**17.97**	**31.18**
Communications	100.29	54.34	25.92	79.99	18.53	25.74
Utilities	104.67	43.43	38.09	44.18	17.47	36.14
All Trade*	**63.99**	**41.17**	**37.68**	**39.89**	**14.36**	**25.04**
Wholesale Trade	77.40	38.58	46.17	63.80	14.62	34.42
Retail Food Stores	39.38	28.83	10.95	15.29	0.61	9.35
Eating & Drinking Places	84.14	39.04	46.62	51.01	9.25	23.58
Retail, N.E.C.	62.14	46.34	38.87	38.66	20.51	26.42
All Services*	**52.39**	**76.67**	**62.96**	**19.14**	**38.65**	**41.86**
Finance, Insurance & Real Estate	74.76	104.02	48.81	30.62	44.01	37.20
Lodging & Personal Services	37.07	47.60	41.14	10.14	7.06	19.06
Household Services	−8.01	66.47	−42.08	−29.86	20.02	42.28
Business & Repair Services	74.27	56.75	56.05	51.34	25.90	44.64
Recreation Services	82.13	47.52	51.65	24.58	4.34	20.57
Professional Services	68.39	91.90	96.40	43.98	62.15	71.47
Public Administration	**116.74**	**64.68**	**35.58**	**69.30**	**30.76**	**26.31**
Total Employment Percentage Change	45.69	42.18	32.42	25.25	14.51	18.62

*Percentage changes in employment for broad industrial categories such as All Manufacturing, All Transportation, etc., are not to be considered as sub-totals because they do not equal the unweighted sum of the relative employment change for the component industries. This is due to the differential importance of these components in the totals.
†N.E.C. = Not Elsewhere Classified.

TABLE 4.2
Representative Counties in the Colorado River Basin by States

Sub-Basin	Colorado	Utah	Wyoming	New Mexico	Arizona	Nevada
I. Upper Main Stem	1. Delta 2. Dolores 3. Eagle 4. Garfield 5. Grand 6. Gunnison 7. Hinsdale 8. Mesa 9. Montrose 10. Ouray 11. Pitkin 12. San Miguel 13. Summit	1. Grand				
II. Green	1. Moffat 2. Rio Blanco 3. Routt	1. Carbon 2. Daggett 3. Duchesne 4. Emery 5. Uintah	1. Lincoln 2. Sublette 3. Sweetwater 4. Uinta			
III. San Juan	1. Archuletta 2. La Plata 3. Montezuma 4. San Juan	1. Garfield 2. Kane 3. San Juan 4. Wayne		1. San Juan		
IV. Little Colorado				1. McKinley	1. Apache 2. Navajo	
V. Gila				1. Catron 2. Grant	1. Cochise 2. Gila 3. Graham 4. Greenlee 5. Maricopa 6. Pima 7. Pinal 8. Santa Cruz 9. Yavapai	
VI. Lower Main Stem		1. Washington			1. Coconino 2. Mohave 3. Yuma	1. Clark 2. Lincoln

area). The Colorado River Basin is defined as 55 "representative counties" distributed over the six states of Wyoming, Colorado, Utah, New Mexico, Arizona, and Nevada. See Table 4.2 for a listing of counties by state and sub-basin.

Certain patterns immediately emerge. For example, while agricultural employment has been declining in all regions and the nation, the rate of decline in the CRB lagged behind that experienced elsewhere between 1940 and 1960. In the last decade, however, agricultural employment in the CRB fell by 31.4 percent, a sharper fall than in the non-CRB region and in the six states.

Mining employment in the CRB, while falling more slowly in the 1940s than in the other western areas shown in Table 4.1, lagged behind them in the two more recent decades of growth. The interaction of the energy problem and the basin's resources of coal, oil shale, and so forth may make this record a faulty predictor, however. Contract construction as a source of jobs in the CRB has out-distanced its growth rate in the other regions since 1950. In the same period, CRB manufacturing employment has grown dramatically, paced by transportation equipment and machinery manufacturing. CRB employment in transportation, communications, utilities, trade and services all grew at more rapid rates than those experienced elsewhere. In the last two decades, public administration jobs also grew more rapidly in the CRB, and within the rapidly growing service sector, CRB employment in the finance group, lodging, business and professional services out-distanced the other regions.

The percentage of total employment represented by each industrial component in each census year is shown in Table 4.3. Looking at the 30-year period, the most noteworthy changes have been the sharp decline in CRB agricultural employment from 27.3 to 4.4 percent; the decline of almost two-thirds in relative importance of mining from 10.7 to 3.8 percent; and the sharp growth rates in the finance and professional services groups. Manufacturing has grown as a provider of jobs in the CRB but still lags behind the nation, representing 12.8 percent of CRB jobs in 1970 compared with 25.7 percent of all jobs nationally. Nevertheless, in 1940, manufacturing jobs represented 7.5 percent of all CRB jobs and 23.9 percent of national employment. Thus the comparative lag in manufacturing employment between the region and the nation has narrowed from roughly three-to-one to two-to-one.

Table 4.4 permits an easy comparison of the changing rank of the nine basic industry groups as providers of jobs in the four areas at the polar points of the survey period. Two clear patterns emerge—convergence of the regions upon the national pattern and growing concentration of employment. With respect to convergence, in 1940 only two out of nine sectors in the CRB shared the same ranks with their national counterparts, services and contract

TABLE 4.3
Percentage Composition of Employment by Industry, 1940–1970

Sector	CRB				Non-CRB			
	1940	1950	1960	1970	1940	1950	1960	1970
Agriculture	27.327	18.404	9.011	4.371	22.469	14.675	7.429	4.412
Mining	10.695	6.968	5.563	3.838	5.213	3.200	3.281	2.727
Contract Construction	5.851	8.252	9.502	7.835	5.547	8.926	7.922	6.328
Food	1.271	1.388	1.651	0.828	2.462	2.341	2.540	1.587
Textile & Apparel	1.749	0.724	0.488	0.630	0.367	0.376	0.394	0.573
Lumber, Wood, Furniture	1.391	1.375	0.941	0.692	0.755	0.757	0.529	0.427
Printing & Publishing	0.769	0.903	1.184	0.943	1.176	1.270	1.439	1.174
Chemical	0.177	0.352	0.381	0.446	0.289	0.503	0.509	0.396
Machinery	0.176	0.373	1.942	4.625	0.423	0.731	1.410	2.350
Transportation Equipment	0.046	0.086	1.076	0.571	0.112	0.173	0.832	1.007
Manufacturing N.E.C.*	1.880	2.465	3.656	4.079	3.934	4.656	6.523	4.849
All Manufacturing†	7.459	7.656	11.320	12.813	9.518	10.807	14.176	12.364
Railroad	4.141	3.985	1.783	0.842	4.786	4.171	2.305	1.166
Trucking & Warehousing	1.034	1.144	1.271	1.001	1.336	1.480	1.723	1.468
Transportation N.E.C.	0.443	0.880	0.901	1.153	0.833	1.258	1.095	1.232
All Transportation†	5.618	6.009	3.955	2.996	6.955	6.909	5.124	3.867
Communications	0.683	1.183	1.474	1.408	1.176	1.524	1.586	1.509
Utilities	1.289	2.142	2.047	2.117	1.246	1.620	1.656	1.722
Communications & Utilities†	1.972	3.324	3.521	3.524	2.422	3.145	3.242	3.231
Wholesale Trade	2.515	3.146	3.284	3.465	3.191	3.857	3.699	4.184
Retail Foods	2.942	2.926	2.709	2.417	3.274	3.090	2.781	2.254
Eating & Drinking	2.854	3.971	3.671	3.880	2.720	3.294	3.281	3.717
Retail N.E.C.	7.975	10.139	10.782	11.438	10.056	10.708	10.891	11.357
All Trade†	16.286	20.182	20.447	21.200	19.240	20.949	20.653	21.512
Finance, Insurance & Real Estate	1.609	2.544	4.465	5.062	2.930	3.282	4.384	4.900
Lodging & Personal Services	4.499	4.981	5.884	6.818	4.698	4.128	3.885	3.777
Household Services	3.613	2.615	2.808	1.234	3.499	2.078	2.515	1.093
Business & Repair	2.079	2.835	2.881	3.847	2.491	2.846	3.256	3.612
Recreation Services	1.037	1.444	1.498	1.638	1.007	1.201	1.230	1.440
Professional Services	8.083	9.593	13.069	18.414	9.423	10.801	14.623	22.283
All Services†	20.919	24.011	30.605	37.013	24.048	24.336	29.893	37.105
Public Administration	3.872	5.194	6.076	6.408	4.588	7.054	8.279	8.454
Totals	100.000	100.000	100.000	100.000	100.000	100.000	100.000	100.000

*N.E.C. = Not Elsewhere Classified.

†Entries will not always add up to sub-totals in broad industrial categories nor will sub-totals always add to 100.000 due to rounding.

SIX STATES				UNITED STATES			
1940	**1950**	**1960**	**1970**	**1940**	**1950**	**1960**	**1970**
23.820	**15.726**	**7.944**	**4.398**	**19.239**	**12.665**	**7.012**	**3.740**
6.737	**4.262**	**4.023**	**3.113**	**2.069**	**1.674**	**1.054**	**0.830**
5.632	**8.736**	**8.436**	**6.852**	**4.661**	**6.217**	**6.152**	**6.044**
2.131	2.072	2.251	1.323	2.492	2.543	2.938	1.809
0.751	0.474	0.425	0.593	4.396	4.144	3.407	2.849
0.932	0.931	0.663	0.519	2.117	2.140	1.721	1.263
1.063	1.167	1.356	1.094	1.425	1.538	1.840	1.561
0.258	0.460	0.468	0.413	0.992	1.186	1.394	1.286
0.354	0.630	1.583	3.140	2.416	3.748 ·	4.926	5.113
0.093	0.149	0.911	0.855	1.988	2.432	2.932	2.845
3.363	4.036	5.591	4.582	8.040	8.527	9.077	9.012
8.946	**9.919**	**13.248**	**12.520**	**23.866**	**26.257**	**28.233**	**25.737**
4.607	4.118	2.135	1.054	2.562	2.496	1.517	0.836
1.252	1.385	1.576	1.306	1.141	1.264	1.469	1.423
0.724	1.152	1.032	1.205	1.223	1.553	1.430	1.470
6.583	**6.655**	**4.744**	**3.565**	**4.925**	**5.313**	**4.417**	**3.728**
1.039	1.428	1.550	1.474	0.888	1.277	1.321	1.401
1.258	1.767	1.783	1.859	1.227	1.412	1.449	1.663
2.297	**3.195**	**3.333**	**3.333**	**2.115**	**2.689**	**2.770**	**3.063**
3.003	3.657	3.564	3.934	2.725	3.564	3.568	4.043
3.181	3.044	2.758	2.311	3.368	3.100	2.724	2.511
2.757	3.485	3.408	3.774	2.525	3.044	2.904	3.026
9.477	10.547	10.856	11.385	8.424	9.326	9.815	10.460
18.419	**20.733**	**20.586**	**21.404**	**17.042**	**19.035**	**19.011**	**20.040**
2.562	3.074	4.411	4.956	3.312	3.454	4.344	5.024
4.643	4.368	4.535	4.833	3.807	3.348	3.130	3.142
3.531	2.230	2.610	1.142	5.265	2.948	3.090	1.504
2.377	2.843	3.134	3.694	1.955	2.362	2.597	3.166
1.016	1.270	1.317	1.509	0.894	0.890	0.811	0.824
9.050	10.460	14.118	20.939	7.505	8.627	12.216	17.659
23.178	**24.244**	**30.125**	**37.073**	**22.737**	**21.629**	**26.188**	**31.319**
4.389	**6.530**	**7.563**	**7.743**	**3.345**	**4.522**	**5.163**	**5.498**
100.000	**100.000**	**100.000**	**100.000**	**100.000**	**100.000**	**100.000**	**100.000**

TABLE 4.4

Ranking of Industries by Percentage of Total Employment Provided, 1940 and 1970

1940

UNITED STATES		SIX STATES		Non-CRB		CRB	
1940 Sector Rank	%	1940 Sector Rank	%	1940 Sector Rank	%	1940 Sector Rank	%
1. All Manufacturing	23.9	1. Agriculture	23.8	1. All Services	24.1	1. Agriculture	27.3
2. All Services	22.7	2. All Services	23.2	2. Agriculture	22.5	2. All Services	20.9
3. Agriculture	19.2	3. All Trade	18.4	3. All Trade	19.2	3. All Trade	16.3
4. All Trade	17.0	4. All Manufacturing	8.9	4. All Manufacturing	9.5	4. Mining	10.7
5. All Transportation	4.9	5. Mining	6.7	5. All Transportation	7.0	5. All Manufacturing	7.5
6. Contract Construction	4.7	6. All Transportation	6.6	6. Contract Construction	5.5	6. Contract Construction	5.9
7. Public Administration	3.3	7. Contract Construction	5.6	7. Mining	5.2	7. All Transportation	5.6
8. Communication & Utilities	2.1	8. Public Administration	4.4	8. Public Administration	4.6	8. Public Administration	3.9
9. Mining	2.1	9. Communication & Utilities	2.3	9. Communication & Utilities	2.4	9. Communication & Utilities	2.0

1970

UNITED STATES		SIX STATES		Non-CRB		CRB	
1970 Sector Rank	%	1970 Sector Rank	%	1970 Sector Rank	%	1970 Sector Rank	%
1. All Services	31.3	1. All Services	37.1	1. All Services	37.1	1. All Services	37.0
2. All Manufacturing	25.7	2. All Trade	21.4	2. All Trade	21.5	2. All Trade	21.2
3. All Trade	20.0	3. All Manufacturing	12.5	3. All Manufacturing	12.4	3. All Manufacturing	12.8
4. Contract Construction	6.0	4. Public Administration	7.7	4. Public Administration	8.4	4. Contract Construction	7.8
5. Public Administration	5.5	5. Contract Construction	6.9	5. Contract Construction	6.3	5. Public Administration	6.4
6. Agriculture	3.7	6. Agriculture	4.4	6. Agriculture	4.4	6. Agriculture	4.4
7. All Transportation	3.7	7. All Transportation	3.6	7. All Transportation	3.9	7. Mining	3.8
8. Communication & Utilities	3.1	8. Communication & Utilities	3.3	8. Communication & Utilities	3.2	8. Communication & Utilities	3.5
9. Mining	0.8	9. Mining	3.1	9. Mining	2.7	9. All Transportation	3.0

construction. By 1970, five sectors had attained the same rank. The same pattern occurs between the CRB and non-CRB portions of the six states with a movement from four to five similar rankings over the same period. Thus, while the actual figures vary between the regions, in terms of ranking of sectors by share of total employment the CRB is increasingly coming to resemble the nation. Ranking stability is shown in Table 4.5.

TABLE 4.5

**Spearman Rank Correlation Coefficients and
Levels of Statistical Significance for Rank of Industry
by Share of Total Employment, 1940–1970**

	Colorado River Basin				Statistical Significance Levels	
	1940	1950	1960	1970	1%	5%
United States	.576	.734	.884	.917	.783	.683
Aggregate of Six States	.967	.950	.984	.917	.783	.683

All of the areas shown in Table 4.4 also demonstrate a growing concentration of employment in a few industries. The top ranking industry in 1940 was agriculture in the CRB and manufacturing in the nation. They accounted, respectively, for 27.3 and 23.9 percent of total employment in each area. By 1970 the service sector ranked first in each area with 37.0 and 31.3 percent, respectively, of their total employment. The top three sectors moved from 64.5 to 71.0 percent of all employment during the period in the CRB and from 65.8 to 76.0 percent in the nation. This phenomenon of growing dependence upon a smaller number of industries for employment may represent an unhealthy trend toward economic instability.

Table 4.6 presents similar data for each of the six sub-basins of the Colorado. The trends here are not as clear. The Gila showed the greatest change over the thirty years with the number of sectors in the same rank as in the nation growing from two to five. Similar rankings grew from two to three each for the upper main stem and San Juan while the lower main stem remained unchanged with but two sectors in the same rank. Two sub-basins, the Green and the Little Colorado, actually moved away from convergence over the period with the number of similarly ranked sectors falling from four to one and four to three respectively.

The trend toward greater concentration also is uneven at the sub-basin level. The three upper sub-basins—the upper main stem, Green and San

TABLE 4.6

Ranking of Industries by Percentage of Total Employment Provided, 1940 and 1970

UPPER MAIN STEM 1940 Sector Rank	%	GREEN 1940 Sector Rank	%	SAN JUAN 1940 Sector Rank	%
1. Agriculture	38.8	1. Agriculture	31.3	1. Agriculture	50.4
2. All Services	17.3	2. Mining	22.7	2. All Services	14.2
3. All Trade	13.7	3. All Services	15.3	3. All Trade	10.4
4. Mining	9.3	4. All Trade	12.4	4. All Manufacturing	9.6
5. Contract Construction	6.5	5. All Transportation	7.5	5. Contract Construction	5.1
6. All Transportation	5.7	6. Contract Construction	3.8	6. Mining	3.6
7. Public Administration	3.4	7. Public Administration	3.3	7. Public Administration	3.3
8. All Manufacturing	3.3	8. All Manufacturing	2.5	8. Transportation	2.2
9. Communication & Utilities	1.9	9. Communication & Utilities	1.3	9. Communication & Utilities	1.2

GILA 1940 Sector Rank	%	LOWER MAIN STEM 1940 Sector Rank	%	LITTLE COLORADO 1940 Sector Rank	%
1. All Services	25.9	1. Agriculture	22.4	1. Agriculture	39.0
2. All Trade	20.0	2. All Services	20.3	2. All Manufacturing	21.7
3. Agriculture	19.0	3. Agriculture	19.0	3. All Services	12.3
4. Mining	10.9	4. Mining	10.9	4. All Trade	8.3
5. All Manufacturing	7.2	5. All Manufacturing	7.2	5. All Transportation	6.8
6. Contract Construction	5.9	6. Contract Construction	5.9	6. Contract Construction	3.8
7. All Transportation	4.7	7. All Transportation	4.7	7. Public Administration	3.7
8. Public Administration	4.0	8. Public Administration	4.0	8. Mining	3.6
9. Communication & Utilities	2.3	9. Communication & Utilities	2.3	9. Communication & Utilities	0.8

[86]

UPPER MAIN STEM

1970 Sector Rank	%
1. All Services	33.9
2. All Trade	21.5
3. Agriculture	9.9
4. Contract Construction	8.7
5. Mining	6.7
6. All Manufacturing	6.5
7. Public Administration	5.8
8. Communication & Utilities	4.3
9. All Transportation	3.3

GREEN

1970 Sector Rank	%
1. All Services	28.1
2. All Trade	19.3
3. Mining	12.9
4. Agriculture	11.0
5. Contract Construction	7.8
6. Public Administration	6.3
7. All Manufacturing	5.5
8. All Transportation	5.3
9. Communication & Utilities	3.8

SAN JUAN

1970 Sector Rank	%
1. All Services	31.5
2. All Trade	20.9
3. Mining	9.5
4. Contract Construction	9.3
5. All Manufacturing	8.4
6. Public Administration	7.0
7. Agriculture	6.8
8. Communication & Utilities	4.0
9. All Transportation	2.5

GILA

1970 Sector Rank	%
1. All Services	35.5
2. All Trade	21.7
3. All Manufacturing	16.3
4. Contract Construction	7.4
5. Public Administration	5.8
6. Communication & Utilities	3.7
7. Agriculture	3.7
8. Communication & Utilities	3.2
9. All Transportation	2.6

LOWER MAIN STEM

1970 Sector Rank	%
1. All Services	46.7
2. All Trade	20.1
3. Contract Construction	8.9
4. Public Administration	7.7
5. All Manufacturing	5.2
6. Communication & Utilities	4.4
7. Agriculture	3.1
8. All Transportation	3.1
9. Mining	0.7

LITTLE COLORADO

1970 Sector Rank	%
1. All Services	35.9
2. All Trade	19.3
3. Public Administration	12.3
4. All Manufacturing	10.9
5. Contract Construction	7.0
6. Transportation	6.9
7. Agriculture	3.2
8. Communication & Utilities	2.8
9. Mining	1.8

Juan—all showed a movement toward greater dispersion rather than concentration with the share of total employment accounted for by the leading industry and the top three all *declining*. The Little Colorado in the lower sub-basin area performed similarly with only the Gila and the lower main stem showing signs of greater concentration of employment. However, they are the regions in the Colorado River Basin with the major concentrations of population and employment distributed around the Phoenix, Tucson, and Las Vegas Standard Metropolitan Statistical Areas (SMSA).

SHIFT/SHARE TECHNIQUE OF REGIONAL EMPLOYMENT CHANGE

A technique for the analysis of regional growth patterns in employment has been developed by private scholars during the last thirty years (Creamer, 1943; Perloff, et al., 1960) and refined by the Office of Business Economics of the U.S. Department of Commerce (Ashby, 1965). It is frequently referred to as the "shift/share" technique. The method attempts to explain regional variations in the rates of growth of employment by breaking down regional growth into its component parts (see, for example, Tables 4.7-4.10).

The initial component is attributed to overall growth in the national economy. It is assumed that in the absence of regional differences, each area would grow at the same rate and thus retain its share of the national total. This element of national growth is calculated by applying to regional employment in each industry in the base year the percentage increase in aggregate national employment which occurred between the base year and the terminal year. During the 1940–50 decade, total employment in the United States increased from 45,375.8 million to 57,474.9 million or by 26.664 percent (Ashby, 1965). If agricultural employment in the State of Colorado had increased by the same proportion, the 1950 figure would have shown an increase of 19,610 persons over the 1940 total of 73,546. Similarly calculated, the national growth component during the 1950–60 decade would have added 11,117 to state agricultural employment. Symbolically,

$$\text{National Growth Share} = E_i \left(\frac{US^*}{US}\right)$$

where: E_i = regional employment in the ith industry at the beginning of the period.

US^* = change in aggregate national employment at the end of the period.

US = aggregate national employment at the beginning of the period.

Of course, regions do not all grow at the same national rate and, hence, it becomes necessary to examine elements unique to the regions. The shift/

share technique explicitly identifies two elements which contribute to a difference between areal and national growth rates. The first of these arises out of the "industry mix" of the region. If a large proportion of an area's economic activity is accounted for by slow-growing industries, its employment will probably expand at a below-average rate. Conversely, an area with many rapidly growing industries most likely will increase its share of national employment.

This industry-mix effect reflects the industrial composition of employment in the area. It is measured by applying to regional employment in each industry in the base year the difference between the national growth rate of employment in that industry and the rate of growth of employment in all industries in the nation. Symbolically,

$$\text{Industry mix} = E_i \left(\frac{US_i*}{US_i} - \frac{US*}{US} \right)$$

where: US_i = national employment in the ith industry at the beginning of the period.

US_i* = change in national employment in the ith industry at the end of the period.

When the specific industry growth rate is larger than the all-industry figure, the ith industry is considered to be a rapid-growth industry which should help boost area employment growth, depending upon the concentration of that industry in the region. Where the reverse situation holds, industry "i" is a slow-growth industry with a resulting dampening effect on the region's employment. The actual effect of industry-mix on agricultural employment in Colorado is shown below. The relative decline in agricultural employment nationally of 17.9 and 38.5 percent respectively (Ashby, 1965), during each decade, accounted for large negative values in the industry-mix component of agricultural employment in Colorado. Thus,

1940–1950	1950–1960
$73{,}546\,(-.179 + -.267)$	$71{,}808\,(-.385 + -.155)$
$73{,}546 \times (-.446) = -32{,}801$	$71{,}808 \times (-.540) = -38{,}776$

Shifts in a regional industry's share of employment in its national counterpart account for the third element in an area's growth. A regional industry that is growing faster (slower) than its national equivalent is assumed to add to (subtract from) total employment in the area. Symbolically,

$$\text{Regional Share} = E_i \left(\frac{E_i*}{E_i} - \frac{US_i*}{US_i} \right)$$

where: E_i* = change in the regional employment in the ith industry at the end of the period.

TABLE 4.7

Shift/Share Analysis and Related Employment Statistics

DATA REGION – COLORADO RIVER BASIN (CRB)

Sector	Employment Figures 1940	1950	Actual Change
Agriculture	68958	68589	−369
Mining	26988	25967	−1021
Contract Construction	14764	30755	15991
Food	3208	5174	1966
Textile & Apparel	4413	2698	−1715
Lumber, Wood, Furniture	3511	5123	1612
Printing & Publishing	1941	3365	1424
Chemical & Allied	447	1310	863
All Machinery	443	1389	946
Transportation Equipment	115	319	204
Manufacturing N.E.C.*	4744	9153	4409
All Manufacturing	18822	28531	9709
Railroad	10450	14851	4401
Trucking	2609	4262	1653
Transportation N.E.C.	1117	3281	2164
All Transportation	14176	22394	8218
Communication	1724	4407	2683
Utilities	3253	7981	4728
Communication & Utilities	4977	12388	7411
Wholesale Trade	6347	11725	5378
Retail Food Stores	7423	10905	3482
Eating & Drinking Places	7202	14798	7596
Retail N.E.C.	20124	37787	17663
All Trade	41096	75215	34119
Finance, Insurance & Real Estate	4059	9480	5421
Lodging Sources	11353	18563	7210
Household Services	9118	9744	626
Business & Repair Services	5246	10566	5320
Recreation Services	2616	5381	2765
Professional Services	20396	35750	15354
All Services	52788	89484	36696
Public Administration	9771	19356	9585
Totals	252340	372679	120339

*N.E.C. = Not Elsewhere Classified.

[90]

BASE – UNITED STATES

Shift/Share Analysis

Proj. Change at Base Rate	Industrial Mix†	Regional Share†
17410	−29512	11733
6814	−6460	−1374
3728	6175	6088
810	82	1074
1114	−317	−2511
886	49	677
490	193	741
113	109	641
112	306	528
29	32	143
1198	360	2851
4752	2363	2594
2638	−334	2098
659	353	641
282	377	1505
3579	1397	3242
435	944	1304
821	616	3291
1257	1691	4464
1602	2447	1329
1874	−739	2347
1818	1855	3923
5081	2700	9883
10376	6018	17726
1025	218	4178
2866	−1715	6059
2302	−5024	3348
1324	1369	2627
660	−17	2122
5149	3820	6384
13328	−3224	26593
2467	4304	2814
63707	−17809	74444

Total Net Relative Change = 56635

†Industrial Mix and Regional Share employment entries for broad industrial categories (such as All Manufacturing, All Transportation, etc.) are not to be considered as sub-totals in this table. These figures do not equal the sum of the entries for component branches of the industries because all entries are based upon unweighted percentage changes in employment.

TABLE 4.8

Shift/Share Analysis and Related Employment Statistics

DATA REGION – COLORADO RIVER BASIN (CRB)

Sector	Employment Figures 1950	Employment Figures 1960	Actual Change
Agriculture	68589	55090	−13499
Mining	25967	34010	8043
Contract Construction	30755	58088	27333
Food	5174	10095	4921
Textile & Apparel	2698	2982	284
Lumber, Wood, Furniture	5123	5755	632
Printing & Publishing	3365	7238	3873
Chemical & Allied	1310	2332	1022
All Machinery	1389	11874	10485
Transportation Equipment	319	6575	6256
Manufacturing N.E.C.*	9153	22352	13199
All Manufacturing	28531	69203	40672
Railroad	14851	10898	−3953
Trucking	4262	7771	3509
Transportation N.E.C.	3281	5508	2227
All Transportation	22394	24177	1783
Communication	4407	9014	4607
Utilities	7981	12511	4530
Communication & Utilities	12388	21525	9137
Wholesale Trade	11725	20077	8352
Retail Food Stores	10905	16561	5656
Eating & Drinking Places	14798	22444	7646
Retail N.E.C.	37787	65916	28129
All Trade	75215	124998	49783
Finance, Insurance & Real Estate	9480	27296	17816
Lodging Sources	18563	35971	17408
Household Services	9744	17169	7425
Business & Repair Services	10566	17612	7046
Recreation Services	5381	9159	3778
Professional Services	35750	79894	44144
All Services	89484	187101	97617
Public Administration	19356	37142	17786
Totals	372679	611334	238655

*N.E.C. = Not Elsewhere Classified.

BASE – UNITED STATES

Shift/Share Analysis

Proj. Change at Base Rate	Industrial Mix†	Regional Share†
9951	**−35054**	**11604**
3767	**−11002**	**15279**
4462	**−371**	**23243**
751	921	3250
391	−549	442
743	−1150	1039
488	756	2629
190	263	569
202	500	9784
46	75	6135
1328	676	11195
4139	**2458**	**34075**
2155	−6668	561
618	794	2096
476	−296	2048
3249	**−4323**	**2857**
639	177	3790
1158	236	3136
1797	**429**	**6911**
1701	13	6638
1582	−1515	5589
2147	−778	6277
5482	2267	20380
10912	**−108**	**38979**
1375	2797	13644
2693	−1383	16098
1414	537	5474
1533	1204	4309
781	−547	3544
5187	17031	21926
12983	**21599**	**63035**
2808	**3145**	**11832**
54068	**−27921**	**212511**

Total Net Relative Change = 184590

†Industrial Mix and Regional Share employment entries for broad industrial categories (such as All Manufacturing, All Transportation, etc.) are not to be considered as sub-totals in this table. These figures do not equal the sum of the entries for component branches of the industries because all entries are based upon unweighted percentage changes in employment.

TABLE 4.9

Shift/Share Analysis and Related Employment Statistics

DATA REGION = COLORADO RIVER BASIN (CRB)

Sector	Employment Figures 1960	1970	Actual Change
Agriculture	55090	37814	−17276
Mining	34010	33205	−805
Contract Construction	58088	67778	9690
Food	10095	7164	−2931
Textile & Apparel	2982	5446	2464
Lumber, Wood, Furniture	5755	5982	227
Printing & Publishing	7238	8161	923
Chemical & Allied	2332	3858	1526
All Machinery	11874	40009	28135
Transportation Equipment	6575	4937	−1638
Manufacturing N.E.C.*	22352	35285	12933
All Manufacturing	69203	110842	41639
Railroad	10898	7287	−3611
Trucking	7771	8662	891
Transportation N.E.C.	5508	9971	4463
All Transportation	24177	25920	1743
Communication	9014	12177	3163
Utilities	12511	18312	5801
Communication & Utilities	21525	30489	8964
Wholesale Trade	20077	29978	9901
Retail Food Stores	16561	20908	4347
Eating & Drinking Places	22444	33564	11120
Retail N.E.C.	65916	98947	33031
All Trade	124998	183397	58399
Finance, Insurance & Real Estate	27296	43794	16498
Lodging Sources	35971	58977	23006
Household Services	17169	10676	−6493
Business & Repair Services	17612	33280	15668
Recreation Services	9159	14167	5008
Professional Services	79894	159298	79404
All Services	187101	320192	133091
Public Administration	37142	55435	18293
Totals	611334	865072	253738

*N.E.C. = Not Elsewhere Classified.

[94]

Shift/Share Analysis

Proj. Change at Base Rate	Industrial Mix†	Regional Share†
10258	**−30496**	**2963**
6333	**−8582**	**1444**
10816	**−1208**	**82**
1880	−4600	−210
555	−579	2488
1072	−1816	972
1348	−1302	878
434	−213	1305
2211	537	25387
1224	−232	−2630
4162	−187	8959
12886	**−7255**	**36009**
2029	−5808	168
1447	−291	−264
1026	181	3256
4502	**−4470**	**1712**
1678	642	843
2330	2192	1280
4008	**2703**	**2253**
3738	3173	2990
3084	−1534	2798
4179	1114	5827
12274	5143	15615
23275	**8029**	**27095**
5083	5072	6343
6698	158	16150
3197	−10455	766
3279	4582	7807
1705	179	3124
14876	42226	22302
34838	**43489**	**54764**
6916	**2855**	**8522**
113832	**751**	**139165**

Total Net Relative Change = 139916

†Industrial Mix and Regional Share employment entries for broad industrial categories (such as All Manufacturing, All Transportation, etc.) are not to be considered as sub-totals in this table. These figures do not equal the sum of the entries for component branches of the industries because all entries are based upon unweighted percentage changes in employment.

[95]

TABLE 4.10

Shift/Share Analysis and Related Employment Statistics

DATA REGION = COLORADO RIVER BASIN (CRB)

Sector	Employment Figures		Actual Change
	1940	1970	
Agriculture	68958	37814	−31144
Mining	26988	33205	6217
Contract Construction	14764	67778	53014
Food	3208	7164	3956
Textile & Apparel	4413	5446	1033
Lumber, Wood, Furniture	3511	5982	2471
Printing & Publishing	1941	8161	6220
Chemical & Allied	447	3858	3411
All Machinery	443	40009	39566
Transportation Equipment	115	4937	4822
Manufacturing N.E.C.*	4744	35285	30541
All Manufacturing	18822	110842	92020
Railroad	10450	7287	−3163
Trucking	2609	8662	6053
Transportation N.E.C.	1117	9971	8854
All Transportation	14176	25920	11744
Communication	1724	12177	10453
Utilities	3253	18312	15059
Communication & Utilities	4977	30489	25512
Wholesale Trade	6347	29978	23631
Retail Food Stores	7423	20908	13485
Eating & Drinking Places	7202	33564	26362
Retail N.E.C.	20124	98947	78823
All Trade	41096	183397	142301
Finance, Insurance & Real Est.	4059	43794	39735
Lodging Sources	11353	58977	47624
Household Services	9118	10676	1558
Business & Repair Services	5246	33280	28034
Recreation Services	2616	14167	11551
Professional Services	20396	159298	138902
All Services	52788	320192	267404
Public Administration	9771	55435	45664
Totals	252340	865072	612732

*N.E.C. = Not Elsewhere Classified.

[96]

Shift/Share Analysis

Proj. Change At Base Rate	Industrial Mix†	Regional Share†
48356	**−94510**	**15010**
18925	**−27494**	**14787**
10353	**7452**	**35209**
2250	−1494	3201
3095	−2643	581
2462	−2410	2419
1361	315	4544
313	226	2872
311	841	38414
81	84	4657
3327	976	26238
13018	**2512**	**76310**
7328	−11979	1488
1830	1099	3125
783	384	7687
9941	**−5860**	**7663**
1209	1692	7552
2281	1966	10812
3490	**3796**	**18226**
4451	5220	13960
5205	−3213	11493
5050	2431	18880
14112	8276	56435
28818	**12299**	**101184**
2846	3570	33318
7961	−3375	43039
6394	−11081	6245
3679	5532	18823
1834	−350	10067
14302	46950	77649
37017	**33895**	**196492**
6852	**10697**	**28115**
176951	**−60838**	**496620**

Total Net Relative Change = 435782

†Industrial Mix and Regional Share employment entries for broad industrial categories (such as All Manufacturing, All Transportation, etc.) are not to be considered as sub-totals in this table. These figures do not equal the sum of the entries for component branches of the industries because all entries are based upon unweighted percentage changes in employment.

Thus, an agricultural sector in Colorado that out-performed its national parent (in this case by showing a smaller relative loss than agriculture nationally) contributed a positive influence. Colorado agricultural employment fell by $-.024\%$ in the 1940–50 period and by $-.334\%$ in the 1950s.

1940–1950	1950–1960
$[(-.024) - (.179)] \times 73,546$	$[(.334) - (-.385)] \times 71,808$
$(.155) \times 73,546 = 11,400$	$(.051) \times 71,808 = 3,662$

The sum of the three components equals the total employment change actually recorded in Colorado agriculture during each decade.

SHIFT/SHARE RESULTS

The actual results of the shift/share analysis for the Colorado River Basin appear in Tables 4.7–4.10 with the first three tables presenting results for each decade of change and the last covering the entire 30-year span. For the periods shown in the tables the principal task of the analysis was to explain the phenomenon of employment growth. Only one major industry group, agriculture, consistently lost employment over the 30 years. Declines elsewhere were limited to mining in the 1940s and 1960s, textile and apparel manufacture in the 1940s, railroad transportation in the 1950s and 1960s, and food, transport equipment manufacturing and household services in the 1960s.

With respect to the large majority of sectors which experienced employment growth the surprising finding is the nature of the role played by industry mix. Its contribution to employment change over this span of time was negative. Thus of the actual growth in total CRB employment in the 1940–1970 period of almost 613,000, the contribution of industry mix was a negative figure of almost 61,000. A decade-by-decade examination of the industry mix columns shows a growing number of minus signs increasing from 9 in the 1940s to 16 in the 1960s. Thus the employment growth of the CRB was not a result of a heavy concentration of national ''growth'' industries but rather a reflection of the regional vitality of its industry which often out-performed its national counterparts. A sign of possible change in the role of industry mix is its change to a positive value for the 1960–70 decade although the sign of the component was very small, a mere 751.

For comparison the remaining counties of the six states which contain the CRB were aggregated as the non-CRB portion and were separately analyzed by the same techniques. Their shift/share results do not appear here, but a comparison of the overall results for the two regions and for the aggregate six-state area is shown in Table 4.11. While the same general pattern of negative industry mix appears, the differences in magnitude are instructive. Much less of the employment change in the non-CRB portion of the six states

TABLE 4.11
Shift/Share Components of Employment Change
as Percentage of Actual Employment Change

		NATIONAL GROWTH	
Period	Six States	CRB	Non-CRB
1940–1950	55.3%	52.9%	56.2%
1950–1960	34.4	22.7	43.2
1960–1970	57.4	44.9	66.4
1940–1970	40.2	28.9	47.4
		INDUSTRY MIX	
1940–1950	−6.8%	−14.8%	−3.5%
1950–1960	−8.5	−11.7	−6.1
1960–1970	4.7	0.3	7.8
1940–1970	−5.3	− 9.9	−2.4
		REGIONAL SHARE	
1940–1950	51.6%	61.9%	47.3%
1950–1960	74.1	89.0	62.9
1960–1970	37.9	54.8	25.8
1940–1970	65.1	81.1	54.9

is explained by the residual regional share component. Overall national employment change explains a much larger share of employment change in this region and, while also generally negative, the industry mix component is less adverse to the non-CRB region's growth, changing sign and contributing +7.8 percent in the last decade. The non-CRB area contains a larger share of total employment in the six-state area; this larger influence is reflected in the closer resemblance between it and the six-state area.

A rather well-known criticism of the shift/share method of analysis (Houston, 1967) speaks of the method's "...excessive dependence on what happened in the nation as opposed to what happened in the region." This comment raises the question of what constitutes an appropriate base against which the performance of the region under study should be compared. Given the many differences between the industrial structures of the Colorado River Basin and of the nation, such a question might well be raised at this point. In order to determine the influence of using the U.S. as base in the above analysis, the data were recomputed using employment change by industry in the aggregate six-state area in which the CRB lies as the base region. The full shift/share results for the CRB are presented in Tables 4.12–4.15 and a summary for the CRB and non-CRB regions appears in Table 4.16.

TABLE 4.12

Shift/Share Analysis and Related Employment Statistics

DATA REGION = COLORADO RIVER BASIN (CRB)

Sector	Employment Figures 1940	1950	Actual Change
Agriculture	68958	68589	−369
Mining	26988	25967	−1021
Contract Construction	14764	30755	15991
Food	3208	5174	1966
Textile & Apparel	4413	2698	−1715
Lumber, Wood, Furniture	3511	5123	1612
Printing & Publishing	1941	3365	1424
Chemical & Allied	447	1310	863
All Machinery	443	1389	946
Transportation Equipment	115	319	204
Manufacturing N.E.C.*	4744	9153	4409
All Manufacturing	18822	28531	9709
Railroad	10450	14851	4401
Trucking	2609	4262	1653
Transportation N.E.C.	1117	3281	2164
All Transportation	14176	22394	8218
Communication	1724	4407	2683
Utilities	3253	7981	4728
Communication & Utilities	4977	12388	7411
Wholesale Trade	6347	11725	5378
Retail Food Stores	7423	10905	3482
Eating & Drinking Places	7202	14798	7596
Retail N.E.C.	20124	37787	17663
All Trade	41096	75215	34119
Finance, Insurance & Real Est.	4059	9480	5421
Lodging Sources	11353	18563	7210
Household Services	9118	9744	626
Business & Repair Services	5246	10566	5320
Recreation Services	2616	5381	2765
Professional Services	20396	35750	15354
All Services	52788	89484	36696
Public Administration	9771	19356	9585
Totals	252340	372679	120339

*N.E.C. = Not Elsewhere Classified.

[100]

Shift/Share Analysis

Proj. Change At Base Rate	Industrial Mix†	Regional Share†
31504	−34136	2263
12330	−14443	1093
6745	11856	−2610
1466	−128	629
2016	−2374	−1357
1604	−6	15
887	276	261
204	511	148
202	503	241
53	99	52
2167	1383	858
8599	2983	−1872
4774	−1613	1241
1192	403	58
510	960	693
6476	226	1516
788	941	954
1486	1919	1323
2274	2837	2300
2900	2013	466
3391	−467	559
3290	2769	1537
9194	3311	5158
18775	7522	7822
1854	1180	2386
5187	−977	3001
4166	−4895	1356
2397	1500	1424
1195	953	616
9318	4630	1406
24116	3537	9042
4464	6943	−1821
115284	−16889	21950

Total Net Relative Change = 5061

†Industrial Mix and Regional Share employment entries for broad industrial categories (such as All Manufacturing, All Transportation, etc.) are not to be considered as sub-totals in this table. These figures do not equal the sum of the entries for component branches of the industries because all entries are based upon unweighted percentage changes in employment.

TABLE 4.13

Shift/Share Analysis and Related Employment Statistics

DATA REGION = COLORADO RIVER BASIN (CRB)

Sector	Employment Figures 1950	1960	Actual Change
Agriculture	68589	55090	−13499
Mining	25967	34010	8043
Contract Construction	30755	58088	27333
Food	5174	10095	4921
Textile & Apparel	2698	2982	284
Lumber, Wood, Furniture	5123	5755	632
Printing & Publishing	3365	7238	3873
Chemical & Allied	1310	2332	1022
All Machinery	1389	11874	10485
Transportation & Equipment	319	6575	6256
Manufacturing N.E.C.*	9153	22352	13199
All Manufacturing	28531	69203	40672
Railroad	14851	10898	−3953
Trucking	4262	7771	3509
Transportation N.E.C.	3281	5508	2227
All Transportation	22394	24177	1783
Communication	4407	9014	4607
Utilities	7981	12511	4530
Communication & Utilities	12388	21525	9137
Wholesale Trade	11725	20077	8352
Retail Food Stores	10905	16561	5656
Eating & Drinking Places	14798	22444	7646
Retail N.E.C.	37787	65916	28129
All Trade	75215	124998	49783
Finance, Insurance & Real Estate	9480	27296	17816
Lodging Sources	18563	35971	17408
Household Services	9744	17169	7425
Business & Repair Services	10566	17612	7046
Recreation Services	5381	9159	3778
Professional Services	35750	79894	44144
All Services	89484	187101	97617
Public Administration	19356	37142	17786
Totals	372679	611334	238655

*N.E.C. = Not Elsewhere Classified.

[102]

Shift/Share Analysis

Proj. Change at Base Rate	Industrial Mix†	Regional Share†
28932	−48260	5830
10953	−2067	−842
12973	−1503	15864
2182	635	2104
1138	−397	−456
2161	−2093	565
1419	776	1678
553	30	439
586	2987	6912
135	2325	3796
3861	5014	4324
12035	13615	15022
6264	−10169	−48
1798	837	874
1384	−484	1327
9446	−9145	1483
1859	536	2212
3366	100	1064
5225	759	3153
4946	−422	3829
4600	−1455	2512
6242	−464	1868
15939	1571	10619
31727	−757	18814
3999	5862	7955
7830	1005	8573
4110	2367	948
4457	1539	1050
2270	287	1221
15080	17773	11291
37745	30859	29012
8165	4354	5268
157202	−19316	100777

Total Net Relative Change = 81461

†Industrial Mix and Regional Share employment entries for broad industrial categories (such as All Manufacturing, All Transportation, etc.) are not to be considered as sub-totals in this table. These figures do not equal the sum of the entries for component branches of the industries because all entries are based upon unweighted percentage changes in employment.

TABLE 4.14

Shift/Share Analysis and Related Employment Statistics

DATA REGION = COLORADO RIVER BASIN (CRB)

Sector	Employment Figures 1960	1970	Actual Change
Agriculture	55090	37814	−17276
Mining	34010	33205	−805
Contract Construction	58088	67778	9690
Food	10095	7164	−2931
Textile & Apparel	2982	5446	2464
Lumber, Wood, Furniture	5755	5982	227
Printing & Publishing	7238	8161	923
Chemical & Allied	2332	3858	1526
All Machinery	11874	40009	28135
Transportation Equipment	6575	4937	−1638
Manufacturing N.E.C.*	22352	35285	12933
All Manufacturing	69203	110842	41639
Railroad	10898	7287	−3611
Trucking	7771	8662	891
Transportation N.E.C.	5508	9971	4463
All Transportation	24177	25920	1743
Communication	9014	12177	3163
Utilities	12511	18312	5801
Communication & Utilities	21525	30489	8964
Wholesale Trade	20077	29978	9901
Retail Food Stores	16561	20908	4347
Eating & Drinking Places	22444	33564	11120
Retail N.E.C.	65916	98947	33031
All Trade	124998	183397	58399
Finance, Insurance & Real Est.	27296	43794	16498
Lodging Sources	35971	58977	23006
Household Services	17169	10676	−6493
Business & Repair Services	17612	33280	15668
Recreation Services	9159	14167	5008
Professional Services	79894	159298	79404
All Services	187101	320192	133091
Public Administration	37142	55435	18293
Totals	611334	865072	253738

*N.E.C. = Not Elsewhere Classified.

Shift/Share Analysis

Proj. Change At Base Rate	Industrial Mix†	Regional Share†
17859	−32561	−2573
11025	−10191	−1639
18831	−14443	5303
3273	−5510	−692
967	1562	−64
1866	−1656	18
2346	−1849	427
756	−358	1129
3849	15470	8815
2131	−531	−3238
7246	−5342	11030
22434	−5033	24239
3533	−7307	164
2519	−1764	136
1786	1218	1459
7838	−7957	1863
2922	−585	826
4056	710	1035
6978	3	1983
6509	2762	631
5369	−3555	2534
7276	3187	657
21368	4255	7407
40522	6576	11302
8849	4474	3175
11661	3136	8209
5566	−12790	731
5709	4163	5796
2969	1762	277
25900	51116	2388
60654	57141	15296
12041	1176	5077
198182	−3451	59018

Total Net Relative Change = 55567

†Industrial Mix and Regional Share employment entries for broad industrial categories (such as All Manufacturing, All Transportation, etc.) are not to be considered as sub-totals in this table. These figures do not equal the sum of the entries for component branches of the industries because all entries are based upon unweighted percentage changes in employment.

[105]

TABLE 4.15

Shift/Share Analysis and Related Employment Statistics

DATA REGION = COLORADO RIVER BASIN (CRB)

Sector	Employment Figures 1940	1970	Actual Change
Agriculture	68958	37814	−31144
Mining	26988	33205	6217
Contract Construction	14764	67778	53014
Food	3208	7164	3956
Textile & Apparel	4413	5446	1033
Lumber, Wood, Furniture	3511	5982	2471
Printing & Publishing	1941	8161	6220
Chemical & Allied	447	3858	3411
All Machinery	443	40009	39566
Transportation Equipment	115	4937	4822
Manufacturing N.E.C.*	4744	35285	30541
All Manufacturing	18822	110842	92020
Railroad	10450	7287	−3163
Trucking	2609	8662	6053
Transportation N.E.C.	1117	9971	8854
All Transportation	14176	25920	11744
Communication	1724	12177	10453
Utilities	3253	18312	15059
Communication & Utilities	4977	30489	25512
Wholesale Trade	6347	29978	23631
Retail Food Stores	7423	20908	13485
Eating & Drinking Places	7202	33564	26362
Retail N.E.C.	20124	98947	78823
All Trade	41096	183397	142301
Finance, Insurance & Real Est.	4059	43794	39735
Lodging Sources	11353	58977	47624
Household Services	9118	10676	1558
Business & Repair Services	5246	33280	28034
Recreation Services	2616	14167	11551
Professional Services	20396	159298	138902
All Services	52788	320192	267404
Public Administration	9771	55435	45664
Totals	252340	865072	612732

*N.E.C. = Not Elsewhere Classified.

[106]

Shift/Share Analysis

Proj. Change At Base Rate	Industrial Mix†	Regional Share†
120185	−154220	2892
47036	−39820	−999
25732	8772	18510
5591	−3336	1701
7691	−2553	−4104
6119	−4267	619
3383	156	2681
779	739	1893
772	9560	29234
200	2576	2046
8268	4716	17556
32804	**20628**	**38588**
18213	−22106	730
4547	309	1197
1947	2032	4876
24707	**−17829**	**4867**
3005	1982	5466
5670	4265	5125
8674	**6161**	**10676**
11062	5398	7171
12937	−5571	6119
12552	7281	6529
35073	11113	32637
71625	**18267**	**52409**
7074	10403	22258
19787	1278	26559
15891	−16921	2588
9143	7974	10917
4559	3484	3508
35547	73491	29864
92002	**86796**	**88605**
17030	**20484**	**8151**
439793	**−72781**	**245724**

Total Net Relative Change = 172943

†Industrial Mix and Regional Share employment entries for broad industrial categories (such as All Manufacturing, All Transportation, etc.) are not to be considered as sub-totals in this table. These figures do not equal the sum of the entries for component branches of the industries because all entries are based upon unweighted percentage changes in employment.

TABLE 4.16

Shift/Share Components of Employment Change in Absolute Magnitude and as Percent of Actual Employment Change

CRB

Period	Total Actual Change	Total Base Area* Growth Rate	Industry Mix	Regional Share
1940–50	120,339 (100%)	115,284 (95.8%)	−16,889 (14.0%)	21,950 (18.2%)
1950–60	238,655 (100%)	157,202 (65.9%)	−19,316 (8.1%)	100,777 (42.2%)
1960–70	253,738 (100%)	198,182 (78.1%)	− 3,451 (1.4%)	59,018 (23.3%)
1940–70	612,732 (100%)	439,793 (71.8%)	−72,781 (11.9%)	245,724 (40.1%)
		Non-CRB		
1940–50	294,295 (100%)	299,352 (101.7%)	16,899 (5.7%)	−21,940 (7.5%)
1950–60	319,072 (100%)	400,527 (125.5%)	19,325 (6.1%)	−100,762 (31.6%)
1960–70	355,698 (100%)	411,254 (115.6%)	3,464 (1.0%)	−59,003 (16.6%)
1940–70	969,065 (100%)	1,142,003 (117.8%)	72,790 (7.5%)	−245,713 (25.3%)

*Six states

Changing the base region does in fact bring interesting changes in the relative importance of the several components. For example, the broad sweep of employment growth in the western states has clearly carried the CRB along with it, although the effect has not been uniform across all sub-basins of the CRB. Table 4.16 indicates that actual growth rates of total employment in the CRB exceeded those experienced in the total of the six states. On the other hand the non-CRB area failed to match the growth rates of the six-state total. In the non-CRB area industry mix made a modest positive contribution to employment change while regional share was consistently negative. The situation in the CRB was exactly opposite with a rather strong positive regional share component and negative industry mix. Here again, however, while negative in each decade, the industry mix component was declining throughout the period which may presage a switch in its contribution in the future from negative to positive. In summary, the CRB grew more rapidly than did the six-state aggregate despite a negative industry mix, reflecting a strong regional share performance. The non-CRB area lagged behind employment growth in the six-state region with a weakly positive industry mix outweighed by a negative regional share. This would appear to suggest that it was vigorous employment growth in the CRB which sparked growth in the six-state area. A closer look at data for each of the six sub-basins, however, belies that impression. In fact it was only the Gila and lower main stem regions which demonstrated such vigorous growth with the Little Colorado and the three upper sub-basins lagging behind six-state employment growth. Though the magnitudes change, the same pattern at the sub-basin level prevails when the U.S. is used as the base area. This finding is not surprising when the county composition of each sub-basin is studied since the major metropolitan areas of Arizona and Nevada lie within the Gila and lower main stem regions while those of New Mexico, Colorado, Utah, and Wyoming lie outside the remaining sub-basins.

Thus far the influence of changing the geographic base region has been examined in terms of the effect upon the size and sign of the three components of employment change. Another way to trace the consequences of the change in geographic base is to examine the stability in the rank of specific industries in terms of the size of the industry mix and regional share components over time. Table 4.17 contains the results of such an

TABLE 4.17

Spearman Rank Correlation Coefficients and Significance Levels for Employment in Nine Major Industry Groups

Period	Industry Mix Coefficient	Regional Share Coefficient	Significance Level	
			1%	5%
1940–50	.900	.677	.783	.683
1950–60	.958	.800	.783	.683
1960–70	.883	.627	.783	.683
1940–70	.817	.933	.785	.683

analysis. In the first two decades of the period changing the geographic base results in little if any variation in the ranking of industries by the size of the industry mix component. Slightly larger variation occurred in the last decade but all coefficient values were larger than 0.8 and were statistically significant at the 0.01 level. Ranking of industries by regional share component was appreciably less stable in each decade despite a surprisingly high similarity in ranks between the polar years of 1940 and 1970.

TABLE 4.18

Comparison of Industry Groups Appearing in this Chapter,
The Growth Patterns in Employment Volumes, and in the 1970 Census

32—Category Classification (1940, 1950, 1960) Used in OBE (now BEA) Growth Patterns in Employment by County	27—Category Classification Used in this Chapter	39—Category 1970 Classification Used at the County Level, General, Social and Economic Characteristics
1. Agriculture	1. Agriculture	1. Agriculture, Forestry and Fisheries
2. Forestry and Fisheries	2. Mining	2. Mining
3. Mining	3. Construction	3. Construction
4. Construction	4. Food and Kindred	4. Food and Kindred
5. Food and Kindred	5. Textile and Apparel	5. Textile and Fabricated Textile Products
6. Textile Mill Products	6. Lumber, Wood, Furniture	6. Furniture and Lumber and Wood Products
7. Apparel	7. Printing and Publishing	7. Printing and Publishing
8. Lumber, Wood Products, Furniture	8. Chemical and Allied	8. Chemical and Allied Products
9. Printing and Publishing	9. All Machinery	9. Machinery except Electrical
10. Chemical and Allied		10. Electrical Machinery
11. Electrical and Other Machinery	10. All Transportation Equip.	11. Transportation Equipment
12. Motor Vehicles and other Equip.		12. Metal Industries
13. Other Transportation Equip.		13. Other Durable Goods
14. Other and Misc. Manufacturing	11. Manufacturing, N.E.C.	14. Other Nondurable Goods
15. Railroads	12. Railroads	15. Railroads
16. Trucking and Warehousing	13. Trucking and Warehousing	16. Trucking and Warehousing
17. Other Transportation	14. Other Transportation	17. Other Transportation
18. Communications	15. Communications	18. Communications

19. Utilities and Sanitary Services
20. Wholesale Trade
21. Food and Dairy Stores
22. Eating and Drinking Places
23. Other Retail Trade
24. Finance, Insurance, Real Estate
25. Hotels and other Personal Services
26. Private Households
27. Business and Repair Services
28. Entertainment, Recreation Services
29. Medical, Other Professional Services
30. Public Administration
31. Armed Forces (deleted)
32. Industry Not Reported (Distributed Proportionately over Categories 1–30)

16. Utilities and Sanitary Services
17. Wholesale Trade
18. Food and Dairy Stores
19. Eating and Drinking Places
20. Other Retail Trade
21. Finance, Insurance, Real Estate
22. Lodging, Other Personal Services
23. Household Services
24. Business and Repair Services
25. Entertainment, Recreation Services
26. Professional Services
27. Public Administration

19. Utilities and Sanitary Services
20. Wholesale Trade
21. Food, Bakery, Dairy Stores
22. Eating and Drinking Places
23. General Merchandise
24. Motor Vehicle Retailing, Service Stations
25. Other Retail Trade
26. Banking and Credit
27. Insurance, Real Estate, Finance
28. Lodging and Other Personal Services
29. Private Households
30. Business and Repair Services
31. Entertainment and Recreation Services
32. Hospitals
33. Health Services, Except Hospitals
34. Elem., Secondary Schools, Colleges (government)
35. Elem., Secondary Schools, Colleges (private)
36. Other Education and Kindred Services
37. Welfare, Regions, and Non-Profit Organizations
38. Legal, Engineering, and Misc. Professional Services
39. Public Administration

It is difficult to present a satisfactory explanation for these variations in the stability of rankings by industry. On the whole, however, and particularly with respect to the magnitude and sign of the industry mix and regional share components, it appears that changing the base region may produce valuable insights that are veiled when the comparison is limited to the nation. One would expect this conclusion to be particularly applicable when the region under study is as small as a county.

FURTHER EXPLANATION

When this chapter was prepared, the formal shift/share computation of the U.S. Department of Commerce for the 1960–1970 period had not been published. Thus it was necessary to perform the analysis based on the available 1970 Census data. This required various adjustments in the definition of industries to make the earlier data comparable to the 1970 figures. Two consequences worth noting are the elimination of the highly volatile and relatively small "armed forces" sector, and a reduction in the number of separate sectors shown from 32 in the earlier shift/share report of the Commerce Department and 39 in the 1970 Census to the 27 used throughout this chapter. Table 4.18 shows how the three groupings of industries interrelate.

Literature Cited

Ashby, Lowell. 1965. *Growth Patterns in Employment by County, 1940–50 and 1950*–60. Washington, D.C.: U.S. Government Printing Office.

Creamer, Daniel. 1943. Shifts of manufacturing industries. In *Industrial Location and National Resources*, pp. 85–104. U.S. National Resources Planning Board, Washington, D.C.

Houston, David B. 1967. The shift and share analysis of regional growth: a critique. *Southern Econ. J.* 33(4):557–581.

Perloff, Harvey F., Dunn, Edgar S., Lampard, Eric E., and Muth, Richard F. 1960. *Regions, Resources and Economic Growth*. Baltimore: Johns Hopkins Press.

Thompson, Wilbur R. 1973. The economic base of urban problems. In *Contemporary Economic Issues*, ed., Neil D. Chamberlain, p. 1. Revised edition, Homewood, Illinois: Richard D. Irwin, Inc.

5

Policy Goals and Values
in Historical Perspective

HENRY P. CAULFIELD, JR.

The values of a society are reflected in its public policy goals. Obviously, different mixes of these values will appear in different historical epochs. Correspondingly, policy goals will shift. This can be seen in the historical experience of resource use in the Colorado River Basin and more generally, in American natural resources and environmental policy.

THE DEVELOPMENT THRUST

"Politics" is taken to be that of David Easton (1965): the processes by which a society makes authoritative decisions about the allocation of values. Four basic policy thrusts and corresponding values have been manifest generally in the history of federal natural resources politics, including that affecting the Colorado River Basin (Caulfield, 1959, 1961). First, the existence of a development thrust can be asserted which has been manifest in the spirit of the frontier and settlement of the West. Clearly, development has also been equated with "progress" in American society. From the nineteenth century through to the present, development in the economic sense can be said to have been the dominant influence in American society and culture. Prior to 1890, in the development of the far West, mining, lumbering, cattle raising and irrigated farming were the basic dominant modes of livelihood. Development of the West to provide for these modes of livelihood was the predominant concern of western people.

In the context of water and related land policy in river basins of the arid West, irrigated farming is of special importance. Irrigated farming, of course, was present in the West before the white man came and was present in certain parts of the West before the Americans came. In the period after the Civil

War, American settlers fostered irrigation development in the arid West where this could readily be accomplished. In northeastern Colorado, for example, between 1870 and 1890, a land and settlement boom involved the development of irrigation works. The Greeley colony is what Horace Greeley of New York had in mind when he promoted western development by his well-known statement of advice, "Go West, young man, go West." The Mormon settlements in Utah were based upon irrigated agriculture. Likewise, in southern California as well as in the Central Valley of California and in certain locations in Arizona, irrigated agriculture was of great importance as a foundation for economic life.

Two problems became manifest by 1890 with respect to this early irrigated agriculture. With the relatively easy projects already developed, it became clear that greater financial resources than were available locally were needed to undertake the more difficult projects awaiting development. Also, it became evident by 1890 that more technical know-how was needed in the planning, development and management of irrigated agriculture. Few engineers were available to assist in this work and irrigation engineering was a relatively undeveloped field. Soil science and the relationship of soils to ground cover and droughts was just beginning to be understood by technical people exploring the area. *Lands of the Arid Regions of the United States,* the famous report of John Wesley Powell published in 1878, brought national attention to the technical problems of putting western soils to work in irrigated agriculture. The consequences of droughts in terms of soil utilization became known, at least to the technical world.

The first public policy response of the federal government to these problems of finance and technical knowledge involved in the development of irrigated agriculture in the West was the Carey Act of 1894. This act provided for grants of federal public lands to states to assist them in the development of irrigation projects and the sale by the state of such lands to prospective farmers. No technical assistance was provided. Increasingly during this period, however, the views of the leading scientists in the Geological Survey were being heard within the Executive Branch and in Congress. The Geological Survey in those days did not hesitate to voice its policy concerns nor hesitate to make policy proposals. It was an important force in passage of the Reclamation Act of 1902, which, at first, it administered. The 1902 Act provided for the use of technical talent employed by the federal government in the planning, development and initial management of federal irrigation projects on western federal lands and for federal appropriation of funds to finance these activities. In the Townsite Act of 1906, authority was provided for the sale or lease of surplus hydro-electric power from projects of the then Bureau of Reclamation and for the use of the net revenues from such power to subsidize and assist irrigators in their reimbursement of assigned federal costs. Thus, the Federal Reclamation Act of 1902, as supplemented and amended, became a major policy vehicle for development of the West. Some seven projects were undertaken in the early years, including the Salt River Project and the development of Roosevelt Dam above Phoenix, Arizona. The values of irrigated agriculture, not only to the irrigators themselves, but more widely in the interests of economic development of the western United States, were promoted. Political interests of railroads, irrigators, farm suppliers and others were united in support of such publicly financed western development.

THE PROGRESSIVE THRUST

The progressive thrust in American politics is based on the value of individualism. To be sure, economic development in American society also is seen largely as a manifestation of individualism. But the individualism of the progressive thrust carries with it the idea of egalitarianism. As Thomas Jefferson saw it, a nation of yeomen farmers would be the pillar of democracy. Thus the family farm, in contrast to what we would call today corporated farming or then, large-scale plantations, was taken to be the ideal. This ideal of the family farm was embodied in the Homestead Act of 1862. The homesteader received 160 acres of land in return for being a bona fide resident and developing the land for agricultural purposes. Provision for this value was made in the previously noted Carey Act of 1894. It was also embodied in the "excess land law provisions" of the Reclamation Act of 1902, which provided that federally developed water could only be delivered to family farmers residing on their farm lands. In the sale of power from federal projects, the progressive thrust was also manifest in the Townsite Act of 1906, which provided that the sale or lease of power should be, preferentially, for public purposes. Subsequent policy provisions in federal law relating to federally developed hydro-electric power have followed this lead in providing that power shall be sold preferentially to public bodies and rural cooperatives. The thrust of these power policy provisions was "anti-monopoly" and a manifestation of egalitarianism. The progressive saw the private power utilities as a threat to his vision of a good society as well as to his pocketbook.

THE CONSERVATION AND PRESERVATION THRUSTS

The conservation thrust is defined here as that embodied in the traditional Conservation Movement, the coalition of policy forces led by Gifford Pinchot after the turn of the century. For Gifford Pinchot, conservation was not the locking up of resources in the sense of preserving them in their natural state. It was their development and wise use (Pinchot, 1910). Pinchot's conservation aligned itself with the developmental thrust in American politics. But it did so with the constraint that renewable resources should be used only on the basis of "sustained yield." Sustained yield, a concept of the emerging scientific elite, grew out of the development of scientific thought in the latter part of the nineteenth century and was deemed applicable to soil fertility, forests, water, fish, and wildlife. In terms of irrigation development, scientific concern related to soil characteristics in relation to drought possibilities, ground cover and soil erosion, as well as chemical characteristics of water. The traditional Conservation Movement represented a coalition of political forces involving all three of the foregoing policy thrusts. Federal reclamation projects constructed since 1902, an output of that movement, generally have embodied the values of all three thrusts.

The ideology of the traditional Conservation Movement of Gifford Pinchot and his followers was the justification of the main stream of federal natural resources policies that were established up to the early 1960s. By some this ideology was considered the total of conservation concern. But anyone close to the scene was aware of another elite group, less politically visible until the 1950s, which was concerned with the values of undisturbed nature in terms of its natural beauty as well as an ethical concern for undisturbed life other than that of man. The prominence of the preservation thrust should not be thought to reflect its origins as being "in our time." The Sierra Club was founded under the leadership of John Muir in the 1890s. Yosemite Valley was made a California State Park by congressional action in the 1860s. Yellowstone was made a national park in 1872. Yosemite Valley was restored to federal control in 1890 and its boundaries were substantially extended after the turn of the century. The Antiquities Act of 1906 provided the necessary authorization for the subsequent creation by executive action, of national monuments out of federal public domain lands which were subsequently individually authorized by Congress to become, in many instances, national parks.

The sharp split between the preservationists symbolized by John Muir and the traditional conservationists symbolized by Gifford Pinchot, did not clearly manifest itself in public policy until the celebrated Hetch-Hetchy controversy culminating in the Raker Act of 1913. This controversy involved federal authorization to San Francisco to build Hetch-Hetchy reservoir on a river that would flood water back into the extended Yosemite National Park. Pinchot, in accord with his utilitarian philosophy, saw use of this water for municipal purposes and power as an appropriate sustained use of the nation's resources, whereas John Muir saw the desecration of the Tuolumne, a valley somewhat comparable to Yosemite Valley, as being the desecration of nature, which, for Muir, was a manifestation of God. The traditional Conservation Movement won this argument, and others subsequently. But the losing forces were sufficiently strong to obtain creation of the National Park Service by Congress in 1916. Nevertheless, until the Echo Park fight in the 1950s and subsequent development of the New Conservation, or Environmental, Movement, the preservation thrust was no match for the coalition embodied in the traditional Conservation Movement, at least in open political combat. Prior to the Congressional decision deleting Echo Park dam and reservoir, within Dinosaur National Monument in Colorado and Utah, from the Colorado River Storage Project Act of 1956, the intellectual, scientific and social elite which strongly upheld preservational values achieved its then very limited ends through its special access to political power and relatively quiet persuasion.

RELATION TO THE COLORADO

The foregoing analysis of the values and related policy thrusts is pertinent, generally, to the development of natural resources and environmental policy. It provides the context within which value and policy developments specific to the Colorado River Basin have occurred since the turn of the century.

Other chapters in this volume cover in detail public decisions manifesting the development thrust within the Colorado River Basin. Very substantial development and control of the resources of the Colorado mainly in the interests of irrigation, hydroelectric power, flood control, municipal and industrial water and other forms of water development have been the result. A very general sketch of these decisions will suffice. A comprehensive study of the water development potentialities of the Colorado River Basin by E.C. LaRue (1916) of the Geological Survey, and published by the Survey in two volumes in the decade preceding the compact, appears to have set the stage for the subsequent developments. This river basin study, and no doubt other analyses available at the time, established the factual scientific basis for the interstate compact between the upper basin and the lower basin states to which the consent of Congress was granted in 1922. The Boulder Canyon Act of 1928 authorizing construction of Boulder Dam within the lower basin was, in effect, the first development generally consistent with LaRue's comprehensive plan. Major developments in the upper basin were authorized by the Colorado River Storage Project Act of 1956. The Colorado River Basin Project Act of 1968 provided for completing development works in the lower basin through authorization of the Central Arizona Project, and authorized certain further development projects in the upper basin in order to obtain upper basin political support.

The progressive thrust has been manifest in the Colorado River Basin by support for the family farm concept and low-cost public power. Irrigation projects of the Bureau of Reclamation generally incorporated the ''Excess Land Laws'' that sought to perpetuate the concept of the family farm. The most notable exception to this policy, however, occurred within the basin when in the late 1920s the irrigators within the Imperial Valley were exempted from these laws. Although electric power at Boulder Dam is not produced and marketed by the Department of the Interior under the federal public and rural cooperative power policy that has developed since 1928, the policy governing power marketing there can be said to be strongly animated by the progressive spirit. A very substantial allocation of power is produced by generators owned by the Water and Power Department of the City of Los Angeles and marketed within its system.

The conservation thrust in the sense of sustained yield management of water and related land resources, has been increasingly pursued in the Colorado River Basin since the turn of the century. The Bureau of Reclamation has studied many problems of soil and water management and, with increasing skill, it has sought to solve them. Similarly, the land grant colleges and universities of the basin through their extension services have sought to have best available practices used upon basin lands. Research of the U. S. Department of Agriculture has also been brought to bear within the basin in order that agricultural production could be sustained at high levels. Today these agencies and others are increasingly focusing research and actions upon water quality control in the basin.

The development, progressive and conservation thrusts, in the coalition of the traditional Conservation Movement, have dominated public resource decisions in the basin until relatively recently. Nevertheless, for a long time, significant public decisions have been made that manifest an increasing concern with the preservation of the basin's resources in their natural state. In 1908, Grand Canyon National Monument was established by President Theodore Roosevelt under terms of the Antiquities Act of 1906. Sensitivity to preservational interests was manifest by E. C. LaRue. He noted that one of the best dam sites for water and power storage would be at the western extremity of the then Grand Canyon National Monument. But he did not recommend this site because of its effect upon the naturalness of the monument. Instead, he recommended the Bridge Canyon site (now known as the Hualapai site) some 57 miles downstream as the next best site within that reach of the Colorado River. Nevertheless, the major public value placed upon development in the early part of this century, compared to the preservational concerns, is indicated in the terms of the act creating Grand Canyon National Park in 1919. The park was created. But the Act provided that authorization in the future of reclamation dams and reservoirs within the park was not to be considered as precluded. In the establishment of Rocky Mountain National Park and Dinosaur National Monument, similar reservations with respect to potential reclamation projects were included.

The first major development vs. preservation controversy, subsequent to the Hetch-Hetchy controversy discussed above, did not occur until the 1950s, when the Bureau of Reclamation proposed to build Echo Park Dam on a tributary of the Colorado River within Dinosaur National Monument. This proposal created a storm of protest by preservational interests throughout the country, but largely in the West, led by the Sierra Club of California. In the early 1950s, during a critical period of time in the decision process, some 1,000 letters a day were received in the Office of the Secretary of the Interior. Leaders of the Sierra Club testified before Congress not only on the damage to Dinosaur National Monument they saw by the construction of this dam, but on the technical analysis of the Bureau of Reclamation. Their case was substantially aided by the fact that one of their witnesses discovered a major mistake by the Bureau of Reclamation in the analysis of evaporation losses from the reservoir. Also, the Echo Park site was not the same dam site that was reserved in specific terms upon establishment of Dinosaur National Monument. Thus, the Sierra Club argued that there was no reservation in the establishment of Dinosaur National Monument that pertained to the Echo Park site; therefore, the general policy established by Congress that no developments should occur within areas controlled by the National Park Service. The preservational interests won this major fight. Relative values were shifting. The establishment of Canyonlands National Park and other park developments within the basin in the 1960s and since attests to this marked shift.

The change in public values and related public decisions involving development and preservation within the Colorado River Basin is no better seen than in the controversy over Bridge and Marble canyon dams during consideration of the Colorado River Basin Project Act of 1968. The plan proposed by the Department of the Interior included, among other developments, Bridge and Marble canyon dams. Secretary of the Interior Stewart L. Udall was very sensitive to preservational arguments against these proposed dams, but he was very reluctantly persuaded at first that they were necessary to realization of the Central Arizona Project to which, as an Arizonan, he was dedicated. When the practicality of joint public-private steam power development as a means of avoiding the financial necessity of building Bridge and Marble canyon dams for power production to provide irrigation subsidies was made clear to him, his support for these dams quickly vanished. Again, preservational interests led largely by the Sierra Club were behind the movement to delete these two dams from the proposed Colorado River Basin Project Act. Their position prevailed.

Preservational interests had secured language in the authorization in 1956 of Lake Powell to provide protection from having reservoir water backed into Rainbow Bridge National Monument. Efforts were made by the Department of the Interior during the 1960s to develop and propose to Congress structures which would prevent water entering Rainbow Bridge National Monument and thus possibly undermining Rainbow Bridge. But Congress rejected funding these proposals. In subsequent court cases, the conclusion has been reached that Congress's failure to provide for these protective facilities amended, in effect, the Colorado River Basin Project

Act. Thus, as of 1974, water can be lawfully backed up into Rainbow Bridge National Monument. The Sierra Club lost this battle.

These recent manifestations within the Colorado River Basin of the preservational thrust being more or less equal politically to the coalition of forces stemming from the traditional Conservation Movement have their counterpart in the nation as a whole. The New Conservation or Environmental Movement, which began in the early 1960s under the strong, charismatic leadership of Secretary of the Interior Stewart L. Udall, had preservation of natural areas largely in terms of their aesthetic values as its main focus. It gained much stronger political force in the later 1960s and early 1970s when environmental security issues in terms of human preservation from contamination, wildlife preservation, the threat of running out of resources and concomitant interest in no-growth came to the fore. Clearly, economic development, as the chief manifestation of progress, became seriously challenged by the Environmental Movement. The value that the public attaches to economic development, however, is still strong. This is manifest in the setbacks in environmental policy that have occurred as a result of the energy crisis, but they have been less than one might have expected. Much of the progressive thrust in American politics in its traditional egalitarian forms has lost much of its force. But, in the form of protection of consumer interests, civil rights, redistribution of wealth and income, and other common-man interests, it may find a workable alliance with environmental interests. The environmental interests of 1974, of course, are still concerned with conservation in the traditional sense of sustained-yield from use of resources. But, clearly, concerns of the Environmental Movement of the future must be a broader and deeper concern for maintaining a sustainable relationship for man in nature. What is not so clear are the effective political coalitional arrangements which can make the Environmental Movement of the future as sustainable a political force as was the traditional Conservation Movement for, more or less, half a century.

Literature Cited

Caulfield, Henry P., Jr. 1959. The living past in federal power policy. In *Annual Report*. Washington, D. C.: Resources for the Future, Inc.

____. 1961. Welfare, economics and resource development. In *Western Resources Papers*. Boulder: University of Colorado Press.

Easton, David. 1965. *A framework for political analyses*. Englewood Cliffs, New Jersey: Prentice-Hall, Inc.

LaRue, E. C. 1916. *Colorado River and its Utilization*. U.S. Geological Survey Water-Supply paper 395, Washington, D. C.: Government Printing Office.

Pinchot, Gifford. 1910. *The fight for conservation*. Americana Library paperback edition published in 1967. Seattle: University of Washington Press.

Powell, John, W. 1878. *Lands of the Arid Regions of the United States*. Cambridge, Mass.: Belknap Press of Harvard University.

Part II

Future Directions

6

Agriculture and Salinity

B. DELWORTH GARDNER
and CLYDE E. STEWART

The problems of irrigated agriculture and salinity are related in the Colorado River Basin. Complete analysis and resolution of salinity problems requires consideration of sources of salt other than irrigated agriculture. A total resolution of *water quality* problems in the basin would have to deal with other water uses and classes of pollutants. But it is not feasible to proceed that far within the confines of this chapter. Also, we do not presume to argue the merits of irrigation development, although the value of additional irrigation needs to be considered in appraising various salinity control programs.

HISTORICAL PERSPECTIVE OF THE PROBLEM

The Colorado River Basin is divided into upper and lower basins, with Lee's Ferry as the dividing point. Several key reference points are Lee's Ferry, Hoover Dam, Imperial Dam, and the International Boundary.

Salt enters the river from overland runoff and natural or point sources as well as from man-made sources which, in this context, are largely irrigation return flows. A crucial distinction is made between *salt loading,* which refers to increases in quantity of salt entering the stream channel, and *salt concentration,* which is the salt content of the water and is influenced by both salt load and quantities of water in the river system.

Population

As of 1974, 2.5 million people lived in the Colorado River Basin: nearly 90 percent of them in the lower basin. The Water Resources Council (Series C population) projects the population to more than double by 2020 to 5.8

Courtesy of Colorado River Board of California.
Fig. 6.1. Salt on a field in the Imperial Valley.

million, but the upper basin would have only about half a million people, or less than 10 percent of the total, while more than 5 million are projected in the lower basin.

Irrigated Land Use

The total irrigated acreage in the Upper Colorado Basin did not change markedly from 1920 to 1955 and was around 1.4 million acres. During the sixties, however, irrigated land in the upper basin was in the neighborhood of 1.6 million acres, including substantial irrigated non-cropland pasture and some land classed as idle irrigated.

The irrigated cropland acreage increased greatly in the lower basin in the mid-twentieth century to 1,225,000 acres. This increase was mostly in the Phoenix-Tucson area and was based on groundwater development. The irrigated acreage in the State of Arizona more than doubled from 1940 to 1960. Substantial depletions of the groundwater supply are occurring with concurrent deterioration in quality of water.

Much more land could be irrigated in both basins if water were available. The U.S. Soil Conservation Service has classified as land "suitable" for irrigation 38,760,000 acres in the lower basin and 7,058,600 acres in the upper basin.

Irrigated Crops

The upper basin's economy is basically forage-livestock. Relatively small acreages of fruit, sugar beets, beans, and vegetables are important to selected areas. But overall, the irrigated cropland is used mainly for forage crops and feed grains. Nearly one fourth of the "irrigated" land is in non-rotation, permanent pasture. About 1.1 million acres are in harvested crops of which 944,000 acres are in forage crops.

About 58 percent of the irrigated acreage of 1,225,000 acres in the lower basin is devoted to forage and grains. The remainder is highly intensive crops—cotton, vegetables, and citrus; citrus and vegetables have low salt tolerance while cotton is a high tolerance crop.

A spectacular enlargement of the livestock feeding enterprise has occurred in the lower basin in recent years. The average number of cattle on feed in Arizona was 73,000 over the period 1945–54, but by 1972 the number had grown to 600,000 head. This activity is significant from the standpoint of crop production, water use, and water quality.

Salinity Levels

The salt tolerance limits of crops cannot be set in any absolute sense, but the U.S. Salinity Laboratory (1954) established a general classification of the salinity hazard of irrigation water in parts per million (ppm):

Low	100–250
Medium	250–750
High	750–2250
Very High	>2250

Estimates show pristine water quality (i.e., natural water quality without salt contributions from man-made sources) at Hoover Dam at 330 ppm and above Imperial Dam to be 383 ppm. As of 1974, the salt content at this latter location was around 850 ppm, suggesting sharp deterioration from pristine levels.

Salt contributions from irrigation and thirty major point sources in 1965-66 were estimated by the U. S. Environmental Protection Agency in 1971 (Table 6.1).

Total salt load in the upper basin in 1963-64 was 26,160 tons per day, of which 52 percent was contributed by overland runoff and groundwater inflow. Of the remaining 48 percent, 37 percent came from irrigated agriculture, 9 percent from natural point sources, and 2 percent from municipal and industrial uses. This amount of salt was associated with 19,263 cubic feet per second which calculates to be 499 ppm. The contribution of irrigation was about 185 ppm.

In the seven Colorado River Basin states, nearly one-third of the irrigable lands are classified as saline (see Figs. 6.1 and 6.2). That is, the salt content in these soils is sufficiently large to have significant impacts on crop production and incomes.

The estimates at Imperial Dam, shown below, suggest the change that has occurred in the last 50 years. Presumably the changes are attributable to

TABLE 6.1

Daily Salt Contributions to the Colorado River

	Irrigation (tons/day)	Point sources (tons/day)
Green Sub-basin	3,528	363
Upper Main Stem	5,603	2,061
San Juan	518	25
Upper CRB	9,649	2,449
Lower CRB	1,180	1,990
Total	10,829	4,439

increased irrigation and out-of-basin diversions. But some question exists as to whether the changes in ppm were as great as shown.

Pristine	383
1926–35	619
1941–68	751
1958–63	787
1941–68 modified *	865
1965	839

Distribution of sources with some quality impacts of irrigation are shown as follows (U. S. Department of the Interior, 1967):

	Lee's Ferry	Hoover Dam	Imperial Dam
1941–61	544	684	743
1941–61 modified*	580	720	809

Salt pickup rates from irrigation return flows vary substantially among areas. Among the identified high-rate areas is Grand Valley, Colorado, where the rate is estimated at eight tons of salt per year for each irrigated acre (see Fig. 6.3). The Environmental Protection Agency (EPA) projected an average pickup of a little less than two tons per acre per year in the upper basin on several hundred thousand acres of new irrigation projected to 2010. Factors affecting salt pickup rates are being evaluated in a number of ongoing research programs. Figure 6.4 illustrates one such study in Imperial Valley, California.

---

*"Modified" assumes that irrigation projects were operated during the entire period (U.S. Department of the Interior, 1972).

Fig. 6.2. Salinity at tops of furrow irrigation ridges, Imperial Valley, California.

Institutional Framework

The Colorado River Compact, completed November 24, 1922, apportioned in perpetuity to upper and lower basins exclusive beneficial consumptive use of 7.5 million acre-feet each. The Upper Colorado River Basin Compact, finalized October 11, 1948, divided the water among the five upper basin states. The lower basin waters were allocated among the states by the Supreme Court of the United States June 3, 1963, in the case of Arizona vs. California.

The International Treaty of February 3, 1944, between Mexico and the United States guaranteed Mexico an annual quantity of 1.5 million acre-feet of water.

The above compacts and treaty made no references to quality of water. The United States has agreed to furnish water to Mexico that does not exceed the salt concentration at Imperial Dam by more than 115 ppm. Federal legislation and river basin conferences have led to an orientation to 850 ppm at Imperial Dam as a long-time goal.

Costs of Salinity

An indication of the potential magnitude of the problem can be obtained from a study by the Colorado River Basin Water Quality Control Project over the period 1960-1971 (Udis, 1968; Stewart, 1969; U. S. Environmental Protection Agency, 1971).

Fig. 6.3. Grand Valley, Big Salt Wash in Colorado. Salt leaching from natural rocks shows that the river carries a large natural salt load.

This study projected a basin population of 8.5 million by 2010, assumed no augmentation of water supply or shifts of water among basins or areas, and assumed construction of the Central Arizona Project. Further, the study projected an increase by 2010 of 425,000 acres of irrigated land above Hoover Dam. Increased acreages in the lower main stem sub-basin were more than offset by a projected decrease in irrigated acreage in the Phoenix-Tucson area. Acreages of vegetables and cotton were projected to increase relatively and, in the case of vegetables, absolutely.

The above assuption and projections led to the following water quality values in ppm:

	1960	*1980*	*2010*
Hoover Dam	697	876	990
Imperial Dam	759	1056	1223

Penalty costs or effects of yield decrements in agriculture and non-agricultural costs from increased salinity were estimated for the lower basin and for southern California. Total annual direct and indirect (input-output study) costs at 1960 prices by 2010 were estimated at $5.9 million in the lower main stem, and $19.1 million in southern California. More than 80 percent of the costs were in irrigated agriculture. Direct costs were about 60 percent of total costs. These costs are increases over the base period, assuming no remedial programs.

The off-site costs are associated with new development and increased incomes and production in other areas. Their significance attaches (1) as a trade-off with the new development, and (2) to the evaluation of special control measures and programs to reduce the quality impacts of new development.

Over the years, neither the market system nor the institutional environment of the Colorado River Basin were conducive to prevention or retardation of water quality deterioration. In fact, they have been conducive to non-action. Only when the problem became obvious as costs rose, did concern arise.

History has demonstrated a deterioration of quality with additional irrigation development and out-of-basin transfers of water. This lower quality is partly a function of the increased loading of salts and also a function of less water as a result of productive use of water supplies upstream. The total effects have not all been adverse. The costs of more salt in less water downstream are associated with increased production and incomes upstream and in out-of-basin areas.

We proceed next to a conceptualization of that part of the salinity problem which results from irrigation return flows. At least in part, however, the application may be quite appropriate for other non-irrigation sources of the problem.

AN ECONOMIC MODEL OF SALINE RETURN FLOWS

One fundamental reason for the salinity problem is the existence of what economists call "negative externalities" arising out of irrigation itself. Crops valuable to society and to the farmer are produced on saline soils. But as a consequence of irrigation, salts of the soil are dissolved in the water and enter the river as return flows. When this saline water is utilized in other areas downstream higher production costs result for downstream users in the form of reduced yields and changes necessary in cropping patterns to more salt-resistant, but profit-reducing varieties.

The irrigator upstream does not bear the increased costs imposed on downstream users. Although it is not his intention to injure anyone, the water course is a free resource for disposing of the dissolved salts jointly produced with the valuable crops. No legal or market mechanism exists that incorporates the cost of this salt in the price of the upstream crops. The result is that the social value (value to society) of the upstream irrigation is less than the net private benefits from producing crops.

Let MNB (Fig. 6.5) represent the marginal net private benefits of using various quantities of irrigation water to produce crops on a representative acre of saline land. Assume an optimal level of irrigation technology, and optimal cropping patterns and technical production conditions as seen from the view-

Courtesy of Lyman Willardson, USDA ARS.
Fig. 6.4. Experimental salt leaching plots using
Colorado River water, Imperial Valley, California.

point of the irrigator. Variable production costs are subtracted from crop revenues yielding marginal *net* benefits. The function MNB is negatively sloped because of the conventional principle of diminishing marginal returns to increasing quantities of water, assuming adequate water supplies and optimal deliveries over the irrigation season. The function is presented as linear, but in reality it may not be. The logic of the analysis holds as long as it has negative slope.

MED (Fig. 6.5) represents a schedule of marginal *external* damages inflicted on downstream users by diversions of increasing quantities of water upstream. These damages are related to two factors: (1) Irrigation water consumptively used upstream cannot be available to downstream users, and (2) the saline return flows impose higher production costs on downstream users. *A priori,* one would expect that the greater the upstream diversions the greater the damages downstream. MED may not be linear as presented, but a necessary condition to the argument is that it have positive slope.

MED does not enter the decision calculus of the upstream irrigator. If one assumes the attempts to maximize one's own net benefits, one will

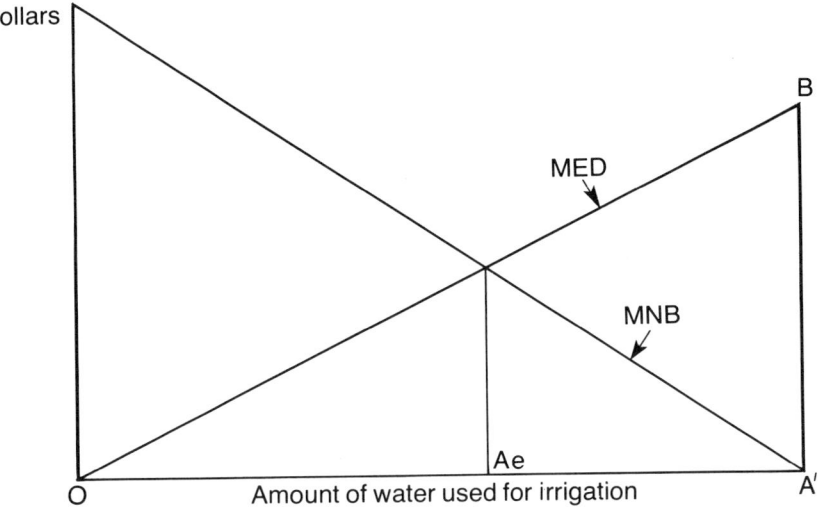

Fig. 6.5. Marginal damages and benefits utilizing irrigation water.

extend water use per acre to OA' where MNB is zero. The total per acre net benefits is the area under the MNB curve. This may be referred to as private water "rent." The external damages at that level of use are A'B.

The socially-optimum position is OA_e where MED = MNB. Before OA_e, the marginal net benefits accruing to the irrigator exceed the marginal external damages imposed on others, and society benefits from expanding per acre water use. After OA_e the reverse is true.

Options for Bringing Salinity Production Toward the Optimum

A basic assumption underlying the following analysis is that the American people wish to have the salinity problem managed so as to maximize the total national economic product from water use. This implies that a social optimum can be defined by an "efficiency" criterion, i.e., an optimum situation will exist when the total value of the economic product along the entire system is maximized. Social welfare may also be influenced by "equity" considerations, however, i.e., how the economic product is distributed among water users along the system. If equity criteria are deemed relevant, social welfare may be enhanced by permitting relatively low-income irrigators to utilize more water, even if marginal external damages exceed marginal net benefits and total economic product is thus below a maximum. Thus, if both criteria are used to arrive at a social optimum, some weighting of "efficiency" and "equity" must be done. Several options for salinity control will be discussed briefly: (1) Investment in water quality improvement, (2) litigation brought by downstream users, (3) imposition of quality standards, and (4) implementation of direct economic incentives to improve quality.

Investment in Quality Improvement. This option is basically different from the others discussed. Investment to improve quality could leave both upstream and downstream users better off, depending on the distribution of the costs of the investment. The other options basically reduce the welfare levels of one region while improving the welfare position of the other.

Investment in quality improvements could take many forms—investment in off-farm desalination plants or other mechanical devices to remove salt, investment in on-farm practices to reduce salt in the return flows or in water saving methods that would reduce needed diversions and thus augment river flow downstream, and/or investment to increase water supplies in the entire system or to alter seasonal deliveries that would minimize the damages of given salt loads. Obviously, investment in quality improvement can apply to all forms and sources of pollution, and may be about the only effective alternative for reducing salinity from overland runoff and natural or point sources.

Off-farm investment to develop new water supplies, or in storage to improve the timing of deliveries, or in reducing evapotranspiration losses by phreatophytes, would shift the MED curve to the right. Each level of water use by irrigators would inflict smaller damages on downstream users. This implies also an increase in optimal water use upstream, and thus both groups could be made better off. These investment alternatives should be evaluated by the economic criterion that if they produce more social benefits than their resource opportunity costs, they should be undertaken. They can only very indirectly contribute to the solution of the problem posed in Fig. 6.5, however. Even though MED shifts to the right and the socially optimum level of water use upstream increases, there will still be a discrepancy between the social optimum and the private optimum at OA's unless the shift is very large. There is no incentive for the upstream irrigator to act in a socially optimum manner. It is likely, therefore, that these investment options can mitigate the damages caused by the externality problem, but cannot solve the problem by inducing the irrigators to use the socially optimum quantity of water.

If the investment is made on-farm, and the irrigator bears the full costs, the expected result would be a leftward shift of MNB as well as a rightward shift of MED. If so, the private optimum will be less than OA', but the social optimum might be more or less than OA_e, depending on the relative shifts of MED and MNB. The result of the investment would probably move the private optimum closer to the social optimum. But the irrigator would have no incentive to make the investment to hurt himself in order to aid downstream users. A good argument exists for public subsidy to pay for the social improvement downstream. Any investment to improve quality, whether on-farm or off-farm, should be undertaken if social benefits exceed resource costs, and the distribution of the costs of this investment should be equitable as possible.

Litigation. In a free society in general, parties injured by actions of other parties can bring damage suits in a court of law to recover their losses. Downstream users could sue upstream irrigators and either prevent them from producing wastes or force them to pay damages. This is seldom done. Although their discussion applies mostly to industrial pollutors, Kneese and Bower's (1968, pp. 85-6) explanation applies remarkably well to irrigation waste as well: "(1) Adversary proceedings are a cumbersome procedure . . . , (2) wide dispersion of damages makes it hard to bring suit for full damages, [wide dispersion of 'damagers' makes it doubly difficult]. . . , (3) waste discharge imposes costs in a highly variable fashion over time . . . , (4) damage may be irreversible. . . , and (5) legal standards of 'reasonable' [damages] are notoriously vague."

It is very difficult to see how an effective damage suit could be brought by so many damaged parties against so many prior users. Besides, as we argued earlier, the damages created by the return flows are not an intentionally harmful act by the irrigators, but an accepted consequence of irrigation that is sanctioned by established water rights, interstate water compacts and international treaty.

Rules and Standards. Crops and saline return flows are produced jointly in almost fixed proportions. Because of the intricate and diverse plant, soil, and water relationships in irrigation, however, return flows are very difficult to monitor and assign to an individual irrigator. Therefore, it is difficult to visualize how meaningful standards relative to return flow could be applied to each irrigator. Perhaps they could be to a given project or area. But then rules would have to be invoked that limited irrigation to certain times when return flows would be less salty or to the least saline soils. Restrictions could also be placed on the crops that could be grown, or the water right altered to reduce quantities diverted. Any of these actions would be costly to irrigators and would reduce their management flexibility. They would also require a strong monitoring and enforcement agency. As still another alternative, irrigators may be forced to impose upon themselves investment in water saving practices or quality improving mechanical devices. In terms of Fig. 6.5, the MNB function would be shifted drastically to the left. In theory, MNB might shift far enough so that the private optimum would be at the previous social optimum OA_e. But the social optimum has also shifted to the left. Many of the private net benefits upstream would be destroyed by employing such a standard, and this option would be highly inequitable and would bear higher costs for upstream users in order to confer benefits downstream.

Economic Incentives. In recent years a sizeable literature has arisen on the issue of whether or not externalities, such as those discussed here, can be optimally regulated by negotiations between the injuring and the injured parties. R. H. Coase (1960) demonstrated in his now classic article that this private bargaining would produce a socially optimal result, providing transaction costs were zero. Potential gainers would pay the polluters to reduce their pollution to levels that would make both groups better off. Transactions costs could be small if the number of negotiating parties is small. They most certainly would not be small in the situation we are considering where the number of irrigators causing the externality and those harmed by it number in the tens of thousands (see Kneese, 1971, for example). There is no conceivable way that they could bargain individually to the socially optimal level of salinity and water use.

Of course, possibilities may exist for representatives of *groups* of individuals (say, state government officials) to bargain for their constituents. These bargaining decisions may well lead toward the socially optimal position if all constituents are fairly represented and a mechanism for making payments to alter pollution behavior could be conceived and implemented (see McKean, 1972, pp. 39-41). History, however, does not provide us with many examples of where this kind of bargaining has been successful. Turning now to penalties that might be imposed on irrigators or others that inflict external costs on others, perhaps the most straightforward type of economic sanction is a user surcharge or tax on water use. If this surcharge were placed on each unit of water diverted, the MNB curve of Fig. 6.5 would shift to the left by the amount of the surcharge. Obviously, if the position of MNB and OA_e were known, the surcharge could be set at the level which would produce a private optimum at OA_e units of water use. The irrigator would respond to this surcharge in those ways which would affect his profit position least unfavorably. His management would be more flexible in responding to a water surcharge than to a water quality standard or quantity quota.

A tax surcharge on water utilized by upper basin irrigators would mean that they would be required to bear the costs for providing benefits to downstream users. This may be inequitable. The beneficiaries may be asked to help pay for these benefits. They could be required to pay a higher price for higher quality water and, with the proceeds, a per-acre lump-sum subsidy paid to the irrigators upstream.

Kneese and Bower (1968) report several instances where user charges have been successfully used to control industrial waste in Winnipeg, Canada; Springfield, Missouri; the Delaware estuary, and the Ruhr industrial area of Germany. They point out that private firms respond to these charges by changing production processes, improving management, and/or treating wastes. There is no reason why irrigators would not behave in the same way.

It may be necessary to tax water at a higher rate if used on highly saline soils, although this would be difficult to determine and enforce. Tietenburg (1973, p. 521) recently argued, however, that an "efficient set of taxes for waste control will change over time and might be different for various firms, not only because the benefits of a unit reduction of pollutant concentrations change over time and the number of sources changes, but also because the relationship among pollutants and waste products is governed by varying and basically uncontrollable elements in nature. Hence, social efficiency will not, in general, be achieved by temporally uniform tax rates." This comment would seem to apply to irrigation waste and suggests that user surcharges (taxes) may need to vary over time as water supply conditions change and over various irrigators if the taxes are to be efficient.

SOME ALTERNATIVE FUTURES

While economics can specify conditions for optimality, the market system does not automatically achieve these conditions. In fact, the return flow problem is a prime example where the market cannot handle the problem. Nor does action to remove point sources of pollution or to alleviate salt concentrations from diffuse sources come about by action in the private market. These are public problems that must be solved by public action.

The alternative futures noted below are program responses arising from public action. The focus is generally on the upper basin. In a general sense, the physical and hydrologic structure of the upper basin is largely one of independent sub-basin entities such that quality output of one sub-basin does not influence another sub-basin. Diversions out of the upper basin are usually in the headwaters so that upper basin export areas are not greatly influenced by quality deterioration from the irrigated area. Rather, the immediate impacts of lowered quality are on the lower basin, southern California, and Mexico. Major U.S. users of water below Hoover Dam are offstream and their actions are largely localized.

The status quo in terms of policy is hardly a viable alternative. Developments in the upper basin will diminish the quantity of water to the lower basin and undoubtedly lower the quality there without control programs. The Central Arizona Project will substantially decrease the supply of water at the International Boundary.

Several studies made since 1960 serve as illustrations. These studies seem to have culuminated in proposed legislation for a single program which presumably has been agreed upon by several federal agencies and by the respective states.

CRB Water Quality Control Project

The assumptions and cost estimates of this comprehensive study were discussed earlier in this chapter. Eight salinity control programs were analyzed: three salt-load reduction programs, four flow augmentation programs, and one program to demineralize water supplies at point of use. This study describes the water quality problem, estimates its present and potential economic dimensions, and locates, describes, and evaluates possible control measures and associated costs. Although a major source of salt, Blue Springs in the Little Colorado sub-basin is a basic element in these program formulations. The Department of the Interior concluded subsequently that the Blue Springs program was not feasible.

The results of this study are a basic information source for consideration of water quality in the Colorado River Basin. The study is also basic to subsequent studies and formulation of programs. However, rather than present more findings and conclusions of the study, attention is directed to subsequent and more current studies and proposals which might be viewed as modifications and applications of the Water Quality Control Project Study.

U. S. Water Resources Council Studies

The U. S. Water Resources Council (1971) conducted comprehensive river basin studies in the upper and lower basin.

The upper basin study group analyzed four alternative projections: (1) Basic Office of Business Economics-Economics Research Service (OBERS) (U.S. Water Resources Council, 1972), (2) regional interpretation (RI) or OBERS, (3) states alternative 1, and (4) states alternative 2. Projected populations (1970 = 346,000) for 2020 ranged from 616,000 to 902,000, respectively. Water depletions (3.4 million acre-feet in 1965) ranged from 5.1 million (OBERS) to 8.1 million (states-2). Only alternatives (3) and (4) assumed oil shale development and less irrigation along with more people. Also, alternative (4) enlarged depletion to 8.1 million acre-feet on the assumption that the Mexican Treaty is a national obligation. Selected acreage, water and quality projections are shown in a later section of this chapter.

Under the OBERS-RI projection of 500,000 acres of new irrigated land, the salt concentration at Lee's Ferry would increase by the year 2020 from 586 ppm to 820 ppm without and 600 ppm with an improvement program. The program viewed most favorably is largely on irrigated land plus one stream diversion and one desalination project. Capital costs by the year 2000 were estimated at $230 million.

As indicated earlier, the OBERS and OBERS-RI projections did not provide for oil shale development. States alternative 1 with 6.5 million acre-feet depletion included one million barrels per day by 2000 in Colorado and

0.5 million barrels by 2020 in Utah. States alternative 2 would include four million barrels per day by 2020.

Indications in the early 1970s were that oil shale development might get underway in a short time. At a magnitude of one to four million barrels per day by 2020, water requirements with present processes would be 200,000-800,000 acre-feet diversions per year with net disappearance of 50 percent.

The lower basin study centered primarily on two projections—OBERS and modified OBERS—with the latter setting the primary focus. Apparently both projections are optimistic and may be unrealistic because augmentation of the present or prospective basin water supply would be necessary. The "modification" is based on an assertion that OBERS did not fully recognize likely new development on Indian lands and in some groundwater areas.

Qualities projected in 2020 without a control program were 1,050 ppm at Hoover Dam and 1,350 at Imperial Dam. While not explicit about a control program, the study showed "with program" salinity levels in 2020 to 850 ppm and 1,030 ppm at these two locations.

A large number of potential improvement measures were listed for the lower basin. But the study group concluded that these measures were mostly infeasible now for institutional, legal, and economic reasons. The most promising measures are (1) reduction of evaporation, (2) vegetative management to increase water yield, and (3) augmentation of supply through importation or desalination.

Proposed Legislative Programs

In 1972, the U.S. Bureau of Reclamation set forth a 10-year water quality program for the Colorado River Basin, and in 1973 this program was introduced in the U.S. Congress as bills 5-1807 and HR-7774. As of early 1974 they were still pending before the Senate Interior Committee. The basic policy incorporated into the legislation was approved by the respective states in April, 1972 and by EPA on June 9, 1972.

The stated objective of this program was to maintain salinity concentrations at or below levels then found in the lower main stem of the Colorado River. This goal viewed the salinity problem as basin-wide. The modified quality at Imperial Dam was 865 ppm.

Comparisons of the Water Resources Council and Bureau of Reclamation studies are shown in Table 6.2. Briefly, the reclamation proposal is for less new land, less water depletions, and high quality downstream.

Senate Bill 1807 would have authorized construction of the LaVerkin Springs, Paradox Valley, and Grand Valley salinity control units as the initial stage (see Fig. 6.6). Completion of planning reports on other point, diffuse, and irrigation units would have been authorized. Irrigation source control

TABLE 6.2
Comparative Data for Two Water Studies

	Water Resources Council		Bureau of Reclamation	
New Irrigated Land (acres)				
Upper Basin	500,000*		350,140	
Lower Basin	358,000†		88,640	
	Present	**New**	**New**	
Water depletions (million acre-feet)				
Upper Basin	6.5	2.1	1.9	
Lower Basin	4.8	0.9	.3	
	Program‡		**Program**	
	W/out	**With**	**W/out**	**With**
Water Quality (ppm of salts)				
Lee's Ferry	820	600	—	—
Hoover Dam	1,050	850	—	—
Imperial Dam	1,350	1,030	1,250§	845‖

*Regional Interpretation—OBERS.
†Modified OBERS.
‡Year 2020.
§Year 2000.
‖Stabilized after 1990.

areas included Lower Gunnison, Uintah Basin, Colorado River Indian Reservation, and Palo Verde Irrigation District plus Grand Valley, which would
be authorized for construction.

Point Source Control Projects. Table 6.3 provides comparative data for
these projects which are all above Hoover Dam.

Blue Springs in Little Colorado, Arizona, is a major source of salt. The
study did not recommend a program there because the springs are (1) a source
of 160,000 acre-feet of water, or half the Little Colorado, (2) inaccessible,
and (3) the subject of Indian folklore.

The total program would reduce salt concentration by 38 ppm at Imperial
Dam. Water losses would also be incurred, however.

Diffuse Source Control. These projects pose difficult data and control
problems, but they were all listed in the proposed authorization act.

The areas listed in Table 6.4 are heavy contributors of salt. To achieve
-32 ppm apparently would require substantial costs and losses of water.

Irrigation Source Control. A summary of point, diffuse, and irrigation
sources included in PL 93-320 (the Salinity Control Act of 1974) and in the
Bureau of Reclamation program emphasizes the importance of agricultural
responses to salinity control in the basin. In order to achieve the development
envisioned, a net reduction of 405 ppm would be necessary by the year 2000.

Courtesy of Colorado River Board of California.

Fig. 6.6. LaVerkin Springs is a point source for salinity in the Colorado River.

Point source projects (-38 ppm) and diffuse source projects (-32) would be smaller elements in the program. The irrigation source control program is scheduled to achieve a reduction of 90 ppm in five areas. The remaining component involves more water through desalting, weather modification, and geothermal sources.

The goal of the scheduling and farm management program would be to reduce salt loading from irrigation return flows. This would be accomplished

TABLE 6.3

Salt Loading and Potential Salt Reduction for Selected Point Sources

	Salt/year (tons)	Reduction (tons/yr.)	Hoover Dam (ppm)
LaVerkin Springs	100,000	100,000	− 6
Little Field Springs	30,000	30,000	− 2
Glenwood-Dolores Springs	500,000	200,000	−15
Paradox Valley	200,000	180,000	−14
Crystal Geyser (oil test well)	4,000	4,000	− 1
Blue Springs	550,000	0	

TABLE 6.4

Comparison of Selected Diffuse Salt Sources

	Flow (ac-ft)	Salt (tons)	Salt Removal (tons)	Water Loss (ac-ft)	Effect at Imperial Dam (ppm)
Price River	74,000	240,000	100,000	25,000	− 8
San Rafael River	95,000	190,000	90,000	30,000	− 7
Dirty Devil River	72,000	200,000	80,000*	NA	− 7
McElmo Creek	31,000	115,000	40,000*	NA	− 3
Big Sandy River	30,000	180,000	80,000*	NA	− 7
Totals	302,000	925,000	390,000		−32

*Method not shown.

by minimizing the quantity of water that enters the groundwater regime where saline formations are contacted.

Irrigation efficiency may be increased by (1) proper and timely irrigation applications without more labor, (2) additional labor inputs, and (3) improved on-farm systems and total distribution systems through capital investment. Improved scheduling and management of farm irrigation would require a substantial increase in informational and operational programs as well.

A major obstacle to this program is the institutional structure of western water laws. "Water savings" may diminish water rights if less water is diverted through time. Increased irrigation efficiency in use of a given quantity diverted not only may reduce water deliveries downstream, but also the return flows may be more saline. Only interstate compacts that specify both quantity and quality of deliveries can adequately protect all water users. As of 1974 there were no such compacts.

Point and natural runoff sources, which are a major segment of quality effects, present a different potential than irrigation return flows in terms of control, solutions, and responsibilities. Point and natural sources of salinity relate generally to the public and public agencies and control thus does not impair private property rights. In contrast, irrigation return flows and associated salinity largely involve private farmers and effective control presents a different set of problems. Most importantly, all salinity sources involve off-site or external relationships and effects, which give rise to extremely complicated questions of responsibility, water rights, net economic and social impacts, and institutional and other arrangements for corrective action.

Costs and Cost-Sharing

In benefit-cost analysis terminology, the problem of off-site or external effects is one of incidence of the benefits and costs of the external effect itself and the corrective action. If the beneficiaries of economically feasible corrective action can be identified, an equitable distribution of the costs would relate cost allocation to benefits received.

Capital expenditures in the 1972 program (U. S. Department of the Interior, 1972) were estimated at $400 to $500 million. Substantial funding would also be needed for investigations, feasibility studies, and educational programs. The report recommended that costs be ''shared by the beneficiaries.'' Differences of opinion no doubt exist as to who the beneficiaries are and how much they benefit.

The proposed legislation ties federal responsibility to Mexico along with federal landownership and federal pollution control policy as argument for allocating 75 percent of the total costs as non-reimbursable. The remaining 25 percent would be allocated between the upper and lower basin development funds with the allocation dependent on benefits, causes of salinity, and availability of revenues in the funds with an upper limit of 15 percent to the upper basin fund.

SUMMARY AND CONCLUSIONS

Clearly the economic or environmental optimum levels of salinity at Hoover Dam or Imperial Dam have not been established. The proposed 850 ppm salinity level at Imperial Dam seems to be more of a historical coincidence than a scientifically based optimum. Apparently this goal was set when someone decided that the level of quality should not deteriorate further at this location.

Quality goals and programs appear to have been settled without adequate research foundations in economic and institutional feasibility analyses. The irrigation return flow problem seems particularly lacking in data, analysis, and synthesis. In our opinion, we simply do not yet know the optimum level of water quality in the basin at various locations. What levels can be achieved? What programs are economically feasible and institutionally possible?

It is clear that if 850 ppm are to be maintained or reduced at Imperial Dam, a heavy burden will be placed on future agricultural development. Corrective action will be especially costly to farmers in upstream areas unless incentives are provided. This control program is also by far the most complex one to implement. Inducements or enforcements leading to farmer and irrigation district action would seem to be necessary.

One major problem is the conflict between developmental and environmental interests, whether real or imagined. The Water Resources Council Principles and Guidelines propose project objectives both to enhance national economic development and to enhance quality of environment. Which is consistent with a quality level of 850 ppm at Imperial Dam? Should this level be the reference point, or should a broader look be taken? What are the trade-offs between these two objectives as the quality level is increased or decreased from this level? This broader look is difficult as long as upper and lower basins are viewed as independent economies. As of 1974 the prevailing view seems to be that new development which deteriorates quality must be offset by quality control programs by the new developers or by their neighbors. It ought to be at least considered that downstream users bear part of the cost of upstream quality control.

There is some hope in some quarters that water importation in major quantities is possible. The Water Resources Council study in the lower basin clearly assumed this, but the more recent Bureau of Reclamation proposal was based on substantially less new irrigation development and water depletions.

Partly because international problems will be discussed later in this program, we have largely avoided the recent water quality agreement between the United States and Mexico. Quite obviously it is a crucial issue in resolution of water quality problems in the Colorado River Basin if it is taken seriously. The basin states are holding to the view that the international agreement is a national problem and that they can proceed independently with development. It is difficult to see how the agreement can be met independently of basin development. The large desalination plant on the Wellton-Mohawk channel (see Fig. 6.7) will provide 140,000 acre-feet per year which mixed with 1,360,000 acre-feet at 1,240 ppm yields 1.5 million acre-feet at 1,124 ppm. This plant will require $67 million of the total $115 million planned. It appears that the remaining measures will have to be more efficient than the desalination plant to achieve 965 ppm.

In many ways we have attempted to report the status of the water quality problem. We have also discussed in principle various policy options that might be considered in coping with the problem. We have attempted to survey the relevant literature. In the end, we are impressed that many more research results are required if our planning and policies are to match our hopes for success.

In the first place, there must be a synthesis of the economics and hydrology of the basin as a whole. We see no way that a socially optimum level of water quality can be estimated without giving empirical content to the marginal net benefit and marginal external damage functions of Fig. 6.4. We have more hydrologic and water quality information than economic. Studies that show the value productivity of irrigation water in the upper basin and the value productivity of various levels of water quality in the lower basin must be given top research priority.

We also need hard economic data on the costs and expected benefits of alternative programs to improve quality. Many government agencies seem to be proceeding with quality improvement as if it were obvious that the

Fig. 6.7. Wellton-Mohawk Drainage Canal outfall drain discharging below Morelos Dam. This canal conveys the highly saline pumped drainage water from the Wellton-Mohawk Project to a point below the last gravity diversion on the Colorado River.

benefits will exceed the costs. Public and Congressional support for these programs would be more easily obtained if it were clear that they are economically feasible.

Finally, there are great gaps in our knowledge in assessing the institutional workability of the various policy options to control salinity. We have suggested alternatives that have been employed elsewhere without firm evidence that they could be successfully applied to the Colorado River Basin. A great deal of attention should be given to organizational arrangements that would permit negotiation between upper and lower basin states. Perhaps some kind of regional authority or basin-wide conservancy district is needed. How the costs of quality-improving investment and institutional improvement should be distributed between upper basin irrigators and lower basin beneficiaries is also a problem of first priority.

Discussion by Harold E. Dregne

Gardner and Stewart presented a careful analysis of the salinity problem in the Colorado River Basin and discussed several proposed solutions. They note that there must be synthesis of the economics and the hydrology of the basin as a whole if a socially optimum level of water quality is to be estimated. Attempts should be made to develop organizational arragements which would permit the upper and lower basin states to negotiate costs for salinity control measures rather than assign all costs to the upper basin states. Furthermore, economic data on costs and expected benefits of alternative quality improvement programs are needed.

The additional alternative of reducing the acreage of low-value forage, pasture, and feed grain crops in the upper and lower basin should be explored. This position is based on data given by Gardner and Stewart.

In addition, a report of the U. S. Department of the Interior (1973) mentions that proposals have been made for about 200,000 presently irrigated acres of land in the upper basin to receive supplemental irrigation water. The report also describes eleven proposed water quality improvement projects for the upper basin. Capital expenditures for the salinity control projects are estimated to be about $400 to $500 million.

In view of the low acre-value of the forage, grain, and pasture crops grown on 90 percent of the land, the high cost of proposed salinity control projects, and the likely increase in demand for water for oil shale and coal gasification projects, consideration should be given to restricting irrigation to higher value crops. Water quality is certain to be improved by the reduction in irrigated acreage, perhaps to the point where no other salinity control measures need to be taken.

A water tax would accomplish the acreage reduction, but would impose hardships on the irrigation farmers growing the traditional low-value crops. A better procedure might be to buy land and the water rights at a price which would be attractive to the farmers. Using only the minimum figure of $400 million for the proposed salinity control projects, it would be possible to pay $800 per acre for 500,000 acres of irrigated land instead of investing that money in projects requiring perpetual maintenance. Since there would be no need for additional irrigation projects, a one-time expenditure of much more than this could well be cheaper than undertaking projects which would exacerbate the problem (more irrigated land) or require maintenance forever (desalting plants, by-pass canals, waste water disposal).

At a time when the demand for non-irrigation uses of water in the upper basin for oil shale and coal development are expected to rise rapidly, the opportunity seems to be at hand to change the water development emphasis from irrigation to industrial and municipal. The latter users are prepared to pay much more for their water than are irrigation users. This may enable the irrigated farmers to sell out at highly favorable prices. The combination, then, of government subsidy for irrigated land retirement and of increased demand for water by the non-agricultural sector could result in a beneficial transfer of water use. Such a transfer could reduce or eliminate the salinity problem in the Colorado River Basin without construction by the federal government of any salinity control devices, while at the same time easing the water supply situation in the upper basin.

Literature Cited

Coase, R. H. 1960. The problem of social cost. *J. Law and Econ.* 3:1-44.

Kneese, A. V. 1971. Environmental pollution: economics and policy. In *Am. Econ. Rev.* 61(2):153-168.

Kneese, A. V. and Bower, B. T. 1968. *Managing Water Quality: Economics, Technology, Institutions.* Baltimore: Johns Hopkins Press.

McKean, Roland N. 1972. American property rights, appropriability and externalities in government. In *Perspectives of Property,* eds. Gene Wunderlick and W. L. Gibson, Jr. Pennsylvania State University.

Stewart, Clyde E. 1969. *Economic Impacts of Water Quality and Quality Constraints on Agriculture in the Colorado Basin–An Interindustry Projections Study.* Logan, Utah: Economic Research Service, U.S. Department of Agriculture.

Tietenburg, T. H. 1973. Specific taxes and control of pollution: a general equilibrium analysis. *Quart. J. Econ.* 87(4): 521.

Udis, Bernard, ed. 1968. *An Interdisciplinary Analysis of the Colorado River Basin in 1960 with Projections to 1980 and 2010.* 9 vols. Boulder: University of Colorado.

U. S. Department of the Interior. 1967. *Quality of Water–Colorado River Basin.* Progress Report No. 3. Washington, D. C.

___. 1972. *Colorado River Water Quality Improvement Program.* Washington, D. C.

___. 1973. *Quality of Water.* Colorado River Basin Progress Report No. 6. Washington, D.C.

U. S. Environmental Protection Agency. 1971. *The Mineral Quality Problem in the Colorado River Basin.* 5 volumes. Washington, D. C.

U. S. Water Resources Council. 1971. Upper Colorado River comprehensive framework study. Washington, D. C.

___. 1972. 1972 OBERS Projections. 5 volumes. Washington, D. C.

U. S. Salinity Laboratory. 1954. Diagnosis and improvement of saline and alkali soils. U. S. Department of Agriculture, Agricultural Handbook No. 60. Washington, D. C.: Government Printing Office.

7

Energy Resources Development

JARED CARTER

This chapter will outline U. S. Department of the Interior policy as it relates to energy development in the Colorado River Basin, will present the rationale for that policy, and will discuss a few implications for the future.

POLICY

The Department of the Interior anticipates and plans to encourage substantial development of coal and oil shale in the Colorado River Basin in the years ahead. Through its oil shale and coal leasing programs, various coal research and development projects, and some related past and current studies, the Department is encouraging development of the technology and resource data necessary to facilitate this development. It is also learning whether development can occur at acceptable costs and is developing procedures and technology to assure that environmental damage is kept to the lowest practicable level.

The facts believed to be most relevant to energy policy and some of the key issues involved will be discussed in the sections that follow. While ultimate precision and certainty cannot be ascribed to all the data used, they are believed to be accurate enough to support the arguments made.

In 1972, total U. S. consumption of energy was the equivalent of about 32 million barrels of oil per day from sources indicated in Table 7.1. We note that other sources, such as geothermal, supplied minute percentages of our total consumption.

U. S. net energy imports in 1972 were equal to about 4.8 million barrels of oil per day—or about 14% of total national energy consumption. This includes importation of 30% of the petroleum.

Until the early 1970s, projections of domestic consumption were, roughly, that by 1980, overall U. S. energy demand would be the equivalent of about 45 million barrels of oil per day. The Bureau of Mines predicted that this demand would be met from sources shown in Table 7.1. By 1980 other

TABLE 7.1
Sources of U.S. Energy for 1972 and Projected for 1980

Source	Percent of Total	
	1972 Actual	1980 Projected
Oil	42	44
Natural Gas	36	28
Coal	17	17
Hydro	4	4
Nuclear	1	7
	100%	100%

sources hopefully will be making a larger contribution than they are today, but they will not be greatly significant.

To meet demand in 1980, the Department estimated that at least nine million barrels of oil per day would have to be imported, i.e., about 20% of our total energy requirements and 44% of our total petroleum or an increase of almost four million barrels of oil per day since 1972.

Recent events have led most Americans to the conclusion that it would be not only very costly, but also against the nation's foreign policy interests, to become this dependent upon insecure foreign sources of supply. President Nixon announced Project Independence, with the objective of increasing domestic sources of supply sufficiently by 1980 to avoid this result.

This chapter will not go into the details of Project Independence. However, two additional points about the 1980 supply/demand projections underlie and support adoption of Project Independence and relate to the need for future development of energy resources in the Colorado River Basin. First, these projections assumed a continued and substantial program of domestic exploration and development for oil and gas, coal and nuclear energy. They also assumed a reasonable level of research and development (R and D) on other energy sources as well as reasonable R and D on the technology of energy production, transmission and utilization—with the objective of increasing efficiencies. Second, even a greatly increased research and development program on new, clean energy sources, such as solar and geothermal, coupled with a maximum energy conservation program, would not alter our supply/demand situation sufficiently to avoid greatly increased dependence upon foreign energy sources by 1980.

The second point is not universally accepted. Many honest, well-intentioned people believe that sooner or later U. S. residents must drastically alter their lifestyles, because fossil fuel energy sources are finite and that alteration of lifestyles must begin now rather than later in order to ease the pain of future generations and protect our environment now.

The maximum possible amount of energy consistent with opportunities for all our citizens should be conserved and a commitment should be made to the maximum practical research and development effort to develop clean, plentiful energy sources from such inexhaustible sources as the sun. However, there is very broad acceptance of the conclusion that the nation simply cannot solve our energy supply/demand dilemma for the next decade or two if its initiatives are limited to a strong conservation program and a strong R and D program for new, clean energy. Under 1970s circumstances, for any administration to assume otherwise and therefore fail to emphasize and support programs that rely upon known resources and technologies to solve our known, near term problems would be irresponsible.

THE RESOURCES OF THE BASIN

The second set of facts to be emphasized relates to the energy resources of the Colorado River Basin.

Oil, gas, and hydroelectric power are all produced in the basin. Possibly, nuclear stimulation or hydro-fraction will increase natural gas production greatly by releasing the large amount of natural gas present in tight formations in some areas of the basin. However, I do not expect any really significant improvement to our supply picture to come from these basin resources in the next several years. Coal is the resource that can greatly increase domestic energy supply in the immediate future. Oil shale holds great promise for the intermediate future. Both resources give rise to environmental concern.

Within the basin there is enough coal and oil shale to supply the nation's projected energy demands for at least two hundred years. There are estimated to be over 230 billion tons of coal recoverable by present underground or surface mining technology. There are an estimated 1.8 trillion barrels of shale oil, at least 600 billion of which are thought to be recoverable by technology that either exists now or is expected to be available in the near future (Fig. 7.1). These resources have not been developed because of the ready

Courtesy of Environment and Man Program, Utah State University.
Fig. 7.1. Oil shale outcrop near Vernal, Utah. Reserves in the Colorado Basin total 1.8 trillion barrels.

availability of cleaner and cheaper fuels. Moreover, there remain serious environmental questions that must be answered before full development is allowed to proceed.

The federal government has taken, and is taking, several steps to produce the information and technology necessary for decisions at all levels about how these resources should be developed. The Southwest Energy Study has been completed and several others are underway that together deal with many of the land, resource and water issues in this area. In 1973, Secretary Rogers Morton announced an interim coal leasing policy pending completion of the environmental and other studies necessary for establishment of a new comprehensive policy and leasing schedule. At about the same time the secretary also announced that stringent environmental controls and stipulations would be included in all departmental programs and agreements related to developmental activities to insure compliance with the spirit of the National Environmental Protection Act. The department's prototype oil shale leasing program should fairly quickly produce technical, economic and environmental information necessary for judgments about the prospects for full development of oil shale in Colorado, Utah and Wyoming. Coal gasification and coal liquefaction projects in the Colorado River Basin are being assisted by federal research and development funds; and hopefully they will provide the basis for future full development of clean, economic fuels from our abundant supplies of coal. Federally assisted and private R and D programs indicate that for both coal and oil shale, in-situ processes could minimize water requirements and reduce environmental damage in some areas.

WATER PROBLEMS

The previous sections have indicated that the federal government is moving carefully, but with dispatch, to facilitate development of large amounts of needed energy from resources in the Colorado River Basin at acceptable economic and environmental costs. Other facts show, however, that decisions about development in this area will not be easy.

Among the most important are those that relate to water availability and quality. Because others have discussed this topic in detail, only a few facts to highlight the issues will be emphasized. Existing and planned projects—agricultural, municipal and industrial—will utilize almost all of the water available in almost all of the basin states. The planned projects do not include those required for full development of the energy resources of the basin. A steam-electric generating plant that produces 5,000 megawatts may evaporate as much as 75,000 acre-feet per year for cooling. The Southwest Energy Study estimated that by 1980 an additional 30,000 megawatts of this type capacity will be required in the Southwest. An oil shale plant producing 100,000 barrels of oil per day will use about 15,000 acre-feet of water per year, depending upon the method of production, the method of spent shale disposal, and the extent of on-site refining that occurs. Ground water may be available at some sites, at least in the short run. A coal gasification plant producing the thermal equivalent of 100,000 barrels of oil per day would require about 20,000 acre-feet per year. A coal liquefaction or hydrogeneration plant producing 100,000 barrels of oil per day would also require about 20,000 acre-feet per year.

MONEY PROBLEMS

Low sulfur, low ash, high Btu coal can be turned into electricity relatively cheaply when compared to low sulfur oil. Hopefully, technological advances in stack gas cleanup will continue and such coal can be used in many areas without violating air quality standards. The other methods of utilizing coal discussed earlier—such as coal gasification and coal liquefaction—and methods of producing oil from shale are very costly and require intensive capital investments. A coal conversion facility that can produce 250 million cubic feet of gas or 100,000 barrels of oil per day will cost from $300 to $400 million. The facilities required to produce 100,000 barrels of oil per day from shale will cost about $400 million. If the energy resources are located at a considerable distance from water, costly water transportation facilities will be required—and the facilities required to transport the energy to population centers where it is used will be costly.

As long as the price of oil is high and the supply of oil and natural gas is inadequate to meet energy demands, private interests will probably be willing to make the large investments necessary to develop oil and gas from coal and shale. But this may be a pretty big "as long as." Many people are concerned that maximum production of even the known world oil reserves would drive down the price of oil to a point where production of shale oil or synthetic fuels from coal would be a losing proposition. Exploration for oil and gas is accelerating worldwide.

The unexpectedly high bids received on January 8 and February 12, 1974, for oil shale lands in Colorado suggest that industry is convinced of the economic viability of shale oil production. But in the longer run, if the price of oil drops, it may be necessary to give serious and favorable consideration to some method of guaranteeing industry a reasonable return on their money in order to insure that the massive amounts of money required for full development of clean fuel from coal and oil shale will be forthcoming.

THE PHYSICAL SETTING

Arid lands characteristically have special wilderness qualities and related beauty which mark them for preservation and recreation, particularly by non-residents. People who treasure these qualities resist industrialization of these lands. This is certainly true of the Colorado River Basin. It is largely open space and relatively unspoiled. The air is clear and pure. A large number of people—particularly people from congested, dirty cities—like to visit this area for its scenic, almost wilderness qualities. And there is no doubt that full development of the energy resources of this area will degrade these qualities. Roads, powerplants, powerlines, pipelines and the other indicators of development will mar part of the landscape. Powerplants and other large plants will degrade the air—at least to some extent. An increased population will have many and varied impacts on the environment of the area.

SOCIO-ECONOMIC CONDITIONS

Arid lands characteristically have few permanent residents and, therefore, a limited tax base and a scarcity of industries to support the economy year around. Generally speaking, the basin is sparsely populated; the lifestyle

is quite different from that in more developed parts of the country; job opportunities are less; incomes are lower; and fewer public services are available to
the population.

Many people treasure these qualities. And there can be very little doubt
that full development of the energy resources of this area will change—
perhaps greatly—the existing way of life. Some economic activities that now
thrive in the basin—for example ranching, farming, and outdoor
recreation—will be harmed in at least some areas, either by the loss of water
or by the sheer physical impact of the men and machines that come with
development. The harm will be greater if other industries that use a great deal
of energy, or rely on raw material produced from fossil fuels, locate in this
area. However, development will also create jobs and increase the tax base of
state and local governments, thereby allowing them to improve their services
to their citizens.

THE WISHES OF THE PEOPLE

Coal mined in the basin could be burned or processed at the point of use
rather than in the basin assuming air quality standards could be met at the
point of use. However, as a practical political matter, the people in the basin
could probably prevent development of coal and oil shale, even though most
Americans think of these resources as national, rather than local, in the same
manner that the beauty of this area is considered a national resource.

These points raise several questions: Why not require the coal to be
slurried or hauled out of the basin to be burned or converted to oil or gas
somewhere else? Why not at least prevent the development of other industries
in the basin that will use the energy and further process products produced
from shale oil or coal? What do the relevant majorities of our society feel
about these issues?

There will be disagreement, but I volunteer the following opinions: Most
of the people in the area of potential development want development of the
energy resources to occur, even at the costs that have been mentioned. They
want their property to go up in value; they want jobs that will raise their
standard of living and keep their children at home. They probably have not
thought very much about whether they want simply to export this energy or to
utilize part of it locally for development of other industries. If they thought
about it, a majority of the people in the areas that will consume this energy
would favor development in the basin. They need the energy; but they cannot
build new, clean generating capacity because the fuel is not available and their
air is already so bad that they cannot stand any additional coal-fired plants in
their area. Many, perhaps a majority, of the people who visit the basin area
from other parts of the country would oppose development, even of the

energy resources, because it will detract from the recreational value of the area. Everyone, except perhaps the developers and energy consumers, would favor the imposition of the maximum practicable controls and conditions upon development in order to protect the physical environment. It is national policy to impose such controls and is being done in all the actions that affect development in that area.

Apart from its general responsibilities, the federal government must also make some important decisions regarding these resources because of its position as a landlord and as trustee for several Indian nations.

At least the following conclusions seem to be justified as of 1974. Far too little effort has been made in the past to treat development of the Colorado River Basin in a systems way, which would be far more likely to site powerplants, transmission corridors, and so forth, in more acceptable fashion and utilize the coal and limited water resources more efficiently. The government is attempting to remedy this. The Southwest Energy Study provides a better perspective of these issues in the Colorado River Basin than previously. The Northern Great Plains Study will greatly improve our understanding of the issues in the Upper Missouri River Basin. The federal government's internal decision making processes are better coordinated than in the past and they are improving. Many state and local groups also are attacking the issues involved.

In debating development choices on scenic arid lands, there has been a tendency by many people to equate change with degradation. This result is not inevitable. Properly planned energy resource development can be fitted into a concept of sequential land use. Aside from reclamation and pollution control problems, which eventually will be overcome, development *per se* will have relatively little long-range impact unless it is accompanied by on-site secondary industrialization and urbanization. A choice exists.

Until the results of on-going studies and programs are available the federal government should lease only such lands as are necessary to meet vital demands and sustain research and development efforts, such as coal R and D and the prototype oil shale leasing program. We should be prepared with leasing programs that will facilitate full development, when it is clear that such development is possible.

In the meantime, the maximum practical environmental protections on all development should be imposed and provision should be made in all leases to allow for imposition of even stricter controls as technology advances. Research and Development in environmental control technology should be encouraged and supported.

In closing, two points are emphasized. The first relates primarily to the Colorado River Basin. Perhaps the nation will not need to develop fully the energy resources of the basin—adequate alternatives may come on line more quickly than is expected—or at least some of these resources may be transported to other areas for use. But the only safe course is to assume that there will be a demand for full development and to plan for this result.

Decision about whether or not full development of energy resources should occur will not be easy. Some overly simplified points need to be emphasized; there is not enough water in the basin for full recreational development, full agricultural development and full energy development; to some extent, money can substitute for water in energy projects—i.e., an air-cooled powerplant is more expensive than a water-cooled one—but energy development costs are already high and they will be higher to the extent that the cost of water for other uses is kept artificially low; our past has produced subsidies for some uses of water—such as agriculture and recreation—and a legal system—from the Colorado River Compact to local laws—that puts constraints on establishment of a free market for this scarce resource. While personally sympathetic with this situation—having grown up on a farm where the price for water from the Central Valley Project was $2/acre-foot—I realize that continuation of the existing system involves the perpetuation of value choices that many people in this country would not now choose if they had a choice.

Whether or not there should be a change in the legal and economic systems that allocate Colorado River water, or in assignment of priorities among various uses is not clear to the writer. This reticence stems primarily from lack of enough personal knowledge to be confident. Moreover, it is not clear who should have the greatest say in determining who gets what amount of Colorado River water and at what price—local, state, or federal officials.

It seems clear, however, that in making decisions that affect the future of the basin, all who have a role to play have the responsibility to reassess the values and priorities established in the past to assure that they are equally valid for the future. If they are not, then they should be changed.

My last point, which is not limited to the basin, is that greater emphasis must be placed on research and development programs and policies designed to reduce the nation's consumption of energy. Great attention is now properly directed toward solving the country's short- and long-term supply problems—particularly our short-term ones. Hopefully, solar energy, fusion, geothermal, and other sources not now even being considered will be developed to meet future demands without fouling the environment or requiring changes in lifestyles or the economic system. But to increase the chances of success in the long term and avoid the great wastes now being experienced, even greater weight must be given to the demand side of the energy equation than to the supply side.

It won't be easy to change the consumption patterns that underlie each one's personal life, the country's industrial and transportation systems, and its building codes. But change is necessary and a full bag of "carrots and sticks" must be developed to assure that this occurs. I am convinced that this effort will succeed because it must.

Discussion by Theodore G. Roefs

These comments address three areas: the range of value sets that might apply to future energy development; a definition of a systems study of such development; and the problem of who uses such a study.

Deputy Undersecretary Jared Carter recognizes that both environmental and developmental values are affected by energy production and transmission developments. Other people with other values are affected as well. Their value sets partially overlap those of people who are development oriented or those who are regarded as environmentalists. Among these groups are:

1. The current users of Colorado River water which would be reallocated to energy development
2. Members of native American cultures who reside in the area
3. Recreational users as distinct from preservationists
4. Members of rural societies in the area of development

The point is that a number of parties are interested in energy development. Each has value sets that are at least partially dissimilar to those of others. Thus, negotiation about decisions on energy development, if negotiation is to occur, should not be regarded as a two-party process, but as a multiple party process.

This fact contributes to the need for systems studies. Among the properties of a systems study are:

1. The inclusion of alternative scenarios of future development
2. The best possible estimates of environmental impacts—both probable impacts and worst case estimates
3. Estimates of economic consequences to the national and regional economies if developments are not undertaken
4. Analysis of the impacts on native American and rural cultures
5. A view of the long term—the question of whether a virtually perpetual use of resources (e.g., agriculture or recreation) is being sacrificed to an endeavor with a three-decade life expectancy
6. A method of presenting hard information rather than advocacy text

The 17-volume Southwest Energy Study released in 1972 by the Department of the Interior did not meet any of these criteria adequately. Only

Fig. 7.2. Four Corners coal-fired electrical generating plant.

one scenario of development was considered. No detailed estimate of effects on groundwater in the Navajo Sandstone was presented. Had economic consequences been considered in detail, the study might have made clear the fact that projections of energy requirements taken together with reasonable projections of economic growth in southern California and the intermountain west imply that industry becomes less efficient in terms of output per unit of energy.

A memorandum outlining the plan of the Northern Great Plains Resource Program in 1973 indicated that this effort might more nearly approximate a systems study. For instance, alternative scenarios of development would have been considered.

If a systems study is done, there should be a plan for its use. Provision of a way in which interested parties with dissimilar value sets can react to the various kinds of impacts of alternative scenarios and provide feedback is important. Perhaps an attempt to provide a forum in which a degree of negotiation on the basis of neutrally estimated information could take place would be possible. The current process is one in which multiple decision makers separately and sequentially propose developments. These proposals are frequently followed by litigation based on advocacy information. This process hardly seems likely to produce decisions which are desirable from any point of view.

Rather than attempt to specify either a better process or the shape of a more effective institutional arrangement, an example of the present process is cited. A coal-fired electrical generating plant called Kaiparowits was included in the recommended Phase II development provided by the Southwest Energy Study (Fig. 7.2). After substantial controversy, it was eliminated. New sites in the same general area were then proposed. Meanwhile, two Arizona utilities began work on two other plants. Arizona Public Service proposed a nuclear plant near Phoenix. The Salt River Project proposed something called the Arizona Power Station—a coal-fired plant to be located somewhere in Arizona. The development and timing of these proposals was, in part, a response to the uncertain status of the Kaiparowits plant. The utilities believe they are required to provide the energy that they project will be needed. Apparently no one has the power to choose either between Kaiparowits and the other two plants or to choose that less energy will be supplied. The best systems study in the world is useless if there is no process by which these complicated choices can be made.

8

The Role of Agriculture

GERALD W. THOMAS

Increasing competition for the resources of the Colorado River Basin results not only from pressures *within* the region as residents seek to satisfy their own objectives for livelihood and well-being, but also from pressures outside the region as needs grow for the resources of the region. With the energy crisis and the increasing demands for recreation and environmental preservation, perhaps no region of the United States will be subjected to more external pressure relative to the allocation of its resources than the Colorado region. While agriculture and grazing have been the major activities in the past, they have had relatively little competition. This will not be the case in the future. Agricultural futures in the basin will depend heavily on external forces. What this future will be cannot be predicted.

Increasing worldwide demands for food and fiber will have an impact on the demand for agricultural land which may be amplified in the United States by energy constraints and increasing reliance on agriculture as a source of foreign exchange. Nearly all of the cropped lands of the basin are irrigated, but national policy seems headed toward reducing, or certainly not expanding, irrigation. As energy resources become more scarce, patterns of energy use by agriculture will change. All of these factors will affect the future of cultivated agriculture in the Colorado River Basin.

In terms of land area, the dominant agricultural use is for grazing or timber. These uses, however, are in competition with recreation, wildlife and wilderness and possibly with mineral extraction. Further, all of these uses are related to the use of uncultivated lands for watersheds. Ecological approaches to rangeland problems are reasonably well developed and constitute the basic ingredient for efficient resolution of multiple use.

The purpose of this chapter is to recount the principal features of the context in which agricultural decisions, private and public, will be made and to relate these, at least in general terms, to the resource base of the region.

WORLD FOOD AND FIBER NEEDS

Food and fiber are a matter of concern partly because of the energy crisis (the nation uses vast amounts of energy to produce, process, package and transport food and fiber), but more appropriately because of concern about all resources involved in agricultural production, particularly land and water. Worldwide demands for food and fiber in the future will place resources for agricultural purposes in a much higher priority. Even a doubling of food production by the year 2000 would only maintain the world's population at 1973 dietary levels (Conservation Foundation Letter, 1973).

As Lester Brown (1973) stated, two major reasons for the increased demand for food are: "During the 1970s, rapid global population growth continues to generate demands for more food; but, in addition, rising affluence is emerging as a major new claimant on world food resources." W. Robert Parks (1973), President of the National Association of State Universities and Land-Grant Colleges, added a third factor to the new circumstances facing agriculture. He stated that our "bank account" of food technology has been drawn down to a very low level and there is a special need for new breakthroughs in agricultural research. When the United States decided to open up trade with the Communist Nations, another two billion people became potential customers for U. S. agricultural products. The Colorado River Basin, where a wide variety of crops are produced, will be involved in this increased demand for food and fiber. Thus, priorities for resource allocation in the basin may change.

Competition for Farm Land

It is estimated that, with the technology of the developed countries, nearly one acre of cultivated land is needed on the average to provide an "adequate" standard of living for each person. This takes into consideration some variation in land productivity as well as some variation in levels of affluence. Based on the population in the United States in the early 1970s, there are about 1.2 acres of cultivated land per capita and a potential of about two acres per capita if the cultivated area were expanded to the maximum possible. The effects of farm programs to enlarge the acreage under cultivation and of the new higher yields are still unknown. The experience of many countries has shown that most efforts to expand cultivated acreage have resulted in reduced per-acre yields. U.S. farmers are already using the very best lands for cultivation.

The United States is exporting the produce of about one acre out of every four—leaving about 0.9 of an acre per capita for home use. The Far East has less than 0.8 acre of cultivated land per capita with Communist Asia at about 0.4 acre per capita. Latin America with 1.3 acres per person, and Africa with about 2.3 acres per person still have a good cultivated land resource base.

For the world, in the mid-70s, the cultivated area stands at about 3.5 billion acres, or 11 percent of the earth's land surface. The area actually harvested for crops in a given year is considerably less (due to fallow practices and crop failures)—usually about 2.4 billion acres. This means that only about 0.6 acre of usable cultivated land per person in the world is cultivated compared with an estimated need of one acre per person. By the year 2000, the cultivated land base will probably be reduced to about 0.3 acre per capita. Worldwide yields will have to be doubled to maintain food availability at a 1970s rate.

Worldwide and national needs for farm land will influence priorities of land use within the basin. The basin itself has a surplus land base for farming—about eight acres of cultivated land per capita in the upper Colorado (Upper Colorado River Basin Group, 1971) and about seven acres per capita in the lower Colorado (Lower Colorado Region Group, 1971).

Irrigation Policy and Potential

Water is probably the most important—and most limiting—factor in the growth and development of the Colorado River Basin. A high percentage of the cropland in the basin is irrigated and the potential for irrigation is still greater. One important consideration will be national policy on irrigation. The National Water Commission does not view expansion of irrigation as desirable. One conclusion from the Commission's Report is as follows (Heady, et al., 1972):

> Land will not be scarce by 2000. . . . Output from U. S. farm and rangelands, including lands now set aside in government programs, will be adequate to meet projected food demands even at the high level that would be expected if population increases to 325 million persons and some food exports grow to about twice their 1967-69 levels.

Furthermore, the report concluded that:

> Expansion of irrigation is not needed to meet future food needs. Quite to the contrary. The most efficient pattern of production at most projected demand levels would be achieved with a *reduction* in the acreage of irrigated land use for annual crops.

I do not agree with these statements. A good review of the National Water Commission Report, prepared by the Western Agricultural Economics Research Council (Anderson, 1973), challenged some assumptions of the "Heady Report," on which the Commission conclusions were based. In addition, the world outlook has changed with the emergence of the energy crisis and the opening of markets for American food and fiber in the Communist nations. U. S. agricultural exports in 1973 were about $12.9 billion. The U. S. Department of Agriculture Economic Research Service (1974) estimated that total agricultural exports could reach $20 billion by 1975. Food has indeed become a potent factor in foreign policy.

Reports by the upper and lower Colorado River Basin inter-agency study groups (1971) on land use, project the need for cropland to increase from 1,816,000 acres in 1965 to 1,852,000 acres by the year 2020 in the lower Colorado River Basin and for the upper Colorado region to increase from 1,621,500 acres in 1965 to 2,625,000 acres in 2020. Most of the increases were projected for irrigated lands.

At the Colorado River Basin Environmental Management Conference held in Salt Lake City in October, 1973 (Thomas, 1974; Johnson, 1974), several concerns were expressed about irrigated farming in the basin: (1) there is a serious problem of groundwater mining (some aquifers now being tapped for irrigation water are not being recharged to offset withdrawals, and the "closed basin" approach has not been adopted); (2) increased salinity and other pollution problems—partly associated with irrigation—present both a regional and an international challenge; (3) the status of Indian water rights continues to be unsettled; and (4) transfers of water rights to summer home development, to industries, and to other uses is continuing to reduce irrigation in the agricultural sector.

Thus, while the debate over worldwide requirements for cultivated land and for irrigation continues, and while certain groups are projecting increased food and fiber needs, the Colorado River Basin is actually experiencing continued losses of farm land to other uses—usually to some form of urban or industrial development. These transfer decisions are being made piecemeal, under economic pressure, and without adequate consideration of land-range needs or systematic planning for optimum land use. No part of the basin has so far been able to zone successfully or to "build in" the necessary legal and economic incentives for the protection of good farm land.

Impact of Energy Needs on Cropped Agriculture

The Colorado River Basin is rich in basic energy sources. Energy shortages and increased energy costs will have a substantial impact on all development in the Colorado basin. One can anticipate increased pressure for exploitation and mining of petroleum and mineral resources, increased research and development activity for alternative sources of energy—geothermal, nuclear, hydropower, wind, and solar energy—and economic and political pressure to compromise standards of environmental pollution. Furthermore, changes in expenditure patterns and consumer demands—that is, food, clothing, forestry and housing—may increase in priority while tourism, recreation and energy luxury activities may diminish. Perhaps, the long overdue reconsideration of lifestyles of the affluent American may begin. All of these factors could impact on development and decision-making in the Colorado River Basin.

The gigantic food and fiber industry, taken in its entirety, uses more petroleum products than any other industry in this country. Large amounts of energy are consumed in the "supply sector" to provide the farmer and rancher with fertilizers, pesticides, machinery, and other inputs as well as in the "production sector" for planting, cultivation, irrigation, and care and harvesting of crops and livestock. The "storage, processing, packaging and distribution sector" also requires large amounts of energy in order to place food on the table and clothes on the backs of 210 million Americans. Estimates of the petroleum products used by agriculture, in the broad sense, vary from 10 to 18 percent of total consumption of petroleum in the United States.

Farmers in the United States have substituted over five million tractors for about 22 million horses and mules. About 72 million acres of land that would have been required to feed the horses and mules may be used for direct food production for humans. In addition, efficiency and output per acre have increased (Thomas, 1972). In 1925 a man with a good team of horses could plow about two acres a day. In 1975 mechanized power makes it possible for him to plow over one hundred times that much. However, as a result of this increase in mechanization on croplands in the United States, energy flow patterns have been changed significantly. Horsepower, mulepower, oxenpower, and manpower operate on the solar energy collected by vegetation—a continuing resource for all practical purposes. Tractors and machinery utilize fossil fuel—a finite and depletable resource.

The trend toward mechanization on farms is not confined to this country. In 1950, the Food and Agriculture Organization estimated that there were about 6.1 million tractors in the world. By 1970, this number had exceeded 15.5 million (Brown, 1973). In addition, world fertilizer use, which is heavily dependent on petroleum, increased from 15.2 million metric tons in 1950 to almost 68 million metric tons in 1970. As the technology associated with the "green revolution" spreads, more energy will be required. Thus, world demands for petroleum for the agricultural sector are increasing at an accelerated rate. Farmers in the Colorado River Basin will also feel these pressures on petrochemicals and fuel supplies.

One implication important to agriculture in the Colorado River Basin is that high energy costs of synthetic fibers should result in a shift toward increased use of natural fibers, specifically wool and cotton. A comparison of energy (fossil fuel) required for cotton, cellulosic and non-cellulosic fiber production is presented in Table 8.1. Energy consumption for finished broadwoven fabric for the synthetic fibers is more than double the amount for cotton. Wool places the lowest demand on fossil fuel. The raw materials for the non-cellulosic fibers are petrochemicals from petroleum and natural gas.

TABLE 8.1

Energy Consumption for Selected Fibers*—
Raw Materials to Finished Broadwoven Fabric
(Kilowatt hours per pound of fiber)

Stage of Production	Cotton	Cellulosic	Non-cellulosic
Raw Materials	0.20	1.61	6.28
Fiber Production	3.55	22.09	11.36
Weaving & Spinning	6.30	7.03	7.03
Finishing Mills	6.98	8.52	8.52
Total	17.03	39.25	33.19

*Adapted from Gatewood, 1973.

Synthetic fibers have been capturing an increasing share of the fiber market. In the early 1970s each person in the United States was using over 20 pounds of synthetic fiber per year, and Resources for the Future predicted that synthetic fibers would capture over 54 percent of the fiber market by the year 2000. These projections may not materialize, however, due to the pressure on petroleum products. Japan and western Europe are already placing more emphasis on natural fibers. It is safe to anticipate increased demands for natural fibers produced in the basin.

COMPETITION FOR UNCULTIVATED LANDS

Uncultivated lands constitute the largest acreage in the Colorado River Basin. Management and land-use decisions on these lands are complicated by three major factors: (1) landownership patterns; (2) extreme variability in climate, soils, vegetation, and topographic conditions; and (3) multiple-use possibilities.

The concept of multiple use of western range resources has been accepted and practiced for many years. That is, these lands have value to the individual and to society for more than one purpose. Although the primary income may be from livestock or forest products, the lands also are important from the standpoint of mineral production, wildlife, recreation, and water yield. Another dimension added in the early 1970s is that of "total environmental enhancement," particularly air and water quality as well as aesthetic or wilderness values.

The pressure from individual interest groups is often so great that commitments are made excluding other uses. This trend is of special concern— particularly when political and legal restraints are imposed during a period of emotionalism or under one of the "crisis situations" which appear to develop frequently in the United States. Following is a brief examination of some of these pressures on uncultivated lands.

Livestock Grazing

Approximately 65 percent of the total land area in the basin is public land. Decisions on the use of public land are becoming more and more "everybody's business." Social welfare weighs heavily against economic value. Environmental concerns are more apt to be considered. The tradition for decision-making on private or corporation land is somewhat different. Right or wrong, the landowner still makes most of the decisions on land use and this prerogative is strongly established by American tradition.

Given this difference in management prerogatives and/or objectives on private vs. public lands, the situation is further complicated by the fact that many ranch operators graze livestock on *both* public and private lands. Thus, land-use decisions on federal or state lands have a significant effect on private lands. A change in the use of one has an immediate impact on the other. This dependence on public lands for forage supplies varies among the states— reaching a high in Nevada where 49 percent of the livestock forage comes from public lands.

There is a national movement sponsored by the Natural Resources Defense Council to force the public land agencies—primarily the Forest Service and the Bureau of Land Management—to close all federal lands to grazing. A task force of the Council for Agricultural Science and Technology (1974) prepared a position paper on the effects of such a ban on the national economy and the environment. This task force, composed of fifteen scientists from the western United States, took a stand in opposition to the elimination of all grazing by domestic livestock from federal lands:

> Eating of plant materials by animals is a natural process in terrestrial and aquatic systems. Thus, with the coming of European man to the West, the introduction of domestic livestock did not constitute an entirely new component in the environment. More realistically, the domestic livestock replaced, or were added to, the wild animals that were already there. Rangeland vegetation,

especially grassland and shrubland, in the western states evolved to withstand grazing to a moderate degree. Without grazing, different vegetation characteristics develop. The range forage that livestock utilize is a renewable natural resource because the forage regrows each year and has done so for many centuries. . . .

The environmental effects of grazing depend upon the kind of range, the intensity of grazing, and the kind of management employed to control livestock on the range.

In a paper presented to the 1973 annual meeting of the Society for Range Management, I concluded:

A careful examination of long-range research can only lead to the conclusion that: (1) on vast areas of public lands, livestock grazing, under proper management, is compatible with other uses; (2) on a limited number of sites, grazing by domestic livestock is detrimental to the resources and competitive with other uses; and (3) on other sites, grazing by livestock can be the most beneficial use to society for economic, social and ecological reasons (Thomas, 1973).

Timber Production

Timber production has been increasing in relative importance because of the high demands for lumber, pulp, and paper products (Fig. 8.1). This situation has become increasingly critical with the energy crisis—and even the most extreme environmental groups are recognizing the need for more forest products. From a recent report by the National Commission on Materials Policy (Cliff, 1973), the following statements are pertinent to the discussion of land-use alternatives:

Three-fourths of the nation's softwood is in the west. . . . Projections to the year 2000 for softwood sawtimber demand, at current prices and the present level of management, would require almost doubling the 1970 domestic production. . . . Even with intensified management, prospects for balancing future supply and demand at 1970 prices appear remote. However, stepped up investment in a variety of forestry activities could produce significant increases in timber production by the turn of the century.

From a recorded high of 507 million acres in 1962 in the U. S., the area of "commercial forest" land is projected by the Materials Policy Commission to drop to 475 million acres in the year 2020. They state that, "Quite possibly, additional areas on national forest lands will be removed from the timber supply base for recreation and environmental protection." Nevertheless, forest production will remain a strong competitor as a major use or as a concurrent use of uncultivated land in the basin.

Fig. 8.1. Forests on the Colorado Plateau. The landscape
breaks abruptly into Zion National Park in the background.

Recreation

Recreation use on all of the range and forest lands in the Colorado River
Basin has continued to rise at a more rapid pace than population. Increased
mobility and affluence of people contribute to this pressure on the resource.

In 1957, the Forest Service estimated that the total recreation visits by
1975 would reach 135 million, but this estimate was reached in 1965. Marion
Clawson of Resources for the Future estimates that the recreation visits to
national forests could reach 400 million by 1980 and more than one billion by
the year 2000. The amount of use of Bureau of Land Management lands is
increasing at an even more rapid rate. It is clear that the National Park Service, like most other public agencies and private observers, has also greatly
underestimated the continued growth in recreation demand in the West.

A Forest-Range Task Force (1972) projected outdoor recreation requirements on U. S. uncultivated lands to increase, in terms of 1965 uses, as
follows: camping—560 percent; picnicking—400 percent; horseback
riding—370 percent; and hiking—300 percent. Their projections for requirements for fishing and hunting were much lower. The National Commission on Population Growth and the American Future states as follows
(Cicchetti, 1972):

During the postwar years, participation in outdoor recreation in the United States has grown by an average annual rate of 10 to 15 percent. During more recent years, a slowdown in this rate has been observed for some specific recreational activities; however, the overall annual rate of growth may still be close to 10 percent.

Wildlife

The increased importance of wildlife production and management on the western range can also be illustrated readily. While livestock numbers on federal lands have been reduced substantially since 1935, the number of big game animals has increased. At the present time, estimates by the Forest-Range Task Force indicate that there are more than five million big game animals on the nation's forest-range lands. On Bureau of Land Management lands big game animals increased from an estimated 600,000 in 1945 to about 1.8 million in 1970 (Howard, 1973). Pressure by the public for hunting and other outdoor recreational opportunities has also opened up new possibilities for economic returns to many private ranching enterprises in the Colorado River Basin. Bird watching and nature photography, while not reported in wildlife-use statistics, also constitute an important part of the use of uncultivated lands.

Wilderness

The two comprehensive framework studies of the Colorado River Basin, from which much of the basic material contained in this chapter has been drawn, estimate that the land set aside for wilderness areas in 1965 was about 2.2 million acres. This acreage has already been increased substantially and hearings are continuing in many parts of the basin to add to the nation's wilderness and primitive area base. Naturally, the ranching industry is concerned about this trend since livestock will be excluded from an increasing area of federal land.

Most people recognize the need for wilderness and primitive areas, but the management of wilderness areas is of special concern to ecologists. D. W. Hedrick (1973) stated recently:

> Many of the wilderness and National Park areas are occupied by fragile ecosystems where human and animal impacts are more crucial and significant than on the bulk of public ranges grazed by livestock.

He expressed special concern over the effects of horses—both riding and beasts of burden. Hedrick also stated:

> It is only a matter of time before our policy on use of wilderness and remote recreational areas is attacked by minority

and low-income groups. Our present policies on the use of wilderness areas is among the most discriminatory followed by public officials.

Watersheds

Watershed values relating to both water yield and water quality of the western range are difficult to evaluate. A review of 39 forest-land watershed experiments throughout the world led to the conclusion that when timber stands are harvested, or sufficiently reduced in density, water yield is increased (Forest-Range Task Force, 1972). The magnitude of change varies over a wide range of climates, forest cover types, and geomorphic situations. The role of vegetation management cannot be underestimated. On many brush-infested range areas there may be as much as 100 tons of water associated with the production of each pound of beef. But even water expenditures for "undesirable" vegetation may not be wasted in terms of oxygen production or environmental enhancement.

Minerals

Mineral production in the Colorado River Basin, particularly on federal lands, is subject to much controversy. The total area under petroleum leases or mineral claims may have stabilized somewhat, but the volume of production and the value of production of many minerals is still rising (Forest-Range Task Force, 1972). Increasing concern about the total environment has reduced some of the speculative and haphazard exploration and/or exploitation, but many problems remain to be faced by this and future generations. Land-use policies, as they relate to mineral production, often have a heavy economic impact on small communities in the West. The energy crisis may force a compromise of present standards on environmental protection—at least for the short term.

For purposes of economic analysis, Gray (1968) has classified multiple use of range resources into three categories: competitive, supplementary, and complementary. The traditional viewpoint of the rancher is that all other uses tend to compete with livestock production. This is certainly true for many ranching enterprises. But, for others, it may be both economically advisable and ecologically sound to consider supplementary or complementary activities such as grazing two or more classes of livestock, producing game, and managing the resource for recreational purposes.

Furthermore, while the rancher, as an individual with a direct economic interest in the range resource, may desire single-use management, the public must always consider multiple-use as the most desirable approach. With increased realization of the impact of man's land-use practices on the total environment, the ecology of multiple use management becomes even more important.

Ecological Considerations

Uncultivated lands present a complicated ecosystem involving the inter-relationships among plants, animals, and environment. Because of the need for correlating and analyzing the many variables, ecology has become the dominant science to bring the purpose of man into harmony with the forces of nature on rangeland areas. Plant physiology, soil science, climatology, hydrology, genetics, forestry, entomology, taxonomy, wildlife biology, recreation management, and animal science are all complementary to ecology.

In any ecological analysis, vegetation is the key, since plants are the first step in energy capture, and are the major factor in ecosystem stability. The traditional approach to vegetation surveys on rangelands is often described as "dynamic ecology." The central concepts of this approach—"succession" and "climax"—were developed by Cowles, Clements, and Cooper early in the twentieth century. Figure 8.2 is a schematic diagram of this concept (Thomas, 1969).

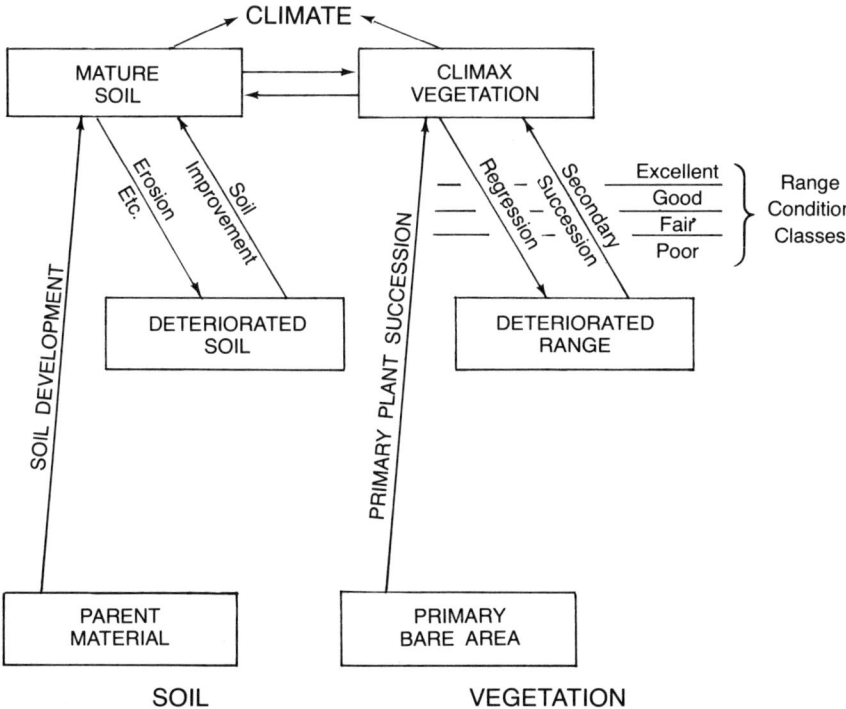

Fig. 8.2. Vegetation succession and regression patterns and the range condition classification system.

Climate is shown as the overall controlling factor in vegetation and soil development. Actually, however, the total environment determines the outcome. On any particular area, vegetation changes with time in a rather systematic pattern (primary plant succession) until a plant community (climax) ultimately appears in equilibrium with the environment. This excludes the influence of man, but includes other natural biotic factors. The climax condition is very dynamic and encompasses "normal" variation in climate.

Man enters the picture and brings about vegetation change (regression) through manipulation of livestock, harvesting of forests, cultivation, recreation, or other disturbance techniques. Man can also bring about improvement by controlled management to hasten "secondary succession." Corresponding changes such as deterioration in physical properties, or erosion, can take place in the soil depending upon the severity of the treatment imposed.

Attempts to quantify the succession-regression patterns were not very successful until the "Range Condition" method was developed following World War II. A major contribution was made by Dyksterhuis (1949) and others when a system of "Range Condition" classes was proposed, based primarily upon the regression sequence using livestock grazing as the disturbance factor. Most federal agencies now are using a modification of this technique in evaluating the effects of grazing. Similar techniques could be used for other "disturbance factors." Unfortunately, few research data are available concerning such man-caused regression effects as the impact of recreational vehicles, uncontrolled hiking, camping, pack horses, etc. Some of the "substitutes" for livestock are probably more detrimental to the ecosystem and more difficult to evaluate and control than cattle or sheep.

Under this ecological approach, vegetation classification "in time" is shown as "range condition," and classification "in space" is determined primarily by soil, topographic, and climatic conditions forming "range sites." Once the boundaries of "sites" are established, the succession-regression patterns are broken into range condition classes: Excellent, Good, Fair, and Poor. These classes represent departures from the climax plant community (Fig. 8.1). All plants on a particular range site are identified as to their response to pressure and their probable place in the climax plant community. Thus, the vegetation survey establishes both present condition and potential productivity. It also reflects "stability" and "diversity" of biological populations.

Reliable soil or site surveys are vital to this system of classification. The spatial pattern of vegetation communities is complex, "a field of phenomena notably lacking in fixed points of reference, lines of division, invariable rules, and easy definitions" (Whittaker, 1953) and determination of the role of soils in this distribution pattern is necessary. For the Colorado River Basin, with its extreme *variability* in rainfall, soils and topographic conditions, the proper identification of "sites" and the proper analysis of vegetation change is critical—not only for evaluating the effects of grazing but also to determine the effects of other man-caused or man-accelerated disturbance factors.

To an increasing extent, values and choices in determining proper land and water use should be based on ecological considerations. Under these conditions, agriculture and grazing will still remain competitive as a part of the economy and development potential in the Colorado River Basin.

Discussion by D. Wynne Thorne

Gerald Thomas has given an excellent overview of the agricultural potential of the Colorado River Basin and some of the constraints on its future. Additional facts could be cited to help clarify the world food situation. Federal and regional land resources may be adequate to compensate for possible reduced crop production in the Colorado River Basin. Potentials for increasing the production of food and fiber for meeting U. S. and world demands are not entirely dismal in the short run. Additional arguments could be marshalled for continued emphasis on livestock in the Colorado River Basin to harvest the marginal fibrous forage plants adapted primarily to the ruminant physiology. However, after assembling some additional data on the quality and present uses of land not now in crop production, on the areas of our wild lands having unique traits for recreation, and on the growing competition for water resources, I would suggest some modifications of the statements made, but would still be in general agreement.

The definition of *agriculture* includes the complex assortment of land, water, climate, and air resources and associated plants and animals and their management for the benefit of man as well as the relative values man attaches to agriculture and these resources and how these values will influence his choice among various alternatives.

Following Gutkind's thesis, man has a long history of gradually shifting values in his relationships to nature. In the beginning, man presumably conceived of himself as the creature of nature and dependent upon it. Nature was often steep topography, far removed from metropolitan centers; a land consecond stage man learned to subdue nature, to clear the forest, to divert water and to alter the environment for his own advantage. The "I-Thou" relationship, while generally dominant, started to crumble.

The third stage is characterized by aggressiveness and conquest. Here man is motivated by the delusion that he is an omnipotent remaker of his environment. Man is the object and end of creation and the earth is his to subdue. But even as man triumphs over nature, he plants the seed of his own defeat. As he destroys plant cover he reaps dust and floods. The revenge of nature from man's mismanagement is bringing man to the fourth stage of cooperation and responsibility toward the land. Here man sees himself as a partner and protector of nature.

While man often perceives himself as having achieved a responsible value system toward agriculture in its broad aspects as depicted in the fourth stage, developments in the Colorado River Basin may well provide a crucial test of his integrity.

The lines of battle are clearly drawn even though all parties to the conflict are not in place nor clearly identified. The issue is energy, the driving force is man's need and greed that might sacrifice nature and agriculture to broad destruction. The potential saving power is man's love and respect for the land and his willingness to invest in its protection.

The Colorado River Basin is a land of rugged beauty, harsh climate, and often steep topography, far removed from metropolitan centers; a land containing this nation's primary energy reserves for at least the next hundred years. Balanced against this are over 150 million acres of land, covered primarily by desert shrubs, some commercial timber, and including less than 3 million acres of irrigated land in a total of 4 million acres of cultivated land. The balance factor between the two is water and man's willingness to proceed with caution and care.

The water requirements for developing these energy resources of coal, oil shale, tar sands and nuclear elements are so large that drastic competition with traditional agriculture seems inevitable. When the water needs for processing coal and oil shale into transportable energy are added to the increased water requirements for the people essential for the industry, a crisis is in the making.

For electric power generation, a 1,000 megawatt plant is reported to consume over 15,000 acre-feet of water per year. Gaseous conversion of coal requires two to three pounds of water for each pound of coal. Oil shale processing would probably require similar large quantities of water and the total requirements may well be greatly underestimated. Substantial increases in population will add 100-150 gallons per day per capita to these requirements.

Obviously, the waters of the Colorado River, now used primarily on 3 million acres of irrigated land, will undergo massive transfers to energy development. The 3 million acres of irrigated land in the basin will not stand high enough in our value system when considered as part of a 40 million-acre irrigated national enterprise to insure their continuance when weighed against our desperate need for energy—energy needed in part to sustain agricultural crop production and to harvest lumber and forage resources from the non-cultivated lands.

The United States does have additional suitable land resources to meet its needs. About 20 percent of the nation's Class I lands are not currently cropped; so also, 35 percent of the country's Class II lands and 52 percent of its Class III lands are still not cultivated! Much of this large land reserve is now used for forests and pasture, and much is irreversibly lost to industry, highways, and urban sprawl so that diversion of large areas to cultivated crops will require a series of land use adjustments. Still adequate adjustments probably can be made to compensate for production lost by diverting water to energy developments.

Our value system related to the Colorado River Basin seems less threatened by major decisions than by small ones. The nation will not debate long about whether energy development in the Colorado River Basin should take priority over continued irrigated agriculture. However, careful consideration needs to be given about how this change is accomplished. Decisions will be made at numerous minor and major points and only a sensitive and alert public can prevent frequent abuses to our environment.

One hundred years or more from now, when the energy resources of the basin are less vital to the well-being of man, people will still need food, and no doubt will look to the Colorado River Basin with increased interest for beauty and recreation. In those days agricultural productivity can perhaps be restored by returning at least part of the diverted water to crop production. Meanwhile, the future welfare of mankind requires that the natural environment of the basin be maintained by reshaping and revegetating disturbed lands, by protecting watersheds and streams, by wisely planning urban developments, and by insisting on a harmonious relationship between man and nature. Public programs must be activated that will reserve part of the returns from exploitation of energy resources to restore and protect the environment, cover costs of adjustments in agriculture both now and in the future when water resources may not be required for energy industries, and build communities and provide social services that will enhance the quality of living in the basin.

Such is the challenge and the test of our values and the wisdom of our choices in relation to agriculture.

Literature Cited

Andersen, Jay C., *et al*. 1973. Review and critique: agricultural water policies and the environment. Unpublished manuscript presented to Western Agricultural Experiment Station Directors Meeting, Newport, Oregon.

Brown, Lester R. 1973. Population and affluence: growing pressures on world food reserves. In *Population Reference Bureau, Inc.* 29(4).

Cicchetti, Charles J. 1972. *Outdoor Recreation and Congestion in the United States*. Report of the Commission of Population Growth and the American Future. Washington, D. C.: Superintendent of Documents.

Cliff, Edward P. 1973. *Timber: The Renewable Material*. Prepared for the National Commission on Materials Policy, August 1973. Washington, D. C.: Superintendent of Documents.

Conservation Foundation Letter. 1973. *How Far Can Man Push Nature in Search for Food.* Washington, D. C.

Council for Agricultural Science and Technology. 1974. *Livestock Grazing on Federal Lands in the Eleven Western States.* Congressional Record, 26 March 1974, pp. 4429-4434.

Dyksterhuis, E. J. 1949. Condition and management of range land based on quantitative ecology. *J. Range Mgmt.* 2:104-115.

Forest-Range Task Force. 1972. *The Nation's Range Resources—A Forest-Range Environmental Study.* USDA Forest Service Report No. 19. Washington, D. C.: Superintendent of Documents.

Gatewood, L. B., Jr. 1973. *The Energy Crisis: Can Cotton Help Meet It?* Memphis, Tennessee: National Cotton Council of America.

Gray, J. R. 1968. *Range Economics.* Ames: Iowa State University Press.

Heady, E. O., *et al.* 1972. *Agricultural and Water Policies and the Environment.* Ames: Iowa State University Press.

Hedrick, D. W. 1973. Grazing on public lands. *Rangeman's News.* Society for Range Management 5(4).

Howard, Paul. 1973. Public attitudes on public land grazing. In *Grazing System Symposium.* Las Cruces: New Mexico State University.

Johnson, Leonard H. 1974. Agriculture's management of the environment in the Colorado River Basin. In *Environmental Management in the Colorado River Basin,* eds., A. B. Crawford and Dean F. Peterson. Logan: Utah State University Press.

Lower Colorado Region Group. 1971. *Lower Colorado Region Comprehensive Framework Study.* Pacific Southwest Interagency Committee Main Report. Washington, D. C.

Parks, W. Robert. 1973. *New Circumstances Facing America's Ability to Meet Expanding Domestic and Foreign Demand for Food and Fiber.* A statement by the National Association of State Universities and Land-Grant Colleges. Washington, D. C.

Thomas, Gerald W. 1969. The western range and the livestock industry it supports. In *Range Research and Range Problems.* Crop Science Society of America.

——. 1972. *Progress and Change in the Agricultural Industry.* Dubuque, Iowa: Kendall-Hunt Publishing Company.

——. 1973. Livestock grazing public lands: unity for political, economic, and ecological reasons. *J. Range Mgmt.* 26(4).

——. 1974. An examination of the environmental carrying capacity concept for the agricultural sector of the Colorado River Basin. In *Environmental Management in the Colorado River Basin,* eds., A. B. Crawford and Dean F. Peterson. Logan: Utah State University Press.

Upper Colorado River Basin Group. 1971. *Upper Colorado Region Comprehensive Framework Study.* Water Resources Council. Washington, D. C.

U. S. Department of Agriculture. Economic Research Service. 1974. *U. S. Agricultural Exports: Commercial and under Government Programs.* Washington, D. C.: Government Printing Office.

Whittaker, R. H. 1953. A consideration of the climax theory: the climax as a population and a theory. *Ecol. Monog.* 23:41-78.

9

Recreation

LAWRENCE E. ROYER and JOHN D. HUNT

The Colorado River Basin has been selected as a case study region with the thought that an assessment of that basin will have applications to other arid lands both in the United States and elsewhere. However, examination shows that the Colorado River Basin cannot be prototypical in regard to future recreation directions on arid lands. It does not provide the appropriate conceptual framework for an examination of recreation values and choices in the arid American Southwest. And in the American West, even the appropriateness of arid geography (rainfall or climate region) as an independent variable can be questioned.

ARIDITY HYPOTHESIS AND THE SOCIOLOGICAL
REGION—AMERICAN SOUTHWEST

One must look to the sociocultural context of recreational use of arid lands in the United States if future directions of values and choices are to be analyzed. If arid lands produce peculiar social patterns, then it would be convenient to interpret recreation in this broader context of aridity's relationship to societal structures and functions. But do arid lands produce their own peculiar societies? Cleland (1966) suggests that there are problems in developing a geographic cohesion between social patterns and aridity. Recognizing oasis-centered social organization as an exception, he emphasizes that the paucity of knowledge about social structures in other arid regions of the earth prevents investigators from utilizing a comparative analytical approach and suggests that a narrower but more rewarding perspective would be the examination of arid geographies in the United States. But even this approach encounters difficulties. Cleland (1966) believed that no "meaningful general work'' had emerged that treated arid lands in the same manner that Kraenzel's *The Great Plains in Transition* (1955) treated semi-arid lands. Perhaps this initial stumbling block has been removed with the availability of D. W. Meinig's *Southwest: Three Peoples in Geographical Change 1600-1970* (1971). The Meinig work is an historical social geography of the arid lands of New Mexico and Arizona. Although not directly comparable in approach and method to *The Great Plains in Transition*, *Southwest* is a ''meaningful general work'' with a number of interpretations that provide insights into recreational choice on arid lands.

Meinig shows that the arid area generally defined by Arizona and New Mexico is dominated by three metropolitan centers at Phoenix, Albuquerque and El Paso. The region exhibits a plural society which results not from a response to aridity but rather from the persistence of immigrant cultures from Texas, Mexico and the Middle West. The development of railroad transportation and mining set the basic economic pattern of this region. Although aridity can influence the space-costliness (emptiness) of an area, Meinig shows that the basic transportation infrastructure is horizontal. It results from the commercial connection of the Midwest and Texas with California (another arid area) and not from the internal connection of central places whose distances could be influenced by aridity. Mining as an internal economic force developed around a random pattern of mineral deposits dependent upon transportation connections to areas outside the region. Again, the development of mining does not indicate a geographic cohesion or economic response to moisture deficiency.

Agricultural cultures, which could be envisioned as containing the most direct adaptations and responses to aridity, seem to be heterogeneous in the Arizona-New Mexico Southwest. And, indeed, they are the result of the persistence of cultural heterogeneity. They do not represent new adaptations to the region's aridity. In a sense, the region's agriculture represents a dichotomy of styles between the more geographically pervasive but less intensive response of the Hispanos of New Mexico, and the Anglo, both Texan and Mormon, intensification and localization of irrigation patterns in the Gila, Rio Grande and Pecos River valleys. The Hispano agricultural tradition is a derivative of Mexico. Intensive irrigation was endemic with the indigenous Hohokam but the Anglo practice represents an American political response with origins from outside the Southwest. As in other regions of the United States, the urban encroachment upon agricultural acreage is a dominating influence upon agriculture in the Southwest. It overrides the influence of aridity.

Meinig contends that the Southwest of Arizona and New Mexico is a viable sociological region in the sense of Cleland's meaning of ''sociological region.'' But the causal relationship between aridity and sociological region seems obscure. Even the Southwest's regional boundaries seem to elude the phenomenon of aridity. The limits of this region are set not by the disappearance of arid lands but rather by the influence of more dominant Texas and California regions, the International Boundary, and the difficult terrain of the Colorado Plateau. Meinig concludes that the social patterns of this arid region will be influenced in the future by the heterogeneity of its cultures. Future values and choices will center on the efforts to cope with this phenomenon and not with a confrontation with aridity. By omission from his discussion, Meinig has also concluded that the Colorado River Basin, which is congruent with some of the major arid portions of Arizona and western New Mexico, has not been influential upon regional development.

The Meinig interpretation of the Arizona and New Mexico Southwest offers a better framework for the description of recreation futures than does aridity. The remaining problem, of course, is that although the Colorado River Basin includes most of Arizona, it only intrudes upon the western edge of New Mexico. Meinig envisions the whole of New Mexico as integral to a sociological region. Thus there is a lack of correspondence between the Colorado River Basin and the Southwest regions.

It is the sociological region that is dominant. In Meinig's interpretation of the Southwest there is ample documentation of the interplay between social development and recreational development in this area. The cultural heterogeneity of the Southwest with its indigenous Indian populations has greatly influenced recreational use of certain portions of the Colorado drainage. The horizontal economic impact of railroads has equally determined touristic development in the basin. The persistence of central places such as Taos and Santa Fe can be attributed to recreational influences. And the contribution of amenities to the urban development of the Southwest in areas such as Tucson is recreational in nature. Elsewhere, the amenities characteristic of the region have created non-urban touristic central places. In fact, the very physiography that fixes the northern boundary of the Southwest is accordant with one of the major recreation-tourism areas of the United States.

ARIDITY HYPOTHESIS AND THE SOCIOLOGICAL REGION—AMERICAN WEST

In a fashion similar to his Southwest study, Meinig (1972) examined the larger West, west of the High Plains. This study is the geographic complement of Walter Prescott Webb's (1957) famous interpretation of the interior West. While both the Webb hypothesis and an interpretation by Wallace Stegner (1953) are founded upon the dual premise of aridity and the oasis as developmental and social determinants, Meinig shows that development directions are essentially independent of aridity. Development has centered upon six nuclear regions whose origins are to be found not in the oasis, but in a diversity of phenomena. Culture, geopolitical patterns, and a horizontal circulation pattern are the variables that explain the ensuing development and refinement of the initial patterns. Webb and Stegner have attempted to explain the social development of a region on the basis of its aridity and montane oases. Meinig's more definitive but introductory study, while directed to the areal components of development, nevertheless refutes the interpretation of Webb and Stegner. In contrast to Kraenzel's and Webb's Great Plains, the heterogeneity of the interior West confounds the search for individual explanatory variables over this large area.

THE COLORADO RIVER BASIN

If one seeks regional formats such as a Colorado River Basin or a Southwest as case studies, then one must be certain of the homogeneity of these regions' social fabrics. If a drainage basin is used as a case for an aridity hypothesis, then the initial requirement of that basin should be that aridity is distributed throughout. Neither of these situations seems to be the case in the Colorado River Basin. The upper portions of the Green River and Colorado River must be eliminated from the basin if any semblance of cohesion between arid geography and drainage geography is to be obtained.

Perhaps the Colorado River Basin embraces two sociological regions and therein could lie the key to describing the recreation future of the basin. There is a contiguous aridity between the western one-half of the Southwest sociological region and the Colorado Plateau physiographic province. The problem is that the Colorado Plateau has not received the attention bestowed upon the Southwest. Durrenberger (1972) provided a holistic interpretation of the plateau, but his study is introductory and cannot rank with the Kraenzel and Meinig studies. We are hesitant to define the Colorado Plateau as a sociological region for the purposes of recreational description. It may acquire this social status in the future as fossil fuel conversions provide identity. But the plateau is a meaningful format for description because of its physiography and emptiness. It is these phenomena, rather than aridity, and perhaps sociological phenomena, that can provide meaningful hypotheses for describing recreational values and choices.

Each area of the Colorado River Basin possesses a unique array of past and present determinants that contribute to future values and choices. Rather than seek a single explanatory hypothesis (such as aridity), we will concentrate upon the description of recreation endemic to the basin focusing upon the identification of the basin as an entity unto itself. We are not optimistic about the Colorado River Basin's prospects as a case study region nor about the correspondence between future choices and aridity in most arid areas.

In 1968, the Committee on Water of the National Academy of Sciences (National Academy of Sciences, 1968) emphasized that the Colorado basin was a unique river basin in the context of future developmental choices. The Committee recognized that the basin exhibited so many developmental fixes that analysis of the basin's future was of less utility than the application of its past experience to less developed river basins. The influence of historical events has drastically narrowed the range of future choice in the basin. This situation is nowhere more evident than in the case of recreational choice, and it gives further cause to question the appropriateness of the Colorado River Basin as a geographic area for the examination of the recreational futures.

RESERVATION OF UNIQUE AREAS
AND RECREATION VALUES AND CHOICES

To explain how the array of future recreation values and choices has narrowed in the basin, we have modified Gilbert White's (1969) perception-and-choice model as a basic analytical framework. The perception-and-choice approach was derived to describe water management, but, with some alterations, the basic format can be utilized to describe recreation values and choices in the Colorado River Basin. The basin's recreational future is analyzed in terms of the arrays of available choices and in terms of the perceptions of the range of choices, supply of recreation, available technologies of recreation management, and spatial linkages. The range of choice and the perception of choice is conditioned by social "guides" and antecedents peculiar to the basin. Thus the organization and geography of social institutions and cultural patterns will be considered as pervasive influences upon range and perception of choice.

Examination of values and choices relevant to recreational opportunities and resources that are characteristic of both the Colorado River and other western drainage basins is not attempted. To do so would negate the objective of utilizing the Colorado River Basin as a case area and for this purpose, the traditional recreational activity spectrum is not an appropriate analytical format. Camping, skiing, hunting and fishing are features of the Rio Grande, Missouri, Snake, and Great basins as well as the Colorado basin. They do not identify recreation values and choices that are characteristic of the Colorado drainage. In contrast, recreational values and choices associated with *amenities* and *landscapes* seem to be peculiar to the Colorado River Basin.

The physical setting of the Colorado basin locates and positions the range of choice possible in the basin (see Fig. 9.1). Physiography and climatic diversity are important physical determinants upon recreation in the sense that they structure the nature of those recreation opportunities that are unique to the basin. They identify the drainage's landscape character, and it is the

Courtesy of the Institute for Study of Outdoor Recreation and Tourism, Utah State University.

Fig. 9.1. Landscape is the most unique recreational resource of the Colorado basin. The Kolob Area in Zion National Park is typical.

domination of landscape quality that distinguishes the Colorado River Basin from all other recreation areas in the United States. In no other geographic area of comparable size can scenic amenities be found in the diversity, areal extent and intensity exhibited in this river basin. In short, the Colorado basin's physiography and climatic diversity are responsible for the best example in the United States of what could be termed an "amenities region."

The physiographic province, because it emphasizes landform and geologic structures, is a good descriptor of landscape types. Fenneman (1931) identified five provinces and nine physiographic sections within the basin. This striking diversity begins at the headwaters of the Green, Colorado, and San Juan rivers which embrace some of the most scenic alpine portions of the middle and southern Rocky Mountains provinces. These alpine situations are shared with the Missouri, Arkansas, and Rio Grande basins. Montane situations of lesser stature are found at the headwaters of the Gila, Yampa, Uinta, and Little Colorado rivers and on the western tier of the High Plateaus of the Utah Section.

Even if the montane headwaters were gerrymandered from the basin to maintain case-study climatic consistency, the Colorado River Basin would still retain its landscape superiority because the Colorado Plateau physical province is clearly the most spectacular arid physical province in the United States. And, with the exception of a portion of the high plateaus of Utah Section, the entire Colorado Plateau is shared with no other drainage. The Colorado River Basin is recreationally dominated by the Plateau Province.

The existence of the Colorado Plateau has extended the range of recreational choice in the basin. In a deterministic sense, it is landform that has indicated an available range of recreation choice. Policy-makers perceived the influence of the Plateau early in this century and the Colorado basin became the United States' "national park" river basin. The National Park Service (1972) announced that the Colorado Plateau was the largest physical region to be "adequately represented" in the natural area category of the National Park System. Congress, by statute, and the Executive, through the Antiquities Act of 1906, have reserved more units and more total acreage for the National Park System in the Colorado River Basin than in any other large drainage. It is important to understand why political decision-makers perceived the recreational resources of the basin in this manner. Furthermore, one should be cognizant of the fact that the basin is no longer perceived politically as a national park region. There is a present disparity between a narrower political perception of recreational range of choice and the recreational consumer's perception of this range.

The basin's archeological wealth has also contributed to the political perception of a national park region. It is not the scientific merit of the Colorado basin archeology which recreationally distinguishes the basin. Rather, it is the fact that the Colorado River Basin possesses scenic archeology unsurpassed by other areas in the United States. As was the case with scenic geology, the scenic archeology endemic to the basin was recognized by the turn of the century and archeological reservations from the public domain became commonplace during the early part of this century.

One of the mechanisms used to establish the National Park System is the Antiquities Act of 1906. This statute is responsible for the national monuments within the system. Its origins can be traced to the archeological resource of the Colorado basin. Prior to the establishment of Mesa Verde National Park in 1906, Congress had become concerned with the vandalism occurring at other scenic archeological areas in the basin. This concern led to the passage of the Antiquities Act. The Act delegated to the President the power to reserve national monuments and this use of executive discretion was initially exercised in the Colorado River Basin. Thirteen of the initial 25 national monuments proclaimed under the Antiquities Act were located in the Colorado River drainage. National monuments representing scenic geology in addition to archeology were proclaimed under the act. Several of these national monuments later became national parks by acts of Congress. The degree of use of the Antiquities Act by the Executive to reserve new units or add large acreages to existing units reflects the range of choice that the President and the Department of the Interior feel is available to them, but the use of the Antiquities Act in this manner has steadily declined.

Congressional interest in the scenic geology manifest on the Colorado Plateau was expressed as early as 1882 when a bill was introduced to create a Grand Canyon park. At that time, Yellowstone National Park was the only national park in the United States. The Grand Canyon was eventually proclaimed a national monument in 1908 and gained national park status in 1919 (Fig. 9.2). The creation of Grand Canyon National Park initiated an era of reservation of those national parks and monuments in the drainage that belong to the natural area category of the National Park System. These reservations continued unabated until the decade of the 1930s. It was as if Secretary of the Interior Franklin Lane's (1918) directive to National Park Director Stephen Mather was being heeded exclusively in the Colorado River Basin.

> In studying new park projects, you should seek to find scenery of supreme and distinctive quality or some national feature so extraordinary or unique as to be of national interest and importance. You should seek distinguished examples of typical forms of world architecture; such, for instance, as the Grand Canyon, for exemplifying the highest accomplishment of stream erosion. . . .

Courtesy of Colorado River Board of California.

Fig. 9.2. Grand Canyon of the Colorado.

Huth (1957), Shepard (1967), and Stegner (1969) have examined the relationship of aridity to geological process and structure in the American West as an influence upon American landscape estheticism and the national park concept. Huth and Stegner emphasize the intellectual role of the Colorado Plateau. Shepard cites other areas of the West as geneses of world architecture concepts. Shepard does, however, emphasize that a later appreciation of color combined with form occurred in the Colorado River Basin where retreating cliff lines exposed a sandstone stratigraphy.

Recognition of the geological wonders did not automatically confer park status. The initiation of an elite non-resident tourist recreational use of the areas in some instances influenced the decisions to reserve those areas. This use was influenced to a large degree by the construction of railroads through the drainage. In particular, the south rim of the Grand Canyon became a tourist use area with completion of a spur from the Santa Fe. Promotion of tourism by the railroads often accompanied the transfer from national monument to park status as was the case with the Union Pacific and Zion and Bryce Canyon national parks. Accompanying the development of tourism in the isolated national park areas was the railroad's exploitation of a Navajo tourist resource in the Gallup area and of the Pueblos in the Rio Grande Basin. Meinig (1971) emphasized the influence of railroads upon socio-economic patterns in the Southwest and the same influence seems to be characteristic of recreational development patterns in the Colorado River Basin.

The political designation of a national park region constitutes a major historical era in the recreational development of the basin. The legacy of this era is evidenced by the present recreational use patterns in the Colorado basin. Non-resident and Phoenix-Tucson visitation to the basin's national parks is motivated by many of the same amenities motivations that spurred the political development of the system. The system's spatial characteristics are determinative upon participant choice and are controlling upon many present day tourist economies in what is a low population density and space costly river basin.

The same influences that ended the historical era of reservation are now beginning to alter the present pattern of recreational choice in the basin and should have dramatic impacts upon the character of future choices available to recreational consumers. The effect is one of a time lag between the narrowing of political options concerning recreational opportunities unique to the basin and the actual range of choice available to recreation users of the endemic resources.

That the period of reservation of unique recreational amenities was to end became apparent in the 1930s. The failure of the 7,000-square mile Escalante National Monument proposal (Richardson, 1965) was symptomatic of that trend. The original proposal in 1935 would have created a unit of the

National Park System in Utah that extended along the Colorado River from the Arizona line to Moab, Utah. The proposal geographically respected the amenities dictates of the Canyonlands section as much as Grand Canyon National Park and Monument respect those of the Grand Canyon section of the Colorado Plateau. Nevertheless, with the defeat of the Escalante National Monument proposal, present day recreationists can avail themselves of only 640 square miles of national parks on the Colorado River in the Canyonlands section. In tracing the history of the Antiquities Act of 1906, Ise (1961, p. 160) shows that western legislators were not initially antagonistic to the act:

> Some years after the Antiquities Act was passed some of the western men saw that the reservation of some of these areas was a threat to the economic development of their states, and turned sour on the reservation policy. Many bills were introduced to amend the act, some of them to abolish the President's power to set aside national monuments, but the attempts always failed.

Ise's "economic development" eventually became competitive with the recreational array of choices. Water can be considered synonymous with the local perception of "economic development." As Richardson (1965) indicated in the case of the Escalante National Monument, by 1940 a "primary interest in resource policy in the Colorado River Basin" had become the "development of mineral, power, and reservoir sites."

It is not water in the sense of rainfall or climatic conditions that requires scrutiny; rather, it is the institutional organization, geopolitical arrangement, and cultural perceptions of the Colorado River itself that demand attention. It is the management of the flow of water in the river that has diminished recreational choice in the Colorado basin and that will narrow choices in the future.

The management of the river has been described on numerous occasions. In *Water and Choice in the Colorado Basin,* the National Academy of Sciences (1968) utilized a perception and choice model to frame its description. The Academy's description is borrowed to describe recreational choice vis à vis choice in river management. The Academy emphasized the constaints upon choice imposed by an evolving law of the river. Several elements seem to have particular relevance to recreation.

The first influence is that of the positioning of the Utah-Arizona border and Lee's Ferry relative to the landforms of the basin. This geopolitical organization of space has had a significant historical influence upon recreation and will dominate the recreation futures in the upper basin. The physical drainage basin does not easily lend itself to analysis. The political facts of the upper and lower Colorado River basins provide a much more suitable analytical framework. The location of the state boundaries within the drainage is another appropriate influence. Southern California as an economic region outside the basin is a third element.

Since the Escalante National Monument proposal, the progressive abandonment of a policy of recreational reservation of areas and opportunities unique to the basin to the water policy arena has been well documented. The Echo Park controversy in the 1950s is instructive. In this case, the positioning of the state boundaries of Wyoming, Utah, and Colorado relative to the amenities of the Green and Yampa rivers canyon country, the perception of water needs held by upper basin decision-makers, the cultural dispositions toward water held by Utahns, and the influence of southern California were significant elements. Although the historical definition of the Dinosaur National Monument recreational resource emerged unscathed, this definition was severely tested. The controversy also elicited the first comprehensive study of the recreational resources of the upper basin (National Park Service, 1950). The study defined the basin's recreation resources in historical fidelity to the nationally significant amenities concept that emerged at the turn of the century. But as Stratton and Sirotkin's study (1959) of the controversy points out, the recreational survey, conducted by the National Park Service, was prompted by the Bureau of Reclamation and was never intended to be an instrument of recreation choice.

The deletion of the Echo Park site from the Upper Colorado Storage Project is often considered to mark a resurgence of the preservation movement in the United States (Nash, 1973). Indeed this did occur, but this perception of a wider range of recreational choice for the Colorado River Basin has always been held by decision-makers, recreational participants, and cultures from outside the basin.

Of greater significance is the fact that the Upper Colorado River Storage Project authorized other sites including the Glen Canyon site. If decision-making relevant to Colorado River flow was to become the dominant decision-making mode in the basin, then the definition of the basin's recreation resource had changed. Recreation would no longer possess a regional identity. Recreation in the drainage was to be developed as one of the outputs of multiple-purpose water development. Under this array of choice, the recreation opportunity spectrum is fixed by the technological capability of large-scale water impoundment and it is indistinguishable from similarly conditioned opportunity spectrums found elsewhere such as the Missouri or Columbia basins. Ingram (1971) explains that the consent building process for water impoundment projects is dependent upon the magnification of benefits. In the Colorado River Basin, recreation has become one of these

Courtesy of U.S. Bureau of Reclamation
Fig. 9.3. Boating on Morrow Point Reservoir, Gunnison River.

"add on" benefits. The policy of distilling the regional essence and potential of recreation that is manifest in the reservation system is not appropriate within this system of water choice.

The reservation of the Grand Canyon of the Colorado River as a national park represents the actual range of recreation choice available in the Colorado

River Basin. The creation of a National Recreation Area in Glen Canyon of the Colorado River is a much narrower interpretation of the drainage's recreational resources. In fact, the National Recreation Area is not a political interpretation of recreational resource potential. Instead, it belongs to the realm of water resource choice.

Recreational choice in the basin is presently dominated by water policy concerning Colorado River flow (Fig. 9.3). The evidence is overwhelming that the broader traditional system of recreational choice is not being employed. The Antiquities Act was last used in the drainage in 1969 to expand Capitol Reef and Arches national monuments and proclaim the Marble Canyon National Monument corridor. At the same time, it was rejected as a mechanism to create a larger Sonoran Desert National Monument from Organ Pipe Cactus National Monument in the lower basin. The Capitol Reef and Arches national monument reservations engendered tremendous opposition from political actors in the basin and these proclamations are perhaps the last time that the statute will be applied to the Colorado River drainage.

Ingram (1969) documented how the geopolitical character of water choice modified a recreational reservation on the Gila River. In this case, controversy over Hooker Reservoir site in the Gila Wilderness Area resulted in the modification of a wilderness reservation. Recreation reservations in the basin generally emanate from interests external to the basin. This type of choice can be disruptive to the process of water choice because it cannot be accommodated within the regionally perceived range of water development choices. However, the Hooker Reservoir experience of the late 1960s indicates that only disruption, not domination, of the water choice process will occur.

There are other recent examples of the erosion of recreational opportunities and resources in the basin. The loss of unique white-water recreational opportunities in the Grand Canyon and Cataract Canyon can be attributed to Lake Powell. The upper Green River has been eliminated from consideration as a wild and scenic river. Trout stream fishing and the present boundary of the High Uintas Primitive Area are threatened by the Central Utah diversion project. The refusal of the Supreme Court to review the intrusion of Lake Powell into Rainbow Bridge National Monument has effectively diminished the boating potential of Lake Powell. As the reservoir reaches full pool, the former dendritic shape and shoreline has been lost to large open bays (Fig. 9.4). In the aggregate these decisions both limit and structure the recreational futures of the basin. And they are decisions of water choice rather than recreational choice.

If the perception of water choices, which dominates recreational choice in the basin, were expanded, would the future of recreation in the basin be correspondingly broadened? We hesitantly conclude, as did the National Academy of Sciences, that expansion of choice over Colorado River flow

Fig. 9.4. Rainbow Bridge. Many natural bridges
characterize the landscape of the Colorado Plateau.

management is unlikely and to paint any other scenario would be an exercise in futility. Nevertheless, there are indications that the range of choice over water flow will be subject to a restructuring in the future. The existence of coal and oil shale fossil fuels in the basin is emerging as a phenomenon to be reckoned with in basin water policy (Baldwin, 1973; U. S. Senate, 1972). One doubts that a separate system of fossil fuels choice will become totally competitive with the traditional system of water management. More likely, water will continue to dominate the decision-making process. But, in the process, the array of alternatives for water management is certain to be restructured and perceived as such by regional water policy actors.

What does this bode for recreational futures? As of 1974, there was only one point of reference. The authors of *Water and Choice in the Colorado Basin* (National Academy of Science, 1968) believed that the Bureau of Reclamation's participation in the Salt River Project constituted an expansion of the perception of water choices in the basin. The use of coal-fired power to pay off the Central Arizona Project was an alternative to the hydroelectric pay-off strategies that had previously dominated water management choices in the basin.

The shift from hydroelectric to thermal-electric power in the planning for the Lower Colorado River Basin Project represents the relaxation of a policy constraint once widely regarded as immutable (National Academy of Sciences, 1968, p.85).

Recognition by basin water managers of the coal-fired generation alternative was spurred by the fact that hydroelectric generation in the Grand Canyon was unacceptable to the nation.

Participation in the Salt River Project involved the construction of the Navajo Power Plant on the shore of Lake Powell, the use of strip-mined coal from the Black Mesa Mine, and the intrusion of transmission lines into the Paria Canyon Primitive Area. Thus this perception of a wider continuum of water management strategies has not enhanced the recreation future of the central portion of the basin. Boating quality, non-resident motor vehicle tourism, and primitive recreational opportunities are diminished by this choice.

A SPACE COSTLY REGION
AND RECREATION VALUES AND CHOICES

The Colorado River Basin is a space costly area, i.e., central places in the drainage are farther apart than in most geographic areas of comparable size. Unlike the Great Plains, the space costliness of the Colorado drainage is not imposed geometrically. Instead, the heterogeneity of the basin's difficult terrain has imposed a more disorganized pattern of distances. Topography as an obstacle remains the basic cause of the basin's space costliness. Aridity does not appear to be a major factor.

The basin has never "filled-in" with central places. Central places exist where they are because of the location of scattered mineral deposits, railroad systems, formally reserved tourist attractions, persistence of Spanish and Mormon settlement efforts, and the objectives of social regions external to the basin. For example, the basic transportation structure of the region resulted from attempts to reach the northern and southern California regions from the Midwest and Texas. Much of the present vertical transportation infrastructure results from attempts by the Wasatch Front Mormon colony to penetrate into the drainage to the Little Colorado region.

Touristic central places were determined and fixed during the first quarter of the twentieth century. The basic pattern was established with the reservation of scenic areas from the public domain. The patterns of recreation consumption and tourist economies have been little modified since this imposition of the National Park System upon the basin. In Utah, Brown (1971) has shown that the pattern of touristic central places involves connecting distances of unusual length. An extreme example in Utah is the proposed San Rafael

Swell National Conservation Area adjacent to a 110-mile section of Interstate 70 devoid of any tourist facilities (Royer and Dalton, 1972). The exceptions to the persistence of this pattern are in the Phoenix-Tucson area and Las Vegas where the basin's larger resident populations reside.

As the motor-vehicle travel mode of tourists began to replace the railroad travel mode, the distances between the drainage's tourist attractions were shortened. The ensuing demands for an improved highway network between local tourism and ranching communities also shortened distances. During the postwar decades of the 1950s and 1960s, non-resident motor vehicle tourism became the characteristic recreational consumption pattern in the drainage. The use of private motor vehicles to reach recreational places enforced the geographic pattern of the recreational economy which had been set much earlier. Automobile tourism was directly responsible for the persistence of local economies which would otherwise have perished as viable communities due to the cost of space. The popularity of motor-vehicle travel thus meant that the basin was no longer space costly in a recreational context. Motor-vehicle travel expanded the available choices of both the recreational consumer and the local entrepreneur.

The persistence of scattered local tourist places was also responsible for the development of a resident cultural disposition toward amenities and other manifestations of the tourist resource. In many cases, the local cultural perception of the recreational resource was not congruent with the perceptions of the non-resident tourist consumers. These cultural perceptions and beliefs about the recreation resource appear to generate local political actors much in the manner that beliefs about water generate local political activists (Ingram, 1971). If they are considered as cultural dispositions toward the environment, both the recreation and water dispositions may belong to the same local belief system.

The Colorado River Basin will become extremely space costly in the immediate future and the recreation consumer's range of choice will be severely curtailed. The extra-basin and intra-basin availability of gasoline to basin-bound non-resident recreationists will be almost deterministic upon choice. If limited gasoline supplies become the norm in the future, then the economic repercussions upon local tourist cultures will also be dramatic. Lacking an economic base, local values and perceptions of the tourist resource will cease to exert a political impact upon the resource. This impact presently influences the reservation of recreation areas removed from the immediate vicinity of the Colorado River and its major tributaries.

The trend of the 1960s toward the "democratization" of the basin's recreational resources will be reversed in the 1970s. The values and choices concerning the recreational resources of the Colorado River Basin had traditionally accumulated political activists external to the basin. If this reversal occurs, then these actors will not be impeded by local perceptions of recreational choice. However, externally derived recreation values will be poised against water values possessing local sources and an emerging energy system of choice with its national origins. We agree with Baldwin's (1973) assessment that these recreation values will not remain competitive in the future.

The recreational phenomenon which distinguishes the Colorado River Basin from other western river basins is that of non-resident tourism to landscape amenity places that have been formally reserved from the public domain. The uniqueness of the National Park System in the basin is the result of the choice to reserve. This process of recreational choice is historical; however, the recreational places that it established in the basin remain determinative upon the present-day recreational participants' range of choice.

The process of reservation as a mechanism for recreation choice has been replaced by a water-choice process which narrows the range of recreation choice in the basin. The decline of the use of the Antiquities Act in the basin is historically congruent with the emergence of water policy domination of recreation choice. The earlier process reflected national recreational values. The water-choice system places recreation in a spectrum of local water values and bestows a degree of ambiguity to the basin's recreational resource.

While the domination of present water choices in recreation does not reflect the innate recreational character of the basin, neither does the expansion of the range of choice in water management expand recreational choices. The blending of energy policy and water policy in the drainage will alter the traditional water choice process in the future. The limited experience with this situation indicates that it diminishes recreation choice.

The National Park System in the basin has created a pattern of touristic central places which is space costly. Motor vehicle tourism in the 1960s diminished the effect of distance and maintained local tourist economies that would not have been viable otherwise. To this pattern accrued local cultural and political dispositions concerning the basin's recreational resources which often were not congruent with the values expressed in National Park reservations or by non-resident tourist consumers. The advent of gasoline shortages will limit recreation participant choice in the future and at the same time eliminate the economic and political activism of the local tourist cultures. Although the lack of local recreational opposition will renew the strength of recreational actors external to the basin, it is doubtful that these actors and their values will be competitive with energy and water actors and values.

Attaching an aridity explanation to the recreation phenomena discussed above does not seem appropriate. If a single natural causal factor could be identified, it would be physiography rather than aridity. Factors other than those discussed here may be influential. For example, Hunt (1971) found that arbitrary geopolitical relationships influenced tourist choice in the basin. Non-resident perception of a state's "image," comprised of impressions of climate, landform, resident population characteristics, and recreation quality, significantly influenced the spatial behavior of tourists to the basin. It is also difficult to generalize to arid regions from a drainage basin that is both humid and arid. Much of the literature of arid areas in the American West laments the fact that the social patterns of these areas do not respect aridity.

The Colorado River Basin possesses a recreational past and future that is unique. The recreational character of the basin will not be duplicated in other drainage basins or in arid areas. Application of the recreational experience at Lake Powell as a case study to situations outside the Colorado River Basin would be difficult. This problem of case study application is typical of most recreational situations in the basin. However, one should not discourage an effort to study recreation values and choices in the Colorado basin for their own utility. Many of the Colorado River Basin's recreation values are too significant to the nation to be abandoned because of academic disinterest.

Discussion by David W. Goodall

Two completely different groups of customers or consumers exist for the recreation potential offered by an area—those living there and the tourists. Royer and Hunt primarily discussed the latter; and in a tourist-oriented society their needs must come first. In some of the Caribbean Islands, for instance, recreation facilities for the local population are relatively unimportant and must take second place to those for tourists except in so far as they can serve both. In most states of this country the reverse applies, and the local population is the main consumer of recreation facilities.

Though tourism is more important in the Colorado basin than in many other parts of the country, in the basin *as a whole* it does not seem to be a major part of the economy. Little more than 10 percent of employment can be related to tourist needs. Consequently, there does not seem strong *economic* justification for putting the tourist's recreational needs ahead of other needs of the local community.

The recreational needs of the local community, apart from those met within the urban framework, are for open-air sports of various kinds. There should be opportunities for hunting, walking, rock-climbing and similar activities, together with camping and picnicking, which require suitable undeveloped areas easily accessible from centers of population. These needs are competitive with urban and agricultural development and to some extent with livestock, but with proper planning from an early stage can be met without serious conflict. Skiing similarly can be provided where suitable terrain exists; and the same applies to snowmobiles, dune buggies and motorcycles as long as they can be restricted to areas where they do not interfere with other uses. Water sports are more difficult. An opportunity for boating and water-skiing is a legitimate wish of the towndweller in an affluent society, even in the desert, as is also the desire to swim in open water and not in a constructed pool. Where there is water development for irrigation or hydroelectric purposes, those demands are easily met, though to meet them otherwise would be prohibitively expensive. Fishing opportunities accompany these, though they also may be developed in more modest ways where topography permits.

Though the tourist will make use of these opportunities if they exist, they are usually not what draws him from his home many miles away. He comes from the eastern states or California, from Chicago or Atlanta, to the Colorado River Basin, not because the skiing is better in Arizona than around San Francisco, say, but because the environment is different from that to which he is accustomed, and has special or unique features. And the two aspects of the Colorado basin that specially attract the tourists are the natural landscapes—canyonlands in the upper and middle basin, deserts in the lower basin—and, in the lower basin, the winter climate.

The natural landscapes of the Colorado River, from the Wyoming border to Las Vegas, are unique in the country and probably in the world. They have an aesthetic appeal that is irreplaceable. And the recreational potential used by these tourists is not available in any other way or any other place. Unique landscapes are akin in this respect to the contents of great art galleries—the Canyonlands and the Grand Canyon take their place alongside the Louvre or the Hermitage as irreplaceable parts of man's aesthetic heritage.

In his demands for such aesthetic experience, the tourist from outside the area may conflict with the resident within the area. The latter may also appreciate the aesthetic value, but other values—for instance, economic values generated by water development—may override it in his appreciation. And to a considerable extent this conflict is reproduced on the political stage. Local and state politicians must reflect the local balance of interests (which may not attach great weight to the tourist's recreational desiderata), while at the national scale these local interests are weighed against the interest of tourists from other parts of the country; and international concern has no real representation in the political process. So the conflict between recreational and other needs is played out, in part, at two different levels of politics.

Even within the field of providing recreational facilities for tourists there may be conflicts between political considerations at different scales. The Canyonlands area could be developed for tourism by providing numerous access roads. Alternatively, by use of helicopters operating along defined access routes to a few limited areas, tourist development could be concentrated in and around a few areas with no access roads whatever. A considerable part of the prospective tourist population would find the latter alternative very attractive; and this approach would minimize the changes in this very special landscape. But one might expect it to be opposed by local interests which would be served by passage of tourists by road.

As in other fields, so in the recreational use of the Colorado River Basin, questions of reversibility and of preserving options should be prominent where any modification of land use is proposed. Irreversibility may often be an argument to be set by recreational interests against water or energy development. Archaeological sites, in particular, are likely to be destroyed or damaged irretrievably by such developments; and archaeological sites may be important to the tourist as well as to the professional. The beauties of the great canyons have been seriously affected by water development, but these changes are not quite so irreversible—a few well-placed bombs and a few decades would probably restore the landscape to something very close to its pristine condition. But, practically speaking, water development has certainly caused the loss of recreational options. And who knows what options may be needed or chosen by our grandchildren? Forty years ago who would have predicted the present interest in snowmobiles as a form of recreation?

Apart from its unique landscape features, what the Colorado River Basin can best offer in recreation to the tourist from other states is space—something already in short supply in the country as a whole, and likely to become shorter. More than almost any other part of the country, one may travel for miles without seeing a man. And to some, this is a most valuable feature. This opportunity—the wilderness experience—must be preserved for those who feel the need of it. But they cannot expect to have all the most exciting scenery preserved for their use—others without the need for solitude may enjoy it too, and should not be deprived of the opportunity—but empty wilderness areas should be preserved wherever they can with relatively little loss to other land uses. And, luckily, the Colorado River Basin has many thousands of square miles which would be so preserved, above and beyond those that have already been so set aside.

For the other tourists—those who are less demanding in space, though perhaps more so in landscape character—the important development question is how to get them to the places they want to visit. I am not prepared to believe, with Hunt and Royer, that energy difficulties will cause a drastic change in the use of individual vehicles. This effect seems likely to be short-lived. But serious consideration should be given to alternatives. In the past, the railway has given thousands of tourists the opportunity of aesthetic satisfaction from outstanding landscapes—one need only take the Grand Canyon South Rim as an example. In Yosemite, good use has been made of buses to reduce dependence on the private car. And, as mentioned, helicopters could be used to avoid even the scarring of landscapes by road development.

In the lower Colorado basin, the tourist potential is different. Here, the tourist is more attracted by the winter climate than the landscape features—attractive as the deserts can be. "Dude ranches" also have their role for the city dweller in some parts of the basin. But in general the winter tourist to the lower basin will go to urban centers, and will expect the usual facilities of urban centers—though preferably without their drawbacks.

Linked with this is a special type of recreational development for those whose main concern is now recreation—the retirement community. Retirement enterprises have developed enormously over the past decade or more in the genial winter climate of Arizona and southeastern California. Their demands for water are substantial, and in other respects they resemble other urban developments. But, in the affluent society of America with its loose family structure, one may expect this demand on what we must regard as the recreation potential of the basin to go on increasing.

Literature Cited

Baldwin, Malcolm F. 1973. *The Southwest Energy Complex: A Policy Evaluation.* The Conservation Foundation, Washington, D. C.

Brown, Perry J. 1971. An exploration of physical attributes in a spatial and contextual analysis of tourist buying behavior. Ph.D. dissertation. Logan: Utah State University.

Cleland, Courtney B. 1966. Do we need a sociology of arid regions. In *Social Research in North American Moisture Deficient Regions,* ed., John W. Bennett. pp. 1-12. Committee on Desert and Arid Zones Research, Southwestern and Rocky Mountain Division, AAAS, Contribution No. 9. Las Cruces, New Mexico.

Durrenberger, Robert. 1972. The Colorado Plateau. *Ann. Assoc. Am. Geog.* 62(2):211-236.

Fenneman, Nevin M. 1931. *Physiography of the Western United States.* New York: McGraw-Hill.

Hunt, John D. 1971. Image—a factor in tourism. Ph.D. dissertation. Fort Collins: Colorado State University.

Huth, Hans. 1957. *Nature and the American: Three Centuries of Changing Attitudes.* Berkeley: University of California Press.

Ingram, Helen M. 1969. *Patterns of Politics in Water Resource Development: A Case Study of New Mexico's Role in the Colorado River Basin Bill.* Institute for Social Research and Development, Publication No. 79. Albuquerque: University of New Mexico.

____. 1971. Patterns of politics in water resources development. *Natural Res. J.* 2(1):102-118.

Ise, John. 1961. *Our National Park Policy: A Critical History.* Baltimore: Johns Hopkins Press.

Kraenzel, Carl F. 1955. *The Great Plains in Transition.* Norman: University of Oklahoma Press.

Lane, Franklin D. 1918. May 13 letter from Secretary of the Interior to Stephen T. Mather, Director, National Park Service. Washington, D. C.

Meinig, D. W. 1971. *Southwest: Three Peoples in Geographical Change, 1600-1970.* New York: Oxford University Press.

——. 1972. American wests: preface to a geographical interpretation. *Ann. Assoc. Am. Geog.* 62(2):159-184.

Nash, Roderick. 1973. *Wilderness and the American Mind.* Revised Edition. New Haven, Conn.: Yale University Press.

National Academy of Sciences, Committee on Water of the National Research Council. 1968. *Water and Choice in the Colorado Basin.* Publication 1689. Washington, D. C.

National Park Service. 1950. *A Survey of the Recreational Resources of the Colorado River Basin.* Washington, D. C.: Government Printing Office.

——. 1972. *Part Two of the National Park System Plan: Natural History.* Washington, D. C.: Government Printing Office.

Richardson, Elmo R. 1965. Federal park policy in Utah: the Escalante National Monument controversy of 1935-1940. *Utah Hist. Quart.* 33(2):109-133.

Royer, Lawrence E., and Dalton, Michael J. 1972. *Land Use in the Utah Canyon Country: Tourism, Interstate 70, and the San Rafael Swell.* Institute for the Study of Outdoor Recreation and Tourism. Logan: Utah State University.

Shepard, Paul. 1967. *Man in the Landscape.* New York: Alfred A. Knopf.

Stegner, Wallace. 1953. *Beyond the Hundredth Meridian.* Boston: Houghton Mifflin.

——. 1969. *The Sound of Mountain Water.* Garden City, N. Y.: Doubleday and Company.

Stratton, Owen, and Sirotkin, Phillip. 1959. *The Echo Park Controversy.* Inter-University Case Program, Inc. No. 36. Indianapolis: Bobbs-Merrill Company.

U. S. Senate, Committee on Interior and Insular Affairs. 1972. *Problems of Electrical Power Production in the Southwest.* Report No. 92-1015. Washington, D. C.: Government Printing Office.

Webb, Walter Prescott. 1931. *The Great Plains.* New York: Grosset and Dunlap.

——. 1957. The American West, perpetual mirage. *Harper's Magazine* 214(1284):25–32.

White, Gilbert F. 1969. *Strategies of American Water Management.* Ann Arbor: University of Michigan Press.

10

Community Development

DEAN E. MANN

An analysis of community development in the Colorado River Basin requires either an encyclopedic knowledge of each and every community in the region or a willingness to rely on intuition and impressionistic accounts. What follows will be an attempt to examine the highly varied urban and rural centers of the Southwest, based in part on data available concerning many of these centers, and especially on data supplied by many of the communities themselves. The focus is on the future—the choices facing these communities and the options they are fashioning for themselves. No attempt is made at prescriptions—the choices must be made by the residents—but some options are presented and some criteria of measurement are suggested.

A discussion of "community development" must necessarily begin with definitions. This is particularly true in the second half of the twentieth century when many of the accepted social definitions have come under attack. For the values that have underlain those definitions either have radically altered or have been called into question by changed environmental conditions, by changes in the relationships among various groups in our society, or by changes in the aspirations of individuals, groups, regions and the nation.

In undertaking research for this chapter I sent letters to state planning officers in the capitals of the states of the Southwest and to planning offices of all of the metropolitan centers and of many of the smaller communities of the region. Considerable response was received from these public agencies, although the utility of the information supplied varied substantially. The planning effort, judging from these documents, often shows considerable imagination but just as often is inadequate or almost non-existent. In some cases—and New Mexico was particularly notable in this respect—the contributions were significant.

Community

What is meant by a "community"? It is surely more than a physical relationship of buildings, communication and transportation systems, and processes of production, distribution and consumption. A community is infused with common understandings, insights, assumptions, and, more generally, values. If this is so, is it meaningful to talk of community in the singular at all? Must we not talk of the plural: communi*ties* of the Southwest? For the Southwest, as repeatedly pointed out by anthropologists, historians and sociologists, consists of many ethnic, racial and national groups that have retained their separate identities despite the forces of the melting pot of American society (see, for example, Spicer and Thompson, 1972). Mexican, Mormon, Indian, Texan, and modern Anglo all inhabit the Southwest and the final resolution of the process of interaction that has been going for the past century and more is entirely uncertain. What is certain is that the value system of each remains strong, making the notion of community stretch almost to the breaking point.

Moreover, the Southwest, like the rest of the United States and the world, is now predominantly urban. It is an oasis culture but the oases now number their residents in the hundreds of thousands and sometimes the millions. Is Los Angeles a community? or Phoenix? or even Albuquerque or Salt Lake City? Perhaps the term applies meaningfully to Kanab, Utah, or Grand Junction, Colorado, or even Santa Fe, New Mexico, or to villages on the Navajo and Hopi reservations. But where relationships have become so anonymous, impersonal and interpersonal, and common symbols so lacking it is difficult to think that the term community even applies. Even the federal government recognizes this distinction in sponsoring community development programs that encourage neighborhoods or ethnic groups in given locations within cities to organize for local improvement.

In many ways the term "community" may transcend geographical and political boundaries. Chicanos in East Los Angeles, for example, may identify far more completely with their ethnic brethren in Phoenix, Tucson and Albuquerque than with their Angeleno neighbors. It has long been said that for Mormons, their principal allegiance is to Salt Lake City, not to the particular towns or states in which they reside.

Development

The second term, "development," equally requires definition. Development has for generations connoted "growth" and the terms might almost have been considered interchangeable. Growth meant additional population, increased economic activity, and the additional variety and choices that came from larger urban areas. It often meant new industry, new farms, new freeways, new subdivisions and new towns. Development was an unmitigated blessing.

While development may continue to connote all of these characteristics of growth, it also connotes other characteristics that formerly were ignored: air pollution, water pollution, urban decay, strip development, inadequate public transportation, rising crime rates, urban sprawl and loss of aesthetic amenities. Can anyone doubt that the residents of Los Angeles, Phoenix or Salt Lake City have not suffered all of these urban "bads" as they have gained the blessings of bigness? There is even evidence that bigness has not paid off economically in that the costs to communities in supporting urban services have often exceeded the benefits flowing from additional jobs and increments to the tax base.

Therefore, should not development connote a qualitative change or improvement rather than growth in size and economic activity? Should it not bespeak an improvement in how people live, their relationships, their opportunities?

The Basin

The third definitional problem relates to the conceptual boundary—the Colorado River Basin. As a natural hydrological phenomenon, there can be little doubt about the boundaries of the Colorado River Basin. But as a man-defined and modified hydrological phenomenon the Colorado River Basin, even hydrologically, spills over into the Rio Grande River Basin, the Platte River Basin, the Great Basin, and the southern coast of California. Furthermore, culturally, economically and politically, the Colorado River Basin lacks any meaning whatsoever. The communities of the Colorado River Basin are not distinguishable by their location either geographically or hydrologically from other communities that exist beyond some watershed. Is Moab, Utah, distinguishable as a community from Spanish Fork, Utah, because the first is within the Colorado River Basin and the second is within the Great Basin? Or does their common Mormon heritage provide the basic similarities and their distinctive setting, not particularly related to their river basin location, supply the differences? Is not Pioche, Nevada—a boom and bust mining town on the Meadow Valley Wash within the Colorado River Basin—more like other mining towns throughout the West than it is similar to other Colorado River Basin communities that share a common hydrologic system?

To make the point positively, the Southwest is a distinguishable region of the United States—distinguishable for its climate, native populations, economy, and perhaps some characteristics of its politics and attitudes—but it should be discussed as the Southwest and not the Colorado River Basin. This discussion, therefore, will be constrained by no such hydrologic boundary.

In this context, one should note that increasingly the Southwest and more broadly the West and their communities are permeated by the same influences that are changing the character of communities throughout the United States

(Mann, 1963, 1969). The larger cities are growing rapidly while the smaller communities have been dying out. Metropolitan areas have been suburbanizing just as the communities of the Middle West, Northeast and South have been doing. Cities of the Southwest have their ghettoes—often Mexican-American rather than black—but ghettoes nevertheless. They suffer the same urban ills of congestion, smog and center-city blight. Despite the occasional effort to continue the romantic notions of the Southwest—some of which may be justifiable—the hard, gut problems confronting policy-makers in the Southwest sound very much like those policy-makers are trying to deal with in Pittsburgh, Atlanta, and St. Paul.

In some ways the Southwest may lead the way. For example, if one examines the statistics on air quality in the Southwest, one discovers that the Southwest has achieved a position of some eminence. Denver in 1968 led such eastern cities as Chicago, Cincinnati, Philadelphia, St. Louis, and Washington, D. C. in the total oxidant concentration of gaseous atmospheric pollutants such as carbon monoxide, nitrous oxides and hydrocarbons (U. S. Environmental Protection Agency, 1972a). Similarly, in 1971 the records of Albuquerque, Denver, and Phoenix were not notably different in concentrations of suspended particulate matter from the records of Chicago, Philadelphia and St. Louis. And the southwestern record was far worse in that category than many other eastern or mid-western cities such as Atlanta and Minneapolis or even western cities such as San Francisco and Seattle (U. S. Environmental Protection Agency, 1972b).

The Southwest and the Nation

In considering choices for urban life in the Southwest one must recognize that the Southwest is far from an autonomous political or economic entity. Perhaps more than any region of the country, it is dependent on decisions made elsewhere. Upwards of 90 percent of the land in some states is owned and managed by the federal government. Decisions made with respect to timber cutting, recreational development, land management practices, military utilization of the land and airspace, all of which are made by federal agencies within the constraints of budgets and the law, have a direct and powerful effect on the quality of life in the region. As revealed by the energy crisis, the Southwest, with its heavy emphasis as a playground for the rest of the country, may be the hardest hit by energy shortages and rising energy costs. Disposable income may be spent at or near home rather than in the communities of the Southwest.

Evidence for this is not hard to come by. The *Los Angeles Times* (1974) reported that many of the principal tourist attractions in California were suffering from the energy shortage. Such popular attractions as Sea World and the zoo in San Diego and Disneyland in Anaheim had suffered declining attendance. The shortage of gasoline in the summer of 1973 had a very serious impact on the tourist industry in Colorado. If these declines become trends as the energy crisis deepens or as energy shortages become a permanent part of American life, the tourist industry on which so many southwestern communities have pegged their future may turn out to be a mirage.

WHAT IS HAPPENING IN THE SOUTHWEST

The issue posed for the people of the Southwest, whether in large or small communities, is the character of the environment in which they will live. Material progress has occurred and the Southwest has offered amenities in abundance. But progress has changed the desert scene often irreversibly. Hikers who long for solitude find their airspace invaded by helicopters and other aircraft. Residents who are inspired by the quiet desert scene often must listen to and see motorcycles, jeeps, trucks, and dune buggies careen across the landscape. The desert atmosphere loses its clarity as it is fouled by automobile exhausts and emissions from the coal-burning steam plants. The Indian who seeks to preserve tradition finds it threatened by the powerful industrial and commercial forces of the modern American economy. "The Southwest shows signs of being overdeveloped, or undercontrolled, or both" (Bakker and Lilland, 1972).

Population Growth and Urbanization

Whether those who presently reside in the Southwest like it or not, rapid growth of population has been the most notable demographic characteristic of the region for the past several decades and is likely to be its most pronounced characteristic in the future. In the decade 1960-1970 the following percentages of population increase were recorded by the U. S. Bureau of the Census (1972).

Arizona	36.1
New Mexico	6.9
Colorado	26.0
Utah	18.9
California (south of the Tehachapi Mountains only)	27.1

Most of that growth took place in the major metropolitan areas, nearly all of which are outside the hydrologic boundaries of the Colorado River Basin (see Table 10.1).

Despite the obvious growth patterns of the past decades, projections of future population growth are extremely risky. The experience of the southern California region is instructive in this regard. Apparently, in-migration now

TABLE 10.1

**Growth of Standard Metropolitan Statistical Areas (SMSA),
Colorado River Basin, 1960–70**

State	SMSA	Percent increase 1960–1970	Percentage of total state population, 1970
Arizona	Phoenix	45.8%	54.6%
	Tucson	32.4	19.9
New Mexico	Albuquerque	20.4	31.1
Colorado	Denver	32.1	55.6
	Colorado Springs	64.2	10.7
Utah	Salt Lake City	24.5	52.6
	Provo–Orem	28.8	13.0
	Ogden	14.0	11.9
Nevada	Las Vegas	115.2	55.9
California	Los Angeles, Long Beach	16.4	35.2
	Anaheim, Santa Ana, Garden Grove	101.8	7.1
	San Bernardino, Riverside, Ontario	41.2	5.7
	San Diego	31.4	6.8

virtually matches out-migration in the Los Angeles area with continued but slower growth in Orange and San Diego counties. The rapid decline in the birthrate has thrown projections even further askew. Furthermore, one can never be certain whether projections are valid predictions of the future or self-fulfilling prophecies and therefore goals to be achieved.

Nevertheless, the southwestern region is expected to experience continued growth and mostly in its urban centers. The U. S. Water Resources Council (1971) projections for the upper and lower Colorado River Basins are instructive. The 1965 population of 366,000 in the upper basin is expected to double by 2020. The lower basin's population is expected to increase nearly four times by the same year. Phoenix planners predict for the Phoenix planning area, the largest segment of the Phoenix Standard Metropolitan Statistical Areas (SMSA), a growth from 596,900 in 1970 to 1,080,000 by 1990 or more than an 80 percent increase (Phoenix, Arizona, Planning Commission,

1972). The State Planning Office for Colorado predicts a population for metropolitan Denver of 2,175,000 by the year 2000, nearly double its 1970 population (Denver Chamber of Commerce). Based on goals set by citizen groups in Albuquerque after considering alternatives, it was estimated by the Albuquerque/Bernalillo County Planning Department (1972) that Albuquerque might reach 500,000 by 1990, up 185,000 from its 1975 figure of 315,000. These may not be accurate projections in view of the dramatic decline in birth rates throughout the nation, but they do suggest the pronounced tendency toward growth.

Rural Communities

At the same time that the major urban centers of the Southwest were growing rapidly, many if not most of the rural counties of the region were losing population in an absolute sense or experiencing a net migration deficit. For the states of Colorado, New Mexico and Utah, figures for the decade 1960-1970 are shown in Table 10.2. Thus, approximately half of all the counties of those three states lost population and many of those that increased in population did so only because natural increase was larger than the deficit in migration. Of the four counties in the Colorado River Basin in Wyoming all suffered a net migration deficit and all but one, Sweetwater, had an absolute loss in population. Half of the Arizona counties had a net migration loss but only one, Greenlee County, had an absolute loss.

TABLE 10.2

Population Change in Counties

State	Total counties	Counties losing population	Counties having net migration deficit
Colorado	63	32	38
Utah	29	13	24
New Mexico	32	17	29

The consequences of absolute declines of population may be severe in an economic sense. Facilities such as schools, churches, libraries, sewage plants may fall into disuse or underuse. If there have been relatively recent investments in public facilities, there may be burdensome public debts that the remaining population must pay off. The inevitable tendency to reduce the cost of public services may result in a general decline in the quality of life that existed in the community.

But the psychological and sociological impact on a particular community may be even more severe. Where there has been a stable and closely-knit social structure, as in Mormon culture, the continued net out-migration signals a loss of young people, the children of the long-term residents of the community. The lack of job opportunities forces the progeny to seek their fortunes in places far removed from immediate family and kinfolk. Unless a new economic base can be established, and this usually means something other than the traditional farming, the tide may seem virtually irreversible. Thus, any spark of interest in investment in the region may be looked upon not only as an economic shot in the arm but also a support for the family as an institution.

Is it therefore surprising that the mayor of Kanab, Utah, should favor construction of the Kaiparowits generating plant on the plateau above Lake Powell? He considered the divergent interests of economic development and environmental protection reconcilable, but his principal concern was the economic well-being of his community:

> I get a little emotional on this because we are living there. I have been president of the school board for 6 years now. We see our population falling off. We are becoming a vicinity where we have elderly people because we have no interest or nothing to keep our youth there. When they finish high school, they must leave our area and they must go away to seek employment. We have no employment that is satisfactory in any way (U.S. Senate, 1971, p. 965).

A similar problem is noted by the Uinta (Wyoming) County Planning and Zoning Commission in their report on population. The County has experienced a high level of net out-migration of people in the 15-24 age group. They are leaving, they assert, to be educated, to join the armed services, or to pursue employment opportunities. Few of those leaving for education return. They comment:

> Migration data is another indicator that can be used to point out needs and problems of the population. If we desire to keep young people or attract those back after they have been educated, it is necessary to plan for economic development to provide necessary jobs to maintain a viable young population (Uinta County Planning and Zoning Commission, 1973, pp. 14-15).

Choices Must Be Made

Despite what may seem inexorable forces controlled by national political and economic demands and propelled by international events such as the embargo on shipments of oil from the Middle East, the people of the Southwest are faced with choices about the kind of environment in which they wish to live. In its most general terms, the choices are between development and growth and conservation and control. Yet between these polar opposites are many intermediate positions in which some growth may be mixed with some levels of conservation or protection and with some measures of control. These various alternatives are further complicated by consideration of the mechanisms by which given ends will be achieved: through essentially market-type operations with little or no public intervention or through comprehensive planning for the achievement of specified goals. Still a further issue must be whether and in what manner the various ethnic, religious and other minority cultures will be treated in this process of choice: whether or to what extent they persist and maintain their specific character in the face of overall planning and market forces. The remainder of this paper is concerned with the manner in which various communities in the Southwest are meeting the challenge of making such choices.

From material supplied by many of the communities of the Southwest it is clear that both planners and members of the communities themselves have become highly sensitized to the choices that lie before them and the necessity to make positive decisions rather than to let the individual choices through market forces determine the character of the communities. Several of these community efforts will be examined to illustrate the manner in which those choices are perceived.

THE PROBLEM OF PERCEPTIONS

Part of the problem is clearly one of perception, for many in urban centers are sufficiently complacent that they fail to see the problems exist now and that they will become exacerbated in the future unless adequate steps are taken. For example, few cities of the United States were as carefully planned as Salt Lake City. Streets were laid out on the grid pattern, wide enough for an ox-cart to make a U-turn without having to back up. Plots of land both within and without the city were apportioned to families in terms of need. The first-time visitor is pleasantly surprised by the abundant vegetation and the magnificent setting. He would not perceive the blight because it is not of the character found in the large urban centers of the east.

> [The blight] has a more unobtrusive form: bad roofs, deteriorating foundations, settlement cracks, leaning porches, poor add-ons and accessory buildings. Structural unsoundness is abetted by environmental difficulties also apt to be dimly perceived at first. Railroad tracks run through residential neighborhoods, parks and play spaces are lacking, yard space and storage areas are insufficient. Single family homes are scattered along heavily traveled commercial corridors and the broad streets, appropriate as major traffic carriers, are inappropriate on a neighborhood scale where they tend to separate communities (Salt Lake City Planning Commission, 1972, p. 3).

Salt Lake City, in common with urban centers throughout the Southwest and the nation, has not been able to avoid the blight of congestion, deteriorating housing and other structures, and an inadequate tax base to support necessary social services. Even the planning process has proved inadequate. The city's 1967 Master Plan was the first effort to bring together disparate elements of a plan since 1943 and the Community Improvement Plan of 1972 was the first to provide detailed and staged accounting of what is necessary to improve conditions in Salt Lake City. The planning staff found the roots of problems in attitudes:

> The first is the city's past complacency about physical blight and its lack of sufficient attention in properly maintaining community facilities and in enforcing existing housing and other municipal codes. The second is a recognition problem. To a majority of citizens, problems simply do not exist and there is a persistent denial of the evidence (Salt Lake City Planning Commission, 1972, p. 90).

Thus, a serious part of the urban problem is a simple willingness to admit that there is a problem and that it is sufficiently serious to warrant public attention and investment.

Alternatives Posed in Albuquerque

Planners of the future metropolitan Albuquerque, in what must be one of the most remarkable planning documents produced by urban planners in the United States, posed the future choices for Albuquerque in terms of modifying the existing trends which are characterized as follows:

> The *TREND* is increasing growth and development. There will be more housing of all types; the form, however, will remain predominantly horizontal, resulting in more sprawl. The scope of education is becoming broader and teaching techniques more innovative, while shorter work weeks will allow more "free" time for recreation, or for drug or alcohol abuse. There will be more cars and more technological means to satisfy desires for individual mobility, status, and accumulation. The negative aspects of these trends will increase proportionately: more development/less open space; more cars/less clean air; more consumption/less resources; etc. (Albuquerque/Bernalillo County Planning Department, 1972, p. 56).

The goals program of the planners produced a sentiment to minimize the negative consequences of growth while strengthening the positive aspects of the area's environment. There are two strategies for doing this:

STRATEGY I is based on modifying the trend. It attempts to fulfill people's needs with a minimum of negative effects on the environment. The suggested policies are structured to accommodate needs with increases and more specific programs. The projection of METROALB—a large, increasingly urbanized center with transportation systems connecting outlying developments. Locations for development would be determined by factors such as soil suitability, desired life-styles and effects on the natural environment. The urbanizing area will be contained by the requirements for open space. The intent is to minimize sprawl and maximize individual opportunity.

STRATEGY II emphasizes action on a different level. The intent is to achieve harmony between people and the rest of the natural environment. Reduction in the need for services is basic to this approach. The idea that there are natural capacities is also essential. This strategy requires determining those capacities and living within their limitations. The projection is PUEBLO METRO—a series of relatively self-sufficient communities designed to provide diversity and opportunity; the common attribute is the recognition of the uniqueness of the area and the commitment to enhance it (Albuquerque/Bernalillo County Planning Department, 1972, p. 56).

The planners thus pose the options: modification of existing trends while endeavoring to protect the quality of the environment for the future or taking a fundamentally different approach—learning to live with less by reducing mobility, discouraging population growth, imposing stricter controls on pollution creating activities, reducing energy use, decentralizing into separate self-contained communities.

In each of these strategies, there was a recognition of the need for protection of the cultural heritage of the area, either through promotion of cultural diversity in Strategy I (protection of historic sites, expanding educational programs about local cultures, displays of artifacts of local culture); or through encouragement of individual identity. In Strategy II: thorough education emphasizing cultural awareness for school children and for social services and law enforcement personnel, discouragement of cultural events by outside groups, and achievement of equal status for minority groups and women.

Throughout 1973 and 1974 Albuquerque planners pursued community reaction to the proposed alternatives by various means: inserts in newspapers describing the options and inviting community responses by questionnaires; public meetings to discuss the alternatives; and scientifically constructed opinion surveys (Albuquerque/Bernalillo County Planning Department, 1974;

Albuquerque Urban Observatory, 1974). The responses from the community and the planning documents prepared in response to community input suggested a number of important preferences which were then expressed to Albuquerque/Bernalillo County citizens:

1. Controlled growth, with emphasis on filling in vacant land in existing developed areas; providing jobs for the disadvantaged population of the area; and protection of the area from pollution produced by job-creating industry
2. Varied developmental patterns, including preservation of historic neighborhoods and planned satellite communities
3. Greater emphasis on public transportation and other alternatives to reliance on the automobile, plus land use planning to permit use of alternative modes
4. Enhancement of recreational opportunities and visual relief from urbanization
5. Various policies to enhance the quality of the air and water and to dispose of solid waste

Tentative conclusions, based on this planning and heavy input from citizen groups, were being prepared for presentation to the County Commissioners with consequences that could not be predicted, either in terms of the reaction of the Commission in the short run or in terms of their ultimate implementation through decisions that must be made in each policy area by public and private agencies.

Attitudes Toward Alternative Patterns of Urban Living: Tucson

If the critical issue facing communities of the Southwest is the choice of the environment or lifestyle the people wish to retain, regain, or create, it is important to know what those people think. It is not enough for planners to impose their definition of the critical issues or their intuition of what an ideal community should look and be like. Some communities have been audacious enough actually to ask their citizens what it is that appeals to them about their communities and what they think are its major problems.

One of these was Pima County and its major metropolitan area, Tucson (City of Tucson, Planning Division, 1973). Using surveys conducted at public meetings and through water-bill mailings, planners obtained a 15 percent response to their inquiries. The validity and reliability of the sample so obtained is difficult to ascertain but it is probably superior to sheer guesses. The results are interesting. Asked why they came to Tucson, 37% said they came for employment or educational reasons. Reasons of health were given by over 30%. Approximately 24% gave reasons of climate; 22% to "escape big cities'' and natural environment. Asked what they liked about living in Tucson, 73% expressed their pleasure in the climate. Ranking far below, but nevertheless second and third were "friendly people" and "natural environment" both between 20% and 25%.

Dissatisfaction with living in Tucson clearly centered on the transportation system, or lack of it, and ills associated with dependence on the automobile:

Traffic	38%
Air Pollution	28
Poor Streets	22
Public Transportation	22

Other responses, when appropriately aggregated, placed emphasis on overpopulation and urban sprawl, 22%; social issues such as drug addiction, crime, housing and poor schools, 24%; and lack of jobs or industry, 24%. Although there is some variation in the responses, with people in the center city area particularly disturbed about low wages, the leading issues of traffic and air pollution were almost uniformly ranked among the first two or three issues throughout the city. Possibly, a more fine-grained analysis would reveal a more pronounced disparity of values among income classes than the data display. And surely the balance between the need for continued economic growth and environmental quality remains the crucial problem for public decision-makers who would retain public confidence.

The result of such surveys and detailed community analyses was the development of "interim concept plans" which were to be used until such time as the comprehensive planning process was completed (City of Tucson, 1973). The planners also developed alternative plans for the future development of Tucson, based on assumptions of continued population growth of 3%–5% per year and certain assumptions with regard to housing construction, imposition of air and water quality standards, reliance on the automobile, and rising costs of energy (*Arizona Daily Star*, 1973). The four plans may be summarized briefly as: (1) *peripheral expansion plan* which is largely an extension of the present trends of development—low density, reliance on the automobile, provision of services as the demand develops; (2) *activity centers plan* which involves clustering activities in "little downtowns," increased housing densities near those centers, increased reliance on public transportation both to and within the centers, long-time horizons for changes in the community; (3) *contained growth plan* which contrasts with peripheral expansion in controlling outward growth and emphasizing the distinctiveness of urban and rural areas. Activities would be centralized with high-speed public access to them and the public sector would play a more vital role in controlling community development; and (4) *satellite cities plan* in which satellite cities would be separated by a distance of 15 to 30 miles through zoning for greenbelts and scenic and recreational easements and with high-speed ground transportation providing easy access to the regional center. Both controls in Tucson and positive incentives in the satellite area would be required to make each a vital place. This plan would require the greatest time, greatest coordination of effort and greatest public investment.

The Tucson planners take explicit note of one problem frequently ignored by those who desire controlled growth, the differential effect that it has on various classes of citizens:

> But the growth controls implicit in the Contained Growth alternative must not be exclusionary for people of lower incomes. To prevent the medium income housing consumer from being frozen out of the market, regulations must be developed [to] insure that some dwelling units in each new development are within the price range of people with moderate incomes. This can be accomplished by affirmative private sector and governmental actions, such as the use of revenue sharing funds for the provision of low and moderate housing, and developing quota systems whereby a certain percentage of total new housing units will be designated for low and moderate income families (*Arizona Daily Star*, Sept. 1973).

Recognizing this potential differential effect is a modest step in the direction of avoiding what most public policy schemes turn out to be—welfare policies for the rich and the middle class and the continued back of the hand to the poor.

Responses to the Interim Concept Plan were obtained through presentations to groups in the community, through questionnaires distributed by newspapers and in public meetings, and through public opinion surveys (City of Tucson, 1973). As one might expect, the responses varied, with some groups such as the Grand Canyon Chapter of the Sierra Club urging the adoption of the contained growth plan, the Chamber of Commerce and others expressing concern about maintaining economic vitality and job opportunities, and the Tucson Regional Plan urging creation of low density peripheral belts and planned desert villages beyond the peripheral belt. In the public opinion survey, of those who knew of the four alternative growth plans, nearly 40 percent favored contained growth with almost equal numbers—17 percent preferring either activity centers or satellite cities. Overall, two-thirds of all respondents preferred either slower growth or none at all. Most expressed a desire for improved transportation networks but balanced between public and private means, as well as other improvements in community amenities. Again, it is not clear what role community input may play in the ultimate decisions that are made or the direction the residents of Tucson may choose to go in creating their future environment.

Population Controls

One of the crucial questions facing communities of the Southwest is whether or not they are going to embark on policies of population limitation and control and what techniques they are going to adopt to avoid severe economic disruptions and constitutional impediments. For, attractive as the idea of population control and limitations may be with an appropriate concern for the environment, the economic and political consequences may be substantial.

The Colorado Environmental Commission (1971), for example, proposed to embark on a policy to control the size of its metropolitan areas, mainly Denver, by imposing a limitation of 1,500,000 persons in any metropolitan area and requiring that there be a distance of 35 miles between a metropolitan area and new cities. As of 1971 55% of the people of the state resided in the Denver metropolitan area; and this was expected to reach 57% to 58% by 1980. The justification of such a policy is found in the following statement:

> Most environmental problems are caused by large concentrations of people. There is no need in the State of Colorado for such a concentration of the population to live in such a disproportionately small area of the state. There is plenty of space in which both people and industry can spread out. There is time to make the plans for a sensible arrangement of our cities. Such plans are indispensable if present trends are to be reversed. All of the amenities of urban environment can be had in a metropolitan area the size of the present metropolitan Denver; there is no need for that metropolitan area to become any larger. In the absence of a state population dispersal policy, the Denver metropolitan area will keep growing indefinitely, with all the resulting pollution and environmental problems (Colorado Environmental Commission, 1971, p. 3).

The goal of the plan was to gain time to plan further but also to obtain specific policy objectives:

> (a) By stabilizing the population level of metropolitan areas, it will stabilize the pollution and environmental problems and thereby render them manageable over time.

> (b) Encouraging the movement of population and industry to other areas of the state which are independent of the Denver metropolitan area will ultimately lead to a resurgence of local government in the state.

(c) Development of other areas and new metropolitan areas will reverse the trend of depression and declining population in many areas of the state.

(d) Colorado will be enabled to retain its independent open-space tradition by having several small metropolitan areas and medium-sized cities instead of one large megalopolis and many depressed small towns.

Conversely, failure to enact a legislative policy simply results in acceleration of the past trends: more people in the metropolitan area, more environment problems in the metropolitan area, higher welfare and other urban costs, more decline in local government in other areas of the state, more depressed areas in other areas of the state (Colorado Environmental Commission, 1971, p. 3).

The problem, however, is to translate the glittering phrases or attractive objectives into realistic policies. Subsequent to proposing such a plan, Colorado officials considered various proposals to control growth or limit population. A Committee on Balanced Population of the General Assembly (Colorado Legislative Council, 1972) considered eight alternative population distribution policies and found support for each but the support "seemed to be based more on hunch and feel rather than on hard data of study" (Colorado Environmental Commission,1971, p. 1). The Committee therefore recommended passage of legislation that would create the capability to gather facts and examine the impact of alternative policy solutions.

Recent information indicates, however, that the various proposals for controlling population and growth have bogged down in controversy among those who would prefer continuation of essentially the same policies of growth under little or no control and those who prefer an array of alternative policies which emphasize population or land-use control.

The challenge facing nearly all proposals for controlling growth or for creating new urban growth centers is that the new urban units must be viable economically. The diseconomies as well as the economies of further growth in a metropolitan area may be obvious but the possibility of inducing private firms to locate in newly created urban locations is not so obvious. As stated by an official of the Colorado Chamber of Commerce and Industry (Colorado Legislative Council, 1972), cities in rural Colorado have found it difficult to persuade industry to locate there unless there were markets in the locality or unless the transportation costs for their goods were an insignificant factor in total costs. He argued that most new industries located in industrial parks in smaller cities were local in origin. Moreover, local growth will continue to stimulate the major metropolitan area:

Even if local communities outside metropolitan areas of the state become more successful in their efforts to obtain new industry, new payrolls and new business, such activities will probably continue to stimulate development in our metropolitan and heavily urbanized areas. The reason is simple: the small communities are dependent upon larger cities for services, goods and financing. It is thus enigmatic that efforts to encourage economic development and population growth in outlying areas of our state will probably continue to develop the very areas we are desirous of bringing under control (Colorado Environmental Commission, 1971, p. 9).

A second factor in growth in the West, decisions by the national government with respect to the location of its facilities, is subject to considerably more control. In Colorado, for example, the decisions to locate the National Center for Atmospheric Research and the Department of Commerce's weather control activities in Boulder rather than Denver and to locate the Air Force Academy in Colorado Springs obviously had an impact on the population distribution in the state. But these are relatively unique activities with the former emphasizing research and therefore not requiring location in an urban center. Other kinds of public activities may require metropolitan location in order to avoid excessive costs that the taxpayer otherwise would have to pay.

Choices for Rural Areas and Smaller Towns: Farmington, New Mexico; Pioche, Nevada; and Southwestern Wyoming

Choices regarding the future quality of life are not limited to major metropolitan areas. Relatively rural and resource-oriented regions are also faced with such choices, although many of the choices already seem to have been made and commitments already have been engendered. Consider, for example, the Four Corners region, the area of northwestern New Mexico, and particularly San Juan County in which Farmington is located. The Four Corners Regional Commission (San Juan Council of Government, 1973a) established as goals for 1980 the creation of 76,000 jobs above the expected level of employment; additional investments of $3.5 billion with $1.1 billion coming from the federal government; added federal expenditures of $500 million; and improvements in employment rates and income levels primarily among those people with the lowest employment rates and income levels, which means particularly Indians and people with Spanish surnames. The strategy for capital investment emphasizes (1) improving accessibility to mineral and other natural resources, to recreation and tourist attractions and to educational and health facilities; (2) improvements in the occupational skill levels of the labor force; (3) stimulation of productive activities such as manufacturing, recreation and tourism, agriculture, or mining potentially offering immediate returns in jobs and incomes; and (4) greatest immediate impact on those areas of high unemployment and low incomes, and especially rural Indians and Spanish surname communities.

Whether these plans will bear fruit remains to be seen. But decisions made to implement these plans may seriously affect the decisions which an individual community must make (San Juan Council of Government, 1973b). Other major decisions lying ahead and having a direct impact on San Juan County are those concerning the construction of coal gasification plants and the development of the Navajo Irrigation Project, both of which will presumably take place on adjacent Indian land. If seven gasification plants and related mining activities are constructed, the 110,000 acre irrigation project is completed, existing industries expand and new ones come in as a result of these projects, there would be an estimated population increase of 190% from 1973 to 1983 with an average annual increase of 5,446 (San Juan Council of Government, 1973b, pp. 7-8).

The impact on the Navajo and surrounding communities would be considerable. On the positive side, it is estimated that more than 50 percent of the permanent employment attributed to the new economic activity would be absorbed by Navajo tribesmen. With a 1973 labor force of 45,000 persons out of a total population of 115,000 and with 20,000 unemployed, these employment opportunities would be a boon. From just one plant, it has been estimated that approximately $10 million annually would be income generated from employment (San Juan Council of Government, 1973b, p. 14).

On the problem side is the necessity to make the necessary community adjustments and improvements to accommodate such an influx of population and economic activity. These include improvement of substandard housing and construction of new housing, development of an adequate transportation system, provision of recreational opportunities for all ages, control of commercial development, particularly to eliminate its present strip character, supply of adequate public utilities, water and sewage services in particular. At this stage of the planning process, judging from documents supplied by San Juan County (San Juan Council of Government, 1973c), responsible officials have identified what is desirable in order for the Farmington area to be a pleasant and attractive community. But the realization of those goals will depend on an ability to control powerful economic forces of the market place, to impose tax burdens on the present members of the community, and to reconcile frequently divergent views of what constitutes a high "quality of life." A degree of humility combined with a large dose of audacity would seem to be necessary equipment to accomplish what seems like an impossible task.

A 1973 article in *The Nevadan* (Lewis, 1973) on Pioche, Nevada, suggests the importance of present and future choices for that town. There are signs of interest in re-opening the mines that gave Pioche its character. Tourism is growing. The federal government is contributing money and land to assist in the process of restoration and development. But there are concerns:

> Many people fear new faces and new ideas, seeing in them a threat to an established way of life. Isolation and unexciting routine mean the comforting reassurance of changelessness.

After citing the virtues of the small community, the author states:

> So it is a valid fear that some of these valued traditions will be lost with the influx of newcomers. In a severely depressed economic area the choice is limited to growth or descent to ghost town classification, a status that Pioche has fought vigorously to avoid.

> Fortunately for the future of Pioche most citizens know that life is change and look forward to the inevitable growth, the seeping out of the rubbish and the brightening of the spirit of the old town. They look forward to more activity for Pioche that now is sadly lacking in entertainment and cultural pursuits (Lewis, 1973, pp. 30-31).

Or consider the other end of the Colorado River Basin, southwestern Wyoming. Between 1960 and 1970, Lincoln, Sweetwater and Uinta counties all had net negative migrations: 18%, 8% and 15% respectively. Only Sweetwater County had an absolute net increase—of 2.6%. Developments in the region involving the construction of a 1500-megawatt power plant near Rock Springs and opening of trona mines near Green River have brought an enormous upsurge in population. A Sweetwater County planner (Young, 1973) states, ''Our population is increasing at such a fantastic rate that no available projections are the least bit accurate.'' Green River was projected by the Mountain States Bell Telephone Company to reach 5,914 residents by 1973 but had between 8,000 and 9,000 residents by early 1974. Rock Springs has between 19,000 and 20,000 rather than the 16,833 projected by the company. A consultant hired by one of the companies operating in the area, after reviewing the newly created problems—housing shortages, lack of public transportation and pressures on schools—perhaps expressed the avowed ethic of America that still receives a highly sympathetic response:

> I won't dwell further on the list of problems, because you all recognize them, and many of you have had a hand in doing what you could to alleviate them. They're not anybody's fault. They've arisen out of recent industrial growth. And that's where your *opportunity* comes in. This area has a great future, if we can

look beyond the immediate difficulties. It seems to be that the rapid development of the trona industry is the first real, positive opportunity for a renaissance of southwest Wyoming since the railroad shut down its mines in the late 1940s. With all its problems, this industrial boom can be made to produce not just payrolls, important as they are, but long-term advantages for this community (Watters, 1973, p. 24).

He did not detail the advantages but urged the creation of a non-profit corporation to plan and coordinate efforts to deal with the problems associated with economic development.

CONCLUSIONS

The notion of "trade-offs" has become commonplace among people who are concerned with planning and decision-making. The term connotes the necessity of "trading off" a benefit for a loss, gaining some advantage while losing another. The term is appropriate in the context of considering choices for communities of the Southwest. At least one state has even elevated the notion to a model "which provides a new method for evaluating the relationships between economic growth and environmental quality"(Myers, n.d.). It is a dynamic simulation procedure which admittedly provides only crude estimates of economic and environmental trade-offs, largely because of the paucity of data.

Whether such economic or social models eventually provide valid and useful information concerning economic and environmental trade-offs remains to be seen. But choices based on available information must be made by individuals in making their private decisions about where to live and work and what kind of lifestyle to adopt, and by communities in their corporate capacity in choosing the kind of environments they wish to fashion. If the past is any indicator, the choice will be made for growth: more economic development and more people with relatively modest concern for the quality of the environment in which they live. These choices will be made by individuals and firms and acquiesced in by communities as political and social units because the ethic of growth remains strong, because the population will continue to grow and because the southwestern region is so vulnerable to economic forces over which the local population has little control.

This is a statement of probability. It is not a prophecy. The people of the communities of the Southwest can influence the quality of life they will have. They have many of the tools but do they have the vision and the desire? The visions presumably will be different and there is no objective standard by which one can judge the "ideal" community.

Larger cities provide greater individual choice, and more varied opportunities while guaranteeing greater privacy. Smaller communities provide less choice but more opportunities for participation and purposeful direction. One political scientist (Dahl, 1967) has argued that the ideal size of a city is between 50,000 and 200,000 inhabitants because of the balance between opportunities and participation. Other measures of the quality of life have to do with levels of unemployment and poverty, quality of housing and health care, public order, racial equality, educational attainment, and levels of public services. But there appears to be no consistent rank ordering of large metropolitan areas along any of these measures (Flax, 1973).

But not even precise measures of quality of life can mechanically substitute for deliberate choices. What is most important is that individuals and groups be provided the means and channels by which those values may be expressed. Some cities and states are opening up those channels and it may well be that the answers they obtain will be as varied in size and quality as the rich and complex heritage of the Southwest itself.

Acknowledgment

I thank those public officials who supplied information for this study. Special thanks go to Alan Wyner for his helpful comments and criticism and to Paul Shoemaker for his able research assistance.

Discussion by Thayer Scudder

I have the unique and dubious distinction of being the only participant in this symposium who has never done research within the Colorado River Basin, or for that matter in any river basin within the United States. If my inclusion is an indication of the poverty of research personnel interested in community development in the Colorado River Basin, then an important point to be made is the need to recruit into the elitist ranks of Colorado basin researchers—and one is intrigued by the "in group" mentality of the participants—more social and behavioral scientists. With that gratuitous suggestion, attention will now be given to some of the issues raised by Dean Mann.

Institutionalizing Community Involvement. There are both contrasts and similarities between the river basins most familiar to me: those of tropical Africa and Egypt, and the Colorado River Basin as described in the preceding chapter by Dean Mann. The contrasts are quite obvious. In the African river basins the large majority of the population is still rural and is intricately articulated to the river systems involved. The basins themselves are underdeveloped, which raises the hope that their development can avoid the mistakes so common within the United States. In contrast, in the over-developed Colorado River Basin, the demographic situation is reversed and Dean Mann quite properly concentrates on the urban as opposed to the rural population.

There is a striking and major similarity, however; namely, the extent to which the impetus and the control for community development lies outside of the communities concerned. Throughout tropical Africa, community development is part of what might be called "the development from above syndrome." This is a tactic introduced by planners from the center to mobilize rural (as opposed to urban) populations to strive toward goals set in national development plans. The local population has virtually no input into the establishment of these goals, nor into the planning or the evaluation of community development programs. Procedures used, by their very nature, narrow down options to those which external planners rather than local citizens consider as appropriate. Interestingly (and depressingly) the situation in the Colorado River Basin is not dissimilar, though historically more complicated. Dean Mann notes that "perhaps more than any region of the country, it [that is, the Southwest] is dependent on decisions made elsewhere." The federal government as landlord and advocate, the market forces and a variety of other powerful pressure groups, like the environmentalists, lurk in the background throughout Dean Mann's paper. Choice alternatives are rarely generated by local communities; they come from without and are based on rather partial information. In terms of values, such a situation cannot be right. For one thing, it makes it easy for outside interests (the Department of the Interior, for example) to claim they have community interests in mind when in fact available information is insufficient to gauge these interests. Though municipal and local councils apparently are changing this to some extent, local communities find themselves responding to options suggested by others; they are not yet an integral part of the planning process via their own interest groups and coalitions.

In sum, something is basically wrong with the present choice structure. Somehow the diverse communities of the region must become more involved in the planning of options processes. Their involvement must be institutionalized and not merely through city councils and town managers. The task, however, is not easy. The growth of the ecology movement, for example, through the proliferation of a large number of small, loosely articulated citizen groups, shows that community pressure groups can develop political clout.

The Content of Choice. Dean Mann does not suggest the choices that such communities should make—he wisely says, "The choices must be made by the residents." Nor does he try to predict what the choices might be though he leans toward polar choice number 1 (business as usual with some environmental safeguards) versus number 2 (paradigm change; that is, a major restructuring of values and lifestyle action based on those values). Prediction of what members of urban communities want is difficult since seldom have residents been asked (even by their own municipal councils and city managers) what they value and what they want. And in a few cases where they *have* been asked, the results are ambiguous, neither supporting nor rejecting the contention that what the citizens really want is more and more job opportunities regardless of environmental costs. The Pima County case that Dean Mann presents is an example:

> Asked why they came to Tucson, 37% said they came for employment or educational reasons. Reasons of health were given by over 30%. Approximately 24% gave reasons of climate; 22% to ''escape big cities'' and natural environment.

Unfortunately, employment and education were lumped together. Obviously people want jobs, but the data also show that the sample population places a high value on the quality of the total ecological context (including sociocultural, biotic and physical environment) in which they live. In other words, the data suggest that these citizens, at least, are not as enthusiastic about growth as earlier authors in this volume have suggested (on the basis, one might add, of inadequate data, scarcely collected according to scientific criteria). But the major point is that regardless of the options chosen, the local people should play major roles in planning the future of the Colorado River Basin. In discussing local involvement, Dean Mann correctly stresses local planning councils and individual choice, but probably gives insufficient attention to local action groups (League of Women Voters, home owners associations and so on) which can bring pressure to bear on elected representatives and on city and regional planners from within the community.

Importance of the River. Throughout his chapter Dean Mann explicitly talks about the Southwest—not about the Colorado River Basin which he intentionally ignores. He is trying to educate us here—to show that sociocultural systems transcend land and water forms. Perhaps he goes too far. The waters of the Colorado River are already oversubscribed. As pressure builds up to develop local energy resources (and especially coal and oil shale reserves) to meet national needs, conflicts over the division of Colorado River waters will become even more acute, and the resolution of these conflicts will have major implications not just for Mexico and irrigation farmers but also for urban communities. In the Colorado case, increased attention needs to be paid to future utilization of river basin waters as it relates to community development.

Literature Cited

Albuquerque/Bernalillo County Planning Department. 1972. *Albuquerque/ Bernalillo County Comprehensive Plan: Metropolitan Environmental Framework,* Work Draft, April 1972, Albuquerque, New Mexico.

___ . 1974. The Albuquerque/Bernalillo County comprehensive plan: input from public meetings, 21 March, 1974. Albuquerque, New Mexico.

Albuquerque Urban Observatory. 1974. Findings of a citizen attitude survey regarding the Albuquerque-Bernalillo County Comprehensive Plan. April, 1974. An untitled report prepared for an 11 October 1974 conference on the comprehensive plan.

Arizona Daily Star. 1973. Tell Tucson where to go! Supplement, September, 1973.

Bakker, Elna, and Lillard, Richard G. 1972. *The Great Southwest: The Story of a Land and Its People.* p. 269. Palo Alto, California: American West Publishing Company.

City of Tucson, Department of Community Development. 1973. *Interim Concept Plan,* Vols. 1 and 2.

Colorado Environmental Commission, 1971. Frequently asked questions and answers regarding proposed population dispersal policy. Denver, Colorado.

Colorado Legislative Council. 1972. *Designing for Growth.* Report to the Colorado General Assembly, Research Publication No. 195, Denver, Colorado.

Dahl, Robert A. 1967. The city in the future of democracy. *Am. Pol. Sci. Rev.* 61(4): 953-970.

Denver Chamber of Commerce. *Denver business,* n. d., p. 3.

Flax, Michael J. 1973. A study in comparative urban indicators: conditions in 18 large metropolitan areas. In *The Quality of Life Concept: A Potential New Tool for Decision-Makers.* pp. II-244-258. The Environmental Protection Agency.

Lewis, Georgia. 1973. *The Nevadan.* pp. 30-31. Las Vegas, Nevada. 9 December 1973.

Los Angeles Times. 1974. 3 February.

Mann, Dean E. 1963. The political and social institutions of the arid region of the United States. In *Aridity and man,* eds., C. Hodge and P. C. Duisberg. Publication American Association for the Advancement of Science, Washington, D. C.

___ . 1969. Political implications of migration to the arid regions of the United States. *Nat. Res. J.* 9:2.

Myers, C. W. Jick, n. d. The Arizona trade-off model. Arizona State Planning Office, Phoenix, Arizona.

Phoenix, Arizona, Planning Commission. 1972. *The Comprehensive Plan 1990, Phoenix, Arizona.*

Salt Lake City Planning Commission. 1972. *Community Improvement Program,* Salt Lake City, Utah.

San Juan Council of Government. 1973a. San Juan County, New Mexico. *Existing Regional Plans.* Prepared by Lewis-Eaton Partnership, Jackson, Mississippi.

___ .1973b. *Land Use Policy Statement.* Prepared by Lewis-Eaton Partnership, Jackson, Mississippi.

___ .1973c. *Goals and Objectives.* Prepared by Lewis-Eaton Partnership, Jackson, Mississippi.

Spicer, Edward H., and Thompson, Raymond H., eds. 1972. *Plural Society in the Southwest.* New York: The Weatherferd Foundation.

Uinta County Planning and Zoning Commission. 1973. *Population Study, Uinta County, Wyoming.*

U. S. Bureau of the Census. 1972. *City and County Data Book.* Washington, D. C.: Government Printing Office.

U. S. Environmental Protection Agency. 1972a. *Air Quality Data for 1968.* p. 164. Office of Air Programs. Research Triangle Park, North Carolina.

——.1972b. *Air Quality Data for Suspended Particulates, 1969, 1970 and 1971.* pp. 38–45. Research Triangle Park, North Carolina.

U. S. Senate, Committee on Interior and Insular Affairs. 1971. *Problems of Electrical Power Production in the Southwest.* Hearings, 92d Congress, 1st Session, Part 3. Washington, D. C.: U. S. Government Printing Office.

U. S. Water Resources Council. 1971. *Water and Land Resources.* An Analytical Summary Report of Framework Studies of Four Regions.

Watters, A. F. 1973. Southwestern Wyoming Development Committee, Rock Springs, Wyoming.

Young, R. Stephen. 1973. Personal communication, 11 December 1973.

11

International Problems

MYRON B. HOLBURT

In August, 1973, the United States and Mexico executed the latest in a series of agreements that have attempted to settle the international problems of the Colorado River. In January, 1974, representatives from the seven Colorado River Basin states introduced H.R. 12165, "The Colorado River Basin Salinity Control Act," to the House of Representatives. This bill would authorize those measures considered necessary by the basin states to implement the latest agreement and also authorize a major basinwide Colorado River Basin Salinity Control Program. Also in January, the administration sent legislation to Congress which contains the measures it considers necessary to implement the 1973 agreement. Commencing March 4, 1974, hearings on both bills were held by the Water and Power Resources subcommittees of the House Interior and Insular Affairs Committee. Major differences between the bill submitted by the basin states and the bill submitted by the administration resulted in continuing controversy. This chapter discusses the Colorado River water quality and quantity problems between the United States and Mexico, considerations that have led to the several agreements between the two countries and possible future actions.

USE OF COLORADO RIVER WATER BY MEXICO

The Colorado River is extremely important to northwestern Mexico. The Mexicali and San Luis valleys occupy the dry delta lands of the Colorado River and are dependent upon the United States, both for the amount of water received and for the usefulness of this water. Colorado River water,

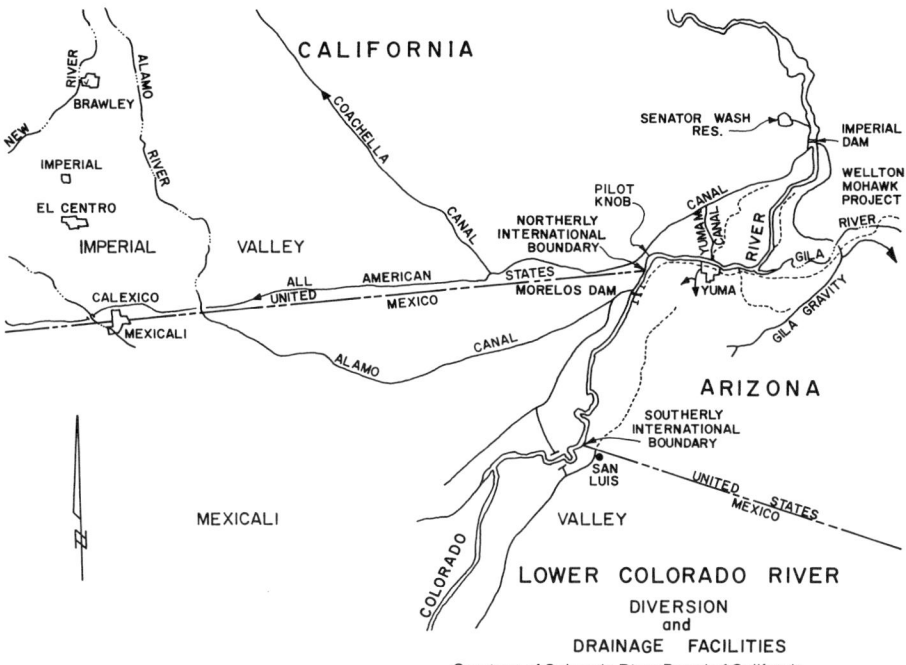

Fig. 11.1. Diversion and drainage facilities on the lower Colorado River.

supplemented by groundwater supplies, has irrigated between 425,000 and 500,000 acres in Mexicali and San Luis valleys. The accompanying map (Fig. 11.1) locates these and other areas (e.g., Morelos Dam, the last diversion dam on the Colorado before it enters Mexico) to be discussed. For many years, the dominant crop was cotton; in 1959, 84% of the irrigated acreage. In recent years, other crops have replaced cotton, as indicated in Table 11.1. The river also supplies water for about 500,000 people in Mexico.

1944 MEXICAN WATER TREATY

Prior to the initial control of the river's flows by Hoover Dam in 1935, the supply ranged from damaging flood flows during the snowmelt season to extremely low flows in the late summer and fall. These natural events acted to limit the use which Mexico was able to make of Colorado River water to an estimated annual maximum historical amount of about 750,000 acre-feet. When negotiations on a water treaty commenced with Mexico in 1928, the United States offered Mexico 750,000 acre-feet a year delivered to meet

TABLE 11.1

1970 Irrigated Acreage in Mexicali and San Luis Valleys

Crop	Area planted (Acres)	(Percent of total)
Wheat	172,140	40.3
Cotton	145,273	34.1
Garden	51,996	12.2
Alfalfa	36,452	8.5
Safflower	20,739	4.9
Total	426,600	100.0

demands, which quantity, by being regulated, was more valuable than the unregulated flows then being received by Mexico. The Mexican negotiators did not accept the principle of a limitation based on maximum historical usage and countered with a request for 3.6 million acre-feet a year (maf/yr). Because of the intractable positions of the United States and Mexico, negotiations were abandoned.

Following control of the river's flows through completion of Hoover Dam in 1935 and the commencement of storage in Lake Mead, Mexico began to expand its usage of Colorado River water. The State Department renewed negotiations for water treaties with Mexico, covering both the Colorado River and the Rio Grande, by a letter on December 29, 1939. The State Department also consulted with the Committee of Fourteen, composed of two representatives from each of the seven Colorado River Basin states, which had been formed in 1938 to consider basinwide problems.

As discussions progressed, the United States presented an offer of 0.9 maf/yr of stored Colorado River water, to be released on demand to Mexican users. Early in 1942, the United States' offer was amended to 1.15 maf/yr, but with the delivery to be from "any source whatsoever." Mexico countered these offers with a demand for 2 maf/yr, and insisted on having one agreement covering both the Colorado River and the Rio Grande, rather than having separate treaties. Since the United States is the basic source of water delivered to Mexico from the Colorado River, while Mexico is the source of a large part of the waters used by the United States in the Lower Rio Grande Valley, Mexico believed she could obtain better terms on the Colorado by considering the two rivers together in one treaty. It was also significant that the Chairman of the Senate Foreign Relations Committee in the 1940s was Senator Tom Connally of Texas, which state would benefit greatly from an agreement with Mexico on the Rio Grande.

After extensive negotiations, an agreement was signed in February, 1944, covering the Colorado and Tijuana rivers and the Rio Grande. The most important provision of the treaty, with respect to the Colorado River, is the allotment to Mexico of a guaranteed annual quantity of 1.5 maf/yr. The State Department defended this agreement by stating that up to 750,000 acre-feet per year would come from irrigation return flows below Imperial Dam and would probably go to Mexico irrespective of any treaty. Another State Department argument for the treaty was that the amount guaranteed to Mexico was less than its Colorado River use in 1943, which was estimated, by the State Department, to be 1.8 million acre-feet.

California, with some support from Nevada, vigorously opposed the Treaty because it gave too much water to Mexico. The upper basin states and Arizona supported the Treaty since they wanted to limit Mexico before her increasing uses invaded their share of compact-apportioned water.

California also made a major issue of water quality during the Senate hearings. Senator Sheridan Downey of California questioned the usability of the supply going to Mexico if the State Department estimates of return irrigation flows were correct (Downey, 1945). He made a remarkably accurate prediction in 1945 by stating that because of the ambiguity in the Treaty concerning water quality, Mexico would come back in 25 or 30 years and demand better quality water. State Department representatives, their consultants, and Senate supporters denied that there was any ambiguity in the Treaty (Tipton, 1945). They stated that water quality was extensively discussed, and that Mexico fully understood that the Treaty required them to take irrigation return flows irrespective of the salinity of those return flows. In response to a question at the hearings before the Senate Foreign Relations Committee, the State Department consultant said the United States could deliver water to Mexico under the Treaty, as much as 500,000 to 750,000 acre-feet per year even if it would not have any value for irrigation purposes (Tipton, 1945). The State Department pointed out that the specific provisions that were included in the Treaty to insure that Mexico must accept return flow and drainage water were in Articles 10 and 11. Article 10 states that Mexico's allotment included water from ''any and all sources'' and would be ''for any purpose whatsoever.'' Article 11 states that ''waters shall be made up of the waters of the said river whatever their origin.''

In testimony before the Mexican Senate, the Mexican negotiators were telling a different story from that told by their United States counterparts. One of the Mexican negotiators said it was understood between the two countries that the water delivered to Mexico must be of good quality (*El Universal*, 1945). He stated that Mexico could demand water similar in quality to that which she was currently using. Recognizing the difficulty of this demand, he said that Mexico would not object to receiving water similar to that used by the United States at Imperial Dam, the last diversion point in the United States. The Mexican Senate unanimously ratified the Treaty in September, 1944 (U. S. Senate, 1946).

In addition to the arguments on the merits of the Treaty, it was apparent that other foreign policy issues were major factors in the Administration's support for the Treaty. On the last day of the Senate hearings, Assistant Secretary of State Dean Acheson put the Treaty in the reference of the larger question of maintaining the Latin America good neighbor policy and world peace (Acheson, 1945). The latter was apparently a reference to the United Nations organizational conference being held in San Francisco. In April, 1945, the Senate ratified the Treaty by a vote of 76 to 10 (*Congressional Record,* 1945). Of the 14 basin state senators, only the two California senators and one Nevada senator voted against the Treaty.

DRAINAGE FROM WELLTON-MOHAWK PROJECT

Between 1945 and 1961, there were no major problems with respect to the river, as the salinity of the water delivered to Mexico at the northerly International Boundary was generally within 100 parts per million (ppm) of the water at Imperial Dam, the last major diversion for users in the United States. In 1947, the Wellton-Mohawk Project in southwestern Arizona was authorized by Congress to deliver water for the irrigation of 75,000 acres. It was constructed by the United States Bureau of Reclamation and completed in 1952. In 1961, the District began to operate a system of drainage wells which discharged saline water into the Colorado River below the last United States diversion but above the Mexican diversion. The drainage water included a substantial proportion of highly saline groundwater that had been concentrated through reuse during the previous half-century. Initially, it had a salinity of around 6,000 ppm. This resulted in a sharp increase in the salinity of the water delivered to Mexico, from an average of around 800 ppm in 1960 to 1,500+ ppm in 1962. Mexico raised strenuous objections to receiving the drainage waters.

Although the Wellton-Mohawk drainage was the primary cause of the increase in salinity, a sharp reduction in river flows to Mexico at around the same time also had an impact. Beginning in 1961, releases into Mexico were sharply reduced in anticipation of storage in Lake Powell behind the newly constructed Glen Canyon Dam. This loss of dilution water can be emphasized by two figures: for the 10-year period from 1951 to 1960 the average delivery to Mexico at the northerly International Boundary was 4.2 maf/yr, while for the succeeding 10-year period from 1961 to 1970, the flow averaged only 1.5 maf/yr.

Although, as previously indicated, the United States intended that Mexico must receive return flows below Imperial Dam under the Treaty, it had not been anticipated that there would be return flows as high in salinity as the Wellton-Mohawk drainage or that there would be such a precipitous rise in the salinity of the waters delivered to Mexico. Consequently, after the winter of 1961-62, the United States undertook certain provisional measures to minimize the impact of the high salinity drainage returns from Wellton-Mohawk. The United States also entered into negotiations with Mexico to arrive at a practical solution. The State Department asked each of the governors of the seven Colorado River Basin states to appoint two members to a reconstituted Committee of Fourteen in order to advise the State Department in connection with the salinity problem.

In the past several years, approximately 525,000 acre-feet per year of the 1.5 maf/yr guaranteed to Mexico has come from drainage water below Imperial Dam. Of this amount, the Wellton-Mohawk Project contributes about 220,000 acre-feet, other projects in the Yuma area contribute 165,000 acre-feet, and 140,000 acre-feet of drainage water from Yuma Valley is delivered at the land boundary near San Luis, Mexico.

INTERIM SALINITY AGREEMENTS

Minute No. 218. Extensive negotiations were conducted between 1962 and 1965 and, in November, 1965, a five-year agreement was incorporated in Minute No. 218. (The "Minute" form is a record of the International Boundary and Water Commission, United States and Mexico.) Under Minute No. 218, the United States undertook the following actions at a cost of $12 million:

1. Constructed an extension of the Wellton-Mohawk Drain so that drainage water could either be bypassed around Morelos Dam or mixed with other Colorado River waters above Morelos Dam, at the option of Mexico (Fig. 11.2; also see Fig. 11.1)
2. Constructed additional drainage wells in the Wellton-Mohawk Project which allowed selective pumping of the most saline waters at times when Mexico would be bypassing Wellton-Mohawk drainage water, and allowed the pumping of higher quality groundwater at times when Mexico would be using Wellton-Mohawk water
3. Replaced a portion of the bypassed Wellton-Mohawk water which resulted in the release of approximately 40,000 acre-feet of mainstream water per year from Imperial Dam in excess of the 1.5 maf/yr guaranteed by the Treaty.

Courtesy of U.S. Bureau of Reclamation.

Fig. 11.2. Morelos Dam in Mexico.

Under the measures taken by the United States, the quality of the water delivered to Mexico was improved from average annual salinity values of about 1,500 ppm in 1962 to 1,240 ppm in 1971.

Minute No. 218 was entered into for a specific period and was to expire in November, 1970. Accordingly, the United States and Mexico commenced negotiations with the purpose of arriving at another five-year agreement, although Mexican officials did not want to enter into a new long-term agreement in November, 1970, since a new administration was taking office in December, 1970.

Negotiations commenced in 1971 with the new Echeverría administration. The United States, supported by the Committee of Fourteen, proposed a new Minute, which would have provided to Mexico Colorado River water having the same salt concentration as would exist were the Wellton-Mohawk Project and all other projects in the United States below Imperial Dam in salt balance. This means that the tonnage of salt in drainage water originating from lands below Imperial Dam in the United States and delivered to Mexico would not exceed the tonnage of salt in the water applied to these lands.

Although the United States negotiators thought they were near agreement with Mexico in November, 1971, Mexico finally rejected the United States' proposals and negotiations were discontinued, pending the results of a forthcoming meeting between Presidents Richard Nixon and Luis Echeverría. In the interim, the two countries agreed to continue operations under Minute No. 218 (Department of State, 1971).

Minute No. 241. On June 15 and 16, 1972, Presidents Nixon and Echeverría met and issued a joint communique dated June 17, 1972. With respect to the Colorado River, President Luis Echeverría gave the essence of the current Mexican position as wanting water under the 1944 Treaty to be the same quality as the water at Imperial Dam. President Nixon replied that "this was a highly complex problem and needed careful examination of all aspects." He said that the United States was prepared to:

1. Undertake certain actions immediately to improve the quality of water going to Mexico
2. Designate a special representative to begin work immediately to find a permanent, definitive and just solution of this problem
3. Instruct the special representative to submit a report to him by the end of the year
4. Submit this proposal, once it has the approval of this government, to President Echeverría for his consideration and approval.

The immediate action referred to by the President was formalized as Minute No. 241 of the International Boundary and Water Commission on July 14, 1972, and replaced Minute No. 218. This agreement was based on the salt balance concept and provided that the United States discharge 118,000 acre-feet per year of Wellton-Mohawk drainage waters below Morelos Dam. On an interim basis the United States agreed to substitute for this water additional Colorado River water releases and waters pumped from wells on the Yuma Mesa so that the total deliveries exceeded the 1.5 maf/yr guaranteed by the Treaty. Actions by the United States under Minute No. 241 included the discharge of 119,490 acre-feet of Wellton-Mohawk drainage water and the substitution by the United States of an equal amount of water during 1972-73. This resulted in reductions in the average annual salinity of waters made available to Mexico from 1,242 ppm in 1971 to 1,140 ppm for the year ending June 30, 1973 (International Boundary and Water Commission, 1973).

Under the Minute, Mexico requested the United States to discharge the balance of the Wellton-Mohawk drainage waters (95,550 acre-feet) below Morelos Dam which was charged to Mexico's 1.5 maf/yr deliveries. As a result, the water diverted at Morelos Dam for the year ending June 30, 1973, had an average salinity of 980 ppm, which was about 130 ppm higher than the mean salinity of water arriving at Imperial Dam for the same period.

A SOLUTION
TO THE INTERNATIONAL SALINITY PROBLEM

On August 16, 1972, President Nixon designated former Attorney General Herbert Brownell, Jr., as his special representative and gave him the task of finding a permanent solution to the Mexican salinity problem.

A federal task force consisting of policy and working level representatives from a number of the major departments of government, including Interior, State, Agriculture, Environmental Protection Agency and Office of Management and Budget, was formed to assist Brownell. The Task Force developed possible solutions, evaluated them and presented the results to Brownell. He also met with the Committee of Fourteen periodically to seek their advice on possible solutions. Brownell submitted his report to the President by December 31, 1972. The Colorado River Basin states supported the concepts of the Brownell report subject to several important conditions.

Several months after receipt of the Brownell report, President Nixon appointed Brownell as a special ambassador and negotiations commenced between him and the Mexican representatives in the spring of 1973. Ambassador Brownell continued to meet and discuss the negotiations with the Committee of Fourteen. Although the committee members supported the thrust of the proposed agreement with Mexico, they continued to state that the agreement would require certain actions on the part of the federal government in order to avoid damage to the basin states. The Committee of Fourteen was unable to receive firm assurances from responsible members of the government that actions satisfactory to the states would be taken.

Senators and congressmen from the seven basin states were also concerned that terms of the agreement could be detrimental to the water interests of the states. On July 20, 1973, all 14 senators and 36 congressmen from the seven basin states signed a letter to President Nixon, asking that final negotiations with Mexico not take place until there was substantial agreement with the basin states on a number of issues. They also asked for a meeting with President Nixon.

Ambassador Brownell met with the Committee of Fourteen on August 18, to discuss the issues that had been raised by the states, senators, and congressmen. Although agreement was not reached with the states on the outstanding issues, Brownell stated that he intended to try to reach an agreement with Mexico. The key unsettled issues will be discussed later in this chapter.

In his negotiations with Mexico, Brownell had to consider a number of conflicting views. The most significant are summarized below:

1. The United States agencies represented on the federal task force had

differing solutions to the problem. For example, the State Department desired a negotiated settlement in order to avoid further conflict with Mexico and the possibility of having to agree to some third party solution, such as the World Court, while the Office of Management and Budget wanted an inexpensive solution.

2. The seven basin states were anxious to work with the federal government to achieve a solution to this vexing problem. They did not object to deliveries by the United States of water in excess of 1.5 maf/yr on a temporary basis in order to have a practical solution to the problem. However, the basin states opposed any permanent commitment to Mexico of water deliveries beyond that required by the Treaty.

3. Mexico's basic position was that it was entitled to the same quality water as is delivered to Imperial Valley just across the border from Mexicali Valley. Mexican representatives also believed that they were entitled to substantial compensation for damages caused by saline Wellton-Mohawk waters since 1961.

After extensive negotiations, agreement was reached between Ambassador Brownell and Secretary of Foreign Relations of Mexico, Emilio O. Rabasa, in the latter part of August, was approved by the two Presidents on August 30, 1973, and was subsequently incorporated in Minute No. 242 of the International Boundary and Water Commission. Minute No. 241 was terminated. Minute No. 242 contained the following major provisions:

1. The United States shall adopt measures to assure that by no later than July 1, 1974, the waters delivered to Mexico upstream from Morelos Dam would have an average annual salinity of not more than 115 ppm ± 30 ppm over the annual average salinity at Imperial Dam. This quality guarantee would become effective upon authorization by Congress of the funds to construct the necessary works. Statements by the administration clarified that Mexico would not receive further improvement in the quality of the water delivered to Mexico by the United States until Congress enacted enabling legislation, even if this did not occur until after July 1, 1974.

2. Until Congress authorized the necessary works to provide the quality guarantee, the United States would continue to bypass Wellton-Mohawk drainage water at the annual rate of 118,000 acre-feet per year and substitute therefor an equal volume of better quality water.

3. The United States would continue to deliver approximately 140,000 acre-feet to Mexico on the land boundary at San Luis, Mexico, with a salinity essentially the same as that of the waters customarily delivered there (approximately 1,550 ppm).

4. The concrete-lined Wellton-Mohawk drain would be extended approximately 53 miles to Santa Clara Slough (on the Gulf of California) with a capacity of 353 cubic feet per second. Construction and operation in Mexico would be performed by the Mexican Government, but at the expense of the United States.
5. Each country would limit pumping of groundwaters in its territory within five miles of the Arizona-Sonora boundary near San Luis to 160,000 acre-feet annually.
6. The United States would support efforts by Mexico to obtain appropriate financing for improvement and rehabilitation in the Mexicali Valley. The United States would also provide non-reimbursable assistance for those aspects of the rehabilitation program relating to the salinity problem, including tile drainage. The extent of this participation was to be negotiated.
7. The new Minute was to be recognized as a permanent and definitive solution to the Colorado River salinity problem.

ANALYSIS OF MINUTE NO. 242

Although not fully spelled out in the Minute, the administration stated that the following "measures" would be undertaken to comply with Minute No. 242:

1. Construction of a major desalting plant and appurtenant works for Wellton-Mohawk drainage waters scheduled to be completed by December, 1978
2. Extension of the Wellton-Mohawk drain by 53 miles to the Gulf of California scheduled to be completed by December 1976
3. Lining or construction of a new Coachella Canal in California scheduled to be completed by April 1977
4. Reduction in Wellton-Mohawk District acreage, and improved Wellton-Mohawk irrigation efficiency scheduled for completion by December 1978

It was expected that the Wellton-Mohawk drainage waters could be reduced to approximately 178,000 acre-feet per year. The desalting plant would treat approximately 143,000 acre-feet per year. The resulting 100,000 acre-feet per year of product water would be mixed with 35,000 acre-feet of untreated Wellton-Mohawk water to produce water equivalent to the salinity of the river at Imperial Dam. The reject stream, which would be carried to the Gulf of California, initially would be approximately 43,000 acre-feet per year and was estimated to have a salinity of 9,600 ppm. Estimated cost of the desalting plant and associated works was $98 million.

It was anticipated that a new concrete-lined canal would be constructed to replace the first 49 miles of the presently unlined portion of the Coachella Canal. This, together with associated works, was estimated to cost $21,450,000.

Mexico received major new benefits from the new agreements. She received a guaranteed quality at Morelos Dam related to the quality of water at Imperial Dam. All of the costs in money or water to achieve this quality guarantee was to be borne by the United States. Mexico had the promise of financial assistance with respect to salinity problems, including tile drainage in Mexicali Valley. In addition, the United States would support efforts by Mexico to obtain favorable financing for the improvement and rehabilitation of Mexicali Valley. Also at the United States' expense, a concrete-lined canal was to be constructed in Mexico to discharge saline water.

The United States negotiators believed there were considerable tangible benefits to the United States. The agreement eliminated the possibility of long years of acrimonious controversy between the two countries. The agreement did not require any payments to Mexico for any past damages. Since the agreement was described as a permanent solution, presumably Mexico waived any future rights to press for monetary damages. Mexico agreed permanently to accept 140,000 acre-feet per year right at the Arizona-Sonora boundary. This is largely drainage water with a salinity considerably higher than that of the Colorado River. Although Mexico accepted this water for years and utilized it, there was apparently no obligation on Mexico's part to accept the waters at this location until this agreement was signed. Although the United States would reduce the salinity of the Wellton-Mohawk drainage waters under the new agreement, Mexico agreed that other drainage waters below Imperial Dam would continue to be accepted as part of the U. S. Treaty obligation. In summary, negotiators believed that the concessions made by the United States were in terms of money, not water. However, this was not entirely true, as the 43,000 acre-feet per year reject stream from the desalting plant would not be included as part of 1.5 million acre-feet obligation and would have to be replaced.

The Colorado River Basin states considered that the agreement with Mexico was entered into largely on the basis of international comity and that the Colorado River Basin states should not be expected to bear any greater burden as a result of the new agreement with Mexico than that to be borne by the rest of the nation. The basin states considered that the following commitments would be necessary in order to prevent any damages to the basin states.

Colorado River Basin Salinity Control Program

Assurances of support from the administration is needed for contemporaneous authorization of a Colorado River Basin Salinity Control Program then before Congress, together with the works that are required specifically for Mexico. This legislation had as its purpose control of the salinity above Imperial Dam for the benefit of users in the United States. It would also benefit Mexico. Brownell stated that unless the United States controlled the salinity at Imperial Dam, there would be a new salinity problem with Mexico (Brownell, 1974).

The Mexican government apparently believed that the United States committed itself to a program to control the river's salinity upstream from Imperial, for Foreign Minister Rabasa made the following statement to the press in Mexico City when the agreement was announced on August 30, 1973:

> The final result will be that the Mexicali farmers will have forever—they and their children and their children's children—water whose average annual salinity will never exceed 1,010 ppm, which is perfectly acceptable (*El Universal*, 1973).

When correction is made for the method of analysis for salinity used in Mexico, which is different from that used in the United States, and the 115 ppm differential between Imperial and Morelos dams is added, the 1,010 ppm referred to above is equivalent to the 1974 salinity at Imperial Dam.

Prompt Construction of Necessary Works

It would be necessary that the United States promptly construct the new Coachella Canal and Wellton-Mohawk desalting plant in order to carry out the intent of the agreement with Mexico. Failure of Congress to appropriate funds or the administration to release any appropriated funds meant that the United States would continue to deliver 118,000 acre-feet per year to Mexico above the 1.5 million acre-feet guaranteed annual quantity. This water would come from the supplies whose rights belong to the seven Colorado River Basin states or to entities within these states.

Replacement of Reject Water Delivered to Gulf

The United States must assume the responsibility for permanent replacement of the reject water from the desalting plant. If these waters are not replaced, it means that there will be permanent additional deliveries to Mexico in excess of the guaranteed 1.5 maf/yr of approximately 43,000 acre-feet per year, which would reduce supplies available to the basin states by a like amount.

Power Source

It was estimated that approximately 35 megawatts would be required to provide the energy needs for the desalting plant and there would be additional power requirements for the protective groundwater pumping program. It was essential that the power necessary for meeting the needs of Minute No. 242 not be provided by taking power away from existing southwest power users and forcing them to obtain replacement power in a difficult market with considerably higher costs. This demand had to be met by development of new power sources.

Groundwater Pumping Near San Luis

Mexico has the capacity to pump approximately 160,000 acre-feet per year of groundwater from wells one to five miles south of the United States-Mexico boundary. Unless the United States installed its own groundwater pumping system, the Mexican wells would pump a substantial quantity of groundwater originating in the United States. Studies made by the United States indicate that it would be necessary to pump approximately 140,000 acre-feet per year in the United States over and above 1974 pumping within five miles of the border in order to protect against the loss of surface and groundwater to Mexico. This development by Mexico was located so as to have the maximum impact on the United States waters. It was reported that in the negotiations, Mexico refused to accept as a charge against its treaty allotment any waters drawn from the United States by the San Luis well field. Mexico also refused to limit the amount of water pumped by its San Luis well field to an amount that would have a negligible impact on the United States water supply. The new Minute allowed for further negotiations between the two countries on the border groundwater problem. Unless some agreement could be reached on one of the two above alternatives, it would be necessary for the United States to install additional groundwater capacity, in order to balance the groundwater pumping by Mexico.

Legal Issues

Questions have been raised with respect to the legality of Minute No. 242. Governor John D. Vanderhoof of Colorado, referring to Minute No. 242 in a speech, stated:

> The agreement contained in this Minute constitutes either an amendment to or an enlargement of the Mexican Water Treaty of 1944. While it may be within the power of the President to modify or enlarge upon an international treaty without the consent of Congress, Section 2 of Article II of the United States Constitution seems to indicate otherwise. That section states that the President

shall have the power to make treaties, but only upon the advice and consent of the Senate. In the case in hand, such consent has not been given (Vanderhoof, 1973).

No Impairment of Colorado River Basin States' Rights

Confirmation was needed from the administration that nothing in the agreement would impair the rights of the Colorado River Basin states to continue to utilize their Colorado River water rights within the constraints of the "Law of the River."

IMPLEMENTATION OF MINUTE NO. 242

The necessary actions with respect to Minute No. 242 were projected by federal officials to occur in the following manner. During the period until Congress enacted enabling legislation for implementation of the quality guarantee portion of the agreement, Mexico would receive about the same quality of water from the United States that it had been receiving since Minute No. 241 was in effect, which required the bypass of 118,000 acre-feet per year of Wellton-Mohawk drainage water and the substitution of water released from Imperial Dam and pumped groundwater from the Yuma Mesa. During this period, Mexico would continue to bypass the remaining part of the Wellton-Mohawk drainage water to obtain even lower salinity.

The intent of the United States-Mexican negotiators was to have the necessary enabling legislation enacted prior to July, 1974. Upon authorization by Congress, the United States would have to begin to meet the 115 ppm salinity differential. This would require that all Wellton-Mohawk drainage water be discharged below Morelos Dam without charge against the treaty to Mexico and that the U. S. would substitute Colorado River water from storage. The quantity was expected to reduce from 220,000 acre-feet the first year to 178,000 acre-feet by the fourth year as Wellton-Mohawk Irrigation District efficiencies increased and return flows decreased. The federal government would borrow water from storage in the Colorado River reservoirs in order to make the required releases.

The replacement of the first 49 miles of the Coachella Canal was scheduled to be completed in three to four years. This would result in a reduction of seepage losses of approximately 132,000 acre-feet per year. The increased storage in Lake Mead resulting from reducing deliveries to Coachella in an amount equal to the salvaged losses would be credited to the United States. This would reduce the draft on storage to 46,000 acre-feet per year assuming that the Wellton-Mohawk irrigation efficiency and acreage reduction programs would be successful in reducing the Wellton-Mohawk outflow to 178,000 acre-feet per year.

The desalting plant was scheduled to be completed within five years. Once this plant is in operation, water accumulating in storage due to water salvage resulting from the new Coachella Canal would be credited to the United States for the purpose of repaying substitution water borrowed in earlier years. In the mid-1980s, it is anticipated that the Central Arizona Project would be completed and that California would be required to reduce its net diversions from the Colorado River from its current 5.1 maf/yr to 4.4 maf/yr. At that time, it is expected that the Coachella Valley County Water District would commence repayment of the unamortized costs of the canal construction to the extent that it benefits from the new canal.

Congressmen from the seven basin states consider it essential that one bill encompass both the measures to implement the new agreement with Mexico and a basinwide salinity control program. It was also considered that hearings would be held early in 1974 to allow time for enactment by July 1, 1974. On January 21, 1974, Congressman Harold T. Johnson of California and eleven other basin state congressmen on the House Interior and Insular Affairs Committee introduced H.R. 12165. Hearings on the bill were set by the Subcommittee on Water and Power Resources of the House Interior and Insular Affairs Committee for March 4, 5, and 8, 1974. This bill included all the items determined necessary by the administration as well as the items determined to be necessary by the states to implement the new agreement with Mexico. Specifically, it provided for:

1. Authorization of construction of a desalting plant and the necessary appurtenant works to reduce the salinity of the Wellton-Mohawk drainage water to the salinity of the water at Imperial Dam
2. Authorization of construction of a lined brine disposal channel, from the desalting plant location to the Gulf of California, to convey the brine from the desalting plant to Santa Clara Slough in Mexico
3. Authorization of construction or lining of the first 49 miles of the Coachella Canal in California
4. Various changes in the Wellton-Mohawk Irrigation District in order to reduce the amount of drainage water from that district; these changes would include a reduction in the lands irrigated from the authorized level of 75,000 acres to a level of 60,000 to 65,000 acres, an improvement in irrigation efficiency, and other necessary activities
5. Authority to obtain a source of energy for the desalting plant that would not diminish the supply of federal power for preference customers in the region
6. Authorization to construct, operate, and maintain a groundwater well field on the Yuma Mesa located within five miles of the Mexican border, and to acquire rights in lands overlying the groundwater areas

to be pumped; this authorization would be exercised only if the comprehensive groundwater agreement referred to in Minute No. 242 was not reached within two years and Congress had determined that the negotiations failed; the estimated cost of this feature was $16 million

7. Recognition as a national obligation the responsibility for the replacement of the brine from the desalting plant wasted to the Gulf of California, with studies to be conducted on how to provide for the replacement water

8. Authorization to construct, operate and maintain the salinity control units of Paradox and Grand valleys, Colorado, Crystal Geyser, Utah, and Las Vegas Wash, Nevada, that would collectively remove approximately 520,000 tons of dissolved salt per year from the river system at a construction cost of $121,000,000

9. Direction to the Secretary of the Interior to expedite completion of planning studies of salinity control projects within the Colorado River Basin

On February 7, 1974, the administration sent its bill to Congress. It differed from H.R. 12165 on a number of major issues. Some of these differences were:

1. The basinwide salinity control program upstream from Imperial Dam was omitted from the administration bill.

2. The Secretary of State was given the key responsibility in the administration, while the Secretary of the Interior was given the key responsibility in H.R. 12165.

3. The authority to install a well field in the Yuma area to balance the groundwater pumping by Mexico, in the event the two countries were unable to reach a groundwater agreement, was omitted from the administration bill.

4. The administration bill made the replacement of the 43,000 acre-feet per year of reject stream from the desalting plant conditional on the augmentation of the Colorado River by 2.5 maf/yr. H.R. 12165 stated that replacement of the 43,000 acre-feet per year is a national obligation.

FUTURE PROSPECTS

Although the title of the latest agreement with Mexico over the Colorado River (Minute No. 242) is "Permanent and Definitive Solution to the International Problem of the Salinity of the Colorado River," it is apparent that many issues remain to be resolved to assure that it is a permanent solution and that new salinity problems do not occur in the future. Briefly stated, it will require: (a) resolution of the differences between the Administration and the Colorado River Basin states; (b) enactment of major complex legislation by Congress; (c) a salinity control program encompassing natural and man-made sources of salinity in the entire Colorado River Basin; (d) authorization and expenditure of hundreds of millions of dollars for works in the United States and Mexico over the next several decades; (e) construction of the world's largest desalting plant; and (f) better irrigation practice in both the United States and Mexico.

Note added in proof. The essential provisions of Minute 242 (discussed by Holburt) have been incorporated and are being implemented in Title I of the 1974 Salinity Control Act. Title II of the Salinity Control Act authorizes the Secretary of the Interior to construct, operate and maintain four salinity control projects (the Paradox Valley Unit, Colorado; the Grand Valley Basin Unit, Colorado; the Crystal Geyser Unit, Utah; and the Las Vegas Wash Unit, Nevada) and to expedite completion of planning reports on twelve other salinity control projects. Seventy-five percent of the estimated $125.1 million cost of these upstream developments will be funded by the federal government. The other twenty-five percent will be paid out of Colorado River Basin funds, which are revenues (principally power revenues) collected in connection with the operation of the completed Colorado River water storage and distribution projects.

Literature Cited

Acheson, Dean. 1945. Statement before the Committee on Foreign Relations, U. S. Senate, 79th Congress, First Session, pp. 1760-79 of Hearings on the Treaty with Mexico relating to the utilization of the waters of certain rivers.

Brownell, Herbert. 1974. Testimony at hearings before the Subcommittee on Water and Power Resources of the Committee on Interior and Insular Affairs, House of Representatives, 93rd Congress, Second Session, on H.R. 12165, March 4, 5, 8, Serial No. 93-45, p. 107.

Congressional Record. 1945. Senate, April 18, p. 3547, Vol. 91, No. 76.

Department of State. 1971. Press release No. 263, November 15.

Downey, Sheridan. 1945. Statement before Committee on Foreign Relations, U. S. Senate, 79th Congress, First Session, pp. 1103-64 of Hearings on the Treaty with Mexico relating to the utilization of the waters of certain rivers.

El Universal. 1945. August 1, p. 1. Mexico City.

——. 1973. August 30.

International Boundary and Water Commission. 1973. Report on operations for solution of the Colorado River salinity problem under Minute No. 241.

Tipton, R. J. 1945. Testimony before Committee on Foreign Relations, U. S. Senate, 79th Congress, First Session, pp. 341-43 of Hearings on the Treaty with Mexico relating to the utilization of the waters of certain rivers.

U. S. Senate. 1946. Document No. 249, 79th Congress, Second Session, p. 4.

Vanderhoof, John D. 1973. Speech at Thirtieth Annual Meeting of the Colorado River Water Users Association, Las Vegas, Nevada, November 26.

12

Navajo Indian Culture and Lands

BAHE BILLY

The energy crisis has brought to a head the need for a reassessment of cultural values, prevailing socio-economic conditions, and man's prospect for the future among all diversified cultural and ethnic groups in America. Such an analysis is especially critical for short- and long-range social planning in the arid regions of the United States, where water is critically short and large acreages of land are required to sustain livestock, and where similar acreages are needed for mining ore, oil, coal, and minerals.

In October of 1973, a conference was held in Salt Lake City, Utah, to discuss the various forces that were influencing socio-economic conditions in the Colorado River Basin (Crawford and Peterson, 1974). The main concern of the scientists at the conference was the economic dependency of this region on the rest of the world, in that this region had to be an exporter and had to import to survive, and human behavior was a critical consideration in formulating energy policies. Other very important factors were the number of jurisdictional lines and the ownership of the natural resources in the basin (Weatherford, 1974).

This discussion should be the first in a series to focus on identifying the issues that confront Indians and others in the use and misuse of their resources and in promoting an awareness to all citizens of the basin region. The first natives of this area were the American Indians, who are now caught along with all others in the prevailing socio-economic conditions and the worsening resource problems.

The plea for such a re-assessment of the Indian situation was related by Billy and Reno (1973) in the following way:

The United States is preparing for the Bicentennial Celebration of 1976, to commemorate 200 years as an independent nation. A nation whose roots are in the Enlightenment [Age] might appropriately devote some of the Bicentennial Celebration to rational analysis. We know of no such official preparation, but at least among the American Indians this is a time of searching questions about the course of the Indian past and the prospects for the Indian future.

It is very difficult for anyone to describe fully and accurately how all the various cultural, social, political and economic forces that have made America the leading industrial society have altered the various Indian cultures and lands in the Colorado River Basin. The effects vary greatly in degree and in complexity. For this reason, the Navajo Tribe is focused on as a standard reference from which to generalize about the Indian situation in relation to culture and lands.

In order to understand the Navajo situation, one needs first to review the historical background, secondly to focus on the prevailing socio-economic conditions, and then to emphasize the cultural and lands status, and finally, from this framework, to project the future of the Navajos in the Colorado River Basin.

HISTORY

According to Navajo mythology, the people, or Dineh, as the Navajo call themselves, came to this earth after having to escape from four underworlds. In each of these four underworlds, adjustment and change had to be made. The Navajos, unable to adjust and adapt, were forced to move on. When they moved into the present world, Gods had to be petitioned to help get rid of the giants that inhabited the earth. When the land was finally taken, the Holy Ones created four mountains known today as Sierra Blanca Peak, Mount Taylor, San Francisco Mountain, and Mount Hesperus. The land between these four mountains is the area the Navajos call home.

Anthropologists say the Navajos descended from bands of hunters and gatherers who migrated during the thirteenth to sixteenth centuries into the Southwest (Kelly, 1968). The people settled and began to learn and remodel their own lifestyle by watching and assimilating other native customs and cultures of the region. Part of this transition was accomplished by intermarriage. The Navajos were also influenced to take up subsistence agriculture and the art of weaving during this period.

In 1626, the Spanish reported that a separate group of the Apache was sighted apart from the Pueblo people. The Pueblo people referred to these people as Apaches de Navajos. Scholars feel that these were the Navajo people.

Sometime during the seventeenth century a transition from subsistence agriculture to pastoral economy occurred and once again the Navajo culture was altered by an increase in the number of horses and the livestock industry. Many authorities feel that this transition was highlighted during the Pueblo Revolt, while the war was occurring between the Pueblo and the Spanish people. The revolt caused the merger of Pueblo and Navajo religious customs and cultures because large numbers of the Pueblo people joined the Navajos. Furthermore, by adopting captured Spanish and Mexicans, the Navajos assimilated and enlarged their religious and cultural traits and broadened their language base.

The Navajos quickly learned to use the horses well for quick striking and mobility to steal livestock, food, and other people's possessions. This became a regular and established practice against settlers and other Indians that settled or lived within striking distance of Navajoland. The art of raiding was a part of Navajo life even before the Spanish settlers came into the Southwest, and the horse enhanced the practice further (Aberle, 1969; U. S. Commission on Civil Rights, 1973). The accumulation of stolen livestock required more range space, so moving with the herd constantly to provide fresh food supply for the livestock was initiated. This cultural pattern persists today; although it is on a very limited scale.

The Navajo people became a part of the American Indians by virtue of a war victory over Mexico in 1848 by the United States. In 1850 a treaty between the United States and Navajos was drawn up. By Treaty, the people were placed under control of the United States, subject to all laws as were all other Indians. This treaty, however, did not insure peace between the two groups because the Navajos continued to take livestock and other possessions. The government, on the other hand, had the responsibility to protect the settlers and their possessions. The clash between the two cultures was inevitable.

The Navajos were rounded up by the army under the leadership of Kit Carson and relocated to Bosque Redondo in eastern New Mexico for four years. Numerous attempts were made to establish farms on which to educate these people to change and to adopt an agricultural economic base. This effort was a failure. In 1868, the government signed a new treaty with the Navajos, allowing them to return to their homeland. The treaty gave the Navajos 3.5 million acres of land to be held in federal trust, just about one-fifth of the land that they previously occupied. The Navajos, upon returning, were encouraged to rebuild their livestock herds as a means of subsistence.

The population and livestock multiplied rapidly and by the early 1930s the reservation had grown from 3.5 million to 15 million acres by executive order land addition, yet these extensions were not enough to fully support the 1.3 million sheep units belonging to an increased population of 45,000 tribal members. The safe capacity of the reservation area had, by this time, fallen to about 600,000 sheep, far less than the 1,300,000 sheep owned by the people and drastic control measures were necessary.

The government and the Navajo people were forced to face the problems created by a steadily growing population and the increasing livestock numbers confined to a fixed land base because of (1) the overgrazing conditions of the reservation, (2) the defeat of the boundary-extension bill in New Mexico, and (3) the passage of the Taylor Grazing Act.

In trying to solve the reservation problems, a number of events occurred which had a tremendous bearing on Navajo adjustment to modern American life. Three major events were: (1) The sheep reduction program; (2) the delayed action in the establishment of a grazing regulation and its enforcement; and (3) the defeat of the Indian Reorganization Act of 1934 by the Navajos.

The Commissioner of Indian Affairs reviewed the overgrazed conditions and proposed the following three alternatives: (1) use his legal authority to reduce the livestock; (2) let the Tribal Council handle the compliance with the government regulations; or (3) let the people participate in the decision making concerning the livestock reduction. Of these, the second alternative was selected and the burden of the livestock reduction was handed to the Tribal Council. The passion stirred by this event embittered the relationship between the Navajos and the federal government and established a very strong traditional regulated land base referred to as customary land use areas. By putting the responsibility of the livestock reduction in the Tribal Council's hands, the commissioner forced the people to turn against their own Tribal Council members and the force of the impact nearly destroyed the effectiveness of the tribal government for all time.

The other related impact was the long delay in the development of the grazing regulation and its enforcement which was not developed until 1937. The enforcement of these regulations unfortunately was not yet begun in the 1940s. In 1956, the council revised these grazing regulations, but their enforcement continues to be difficult. The Navajos' refusal to adopt the Indian Reorganization Act (IRA) on two occasions has given the Navajo Tribe the opportunity to develop its own government system outside of the IRA structure.

Another important event was the start of wage work to replace sheep raising as the major source of Navajo income beginning with the livestock reduction. Through various New Deal programs such as the Civilian Conservation Corps, Works Progress Administration, and the Soil Conservation Service large sums of money were used on the reservation to compensate Navajos for their livestock losses.

Other very important events that occurred earlier in Navajoland were the oil and mineral leases on the reservation in the early 1920s. These leases gave the Tribe the catalyst to organize a Tribal Council. The big drilling companies paid the tribe millions of dollars for rights to its latent oil reserves and for royalties on the oil they pumped to the state of New Mexico and to the West Coast.

With the outbreak of World War II, some 3,500 Navajos joined the Army, Navy, and Marines. Thousands of other Navajos during this same period left to take positions in war industries and some 15,000 Navajos found employment on the railroads (Navajo Parks and Recreation Department, 1973). When the soldiers and others who had gone away from the reservation to work returned home, they had seen too much to go back to the life they once had known. They were psychologically ready to adapt to and adopt new changes that had taken place. The United States government set up the Indian Claims Commission to settle all American Indian claims against the United States and instructed the Navajos to hire themselves an attorney to prepare their cases.

Congress, in the midst of big spending for postwar foreign aid, was apprised of the plight of the Navajo and Hopi Indians at home and the lawmakers voted $88 million in early 1950 for Navajo-Hopi land rehabilitation. In addition, due to the heavy population pressure on the land, a resettlement program was introduced to encourage Navajos to break traditional cultural ties and take up residence where jobs were available. This program was not very successful because most of these Navajos left their new jobs after only a short period of time in the cities. This approach however is still continued by the Bureau of Indian Affairs (BIA), because no program has been created to take its place.

Almost simultaneously with the passage of the Navajo-Hopi Long-Range Rehabilitation Act, additional oil was discovered and by 1960 over $100,000 from oil alone was deposited into the Tribal Treasury. In addition other resources such as natural gas, uranium, and coal were placed on the market for sale. Sizeable sums of money came into the Tribal Treasury from these new sources and many Navajos found opportunities for employment with the companies that were exploiting these resources.

Although the United States committed itself in the 1868 treaty to educate the people, the Navajos largely resisted schooling for their children until World War II. During and after the war, the people's eyes were opened to life outside the reservation and more federal funds were made available for education. As late as 1950, only 12,000 children were in school. As of 1975, practically all Navajo children attend schools on the reservation and its periphery; the tribe today is having a difficult time trying to provide scholarships to post high school students who want to go on for additional training or college work.

Under Presidents Kennedy and Johnson, in the 1950s and 1960s, a few Indian programs were launched under the War on Poverty program. These included Navajo Economic Opportunity (NEO) and Dine Bi Nah Nahiilna Be Agaditahe (DNA) on the reservation. The OEO, another War on Poverty program, did not go very far; President Nixon did not increase the funding but impounded funds for this program. The Navajo OEO and DNA programs were still in operation on the reservation in 1974.

NAVAJO NATURAL RESOURCES

The Navajo Reservation, approximately 15 million acres, extends into Arizona, New Mexico, and Utah. More than half of the land (55%) is classed as desert, about two-fifths (37%) is classed as steppe and about 8% is mountainous country used for lumber production and having attractive potentials for outdoor recreation (Navajo Agency, 1961). Much of the land is dry and barren, with an annual average rainfall of 6 to 12 inches. Almost all of the land area is used primarily for the production of livestock forage, and cattle and sheep are the main sources of agricultural income. Because of the lagging development of other means of livelihood, Navajo rangeland for many years had to support livestock in excess of the numbers that it could actually carry. This along with periodic drought cycles affected the soils drastically. The result is a badly eroded land base with little of its natural grasses and low shrubs still intact and vigorous.

Although the annual precipitation is relatively low in most locations across the reservation, summer storms can cause considerable runoff and erosion. Dams to control these flows are prohibited by the cost due to the excessive numbers that would have to be built. The Navajos are legally entitled to all the water they can use beneficially from the streams or rivers that flow through or border the reservation, yet they use very little of this resource mainly because of the lack of dams and irrigation systems. Canals and irrigation projects are vitally necessary for water use in this region (MacDonald, 1972). The Navajo underground water supply is limited, particularly in the western portion of the reservation, where limestone underlies the surface landforms. In many areas in the eastern portion of the reservation underground water resources exist, but many of these have high salt content and need treatment before they can be used for human consumption. A complete water inventory, surface and underground, for the reservation has been prepared for the Resource Committee of the tribe by the William Brothers firm from Tulsa, Oklahoma (Brown, Vlassis and Bain, 1973).

The arid and rocky Navajo reservation is spectacularly beautiful, especially during the summer, due to the various soil colors that exist in the landforms. Tourism and outdoor recreation have a considerable economic potential, but capital outlays for development and management are limited or lacking completely. Lake Powell and Monument Valley, which are part of the northern front of the reservation, offer the major tourist potential. There are, however, no paved highways on the Navajo side of Lake Powell. All the road networks lead to retail markets, lodges, marinas and camping sites on non-Indian land and tourists naturally spend their money at these non-Indian sites.

Approximately 473,000 acres of ponderosa timber are located on the eastern portion of the reservation along the Arizona border. The tribe, through the Navajo Forest Products Industries—the first Navajo enterprise, operates one of the largest lumber mills in the United States, employing some 500 Navajos, with a payroll of nearly $2 million annually (Navajo Forest Products Industries, 1973).

On tribal lands are sizeable deposits of oil, gas, coal, uranium, and other minerals. Oil leases and royalty revenues have been the major source of operating capital for the tribe but the oil and gas reserves are presently being depleted very rapidly. Coal and uranium, the other energy source, cannot replace the oil and gas revenues, but these reserves, particularly coal, are extensive and are now being utilized. Utah International, Peabody Coal Company, and others are strip mining coal to be used in generating electrical energy for the Southwest. In 1974, the Exxon Corporation signed a contract with the tribe to explore for uranium around the Four Corners region southwest of Shiprock, New Mexico (Farmington Daily Times, 1974a). In the past, the leasing of lands on the reservation for mining purposes was fairly easy to accomplish, but due to the energy crisis, the cost of exploration and mining, cost of tribal operations, cost of reclamation of strip mining, and opposition against mining, the leasing of lands will become difficult to negotiate in the future.

Tribal capital reserves, primarily accumulated from oil and gas revenues, now total less than 50 million dollars. Some of these funds have been invested in different companies to provide returns to the treasury. Because the tribe has been spending more money than its proposed budget each year, tribal funds must be guarded carefully against the time when oil and gas depletion will reduce income below the amount necessary to maintain essential services (Todacheene, 1974).

HUMAN RESOURCES

The reservation, shown in Fig. 12.1, is the home of approximately 130,000 people. Some 128,000 of these now live on or adjacent to the reservation. The others live elsewhere all over the world, but most of these people still claim Navajoland as their home. The birth rate, 31.8 per 1,000 people per year is comparable to India or Mexico (Boyle, 1973). The majority of the reservation residents live in Arizona—73,660 (58%); approximately

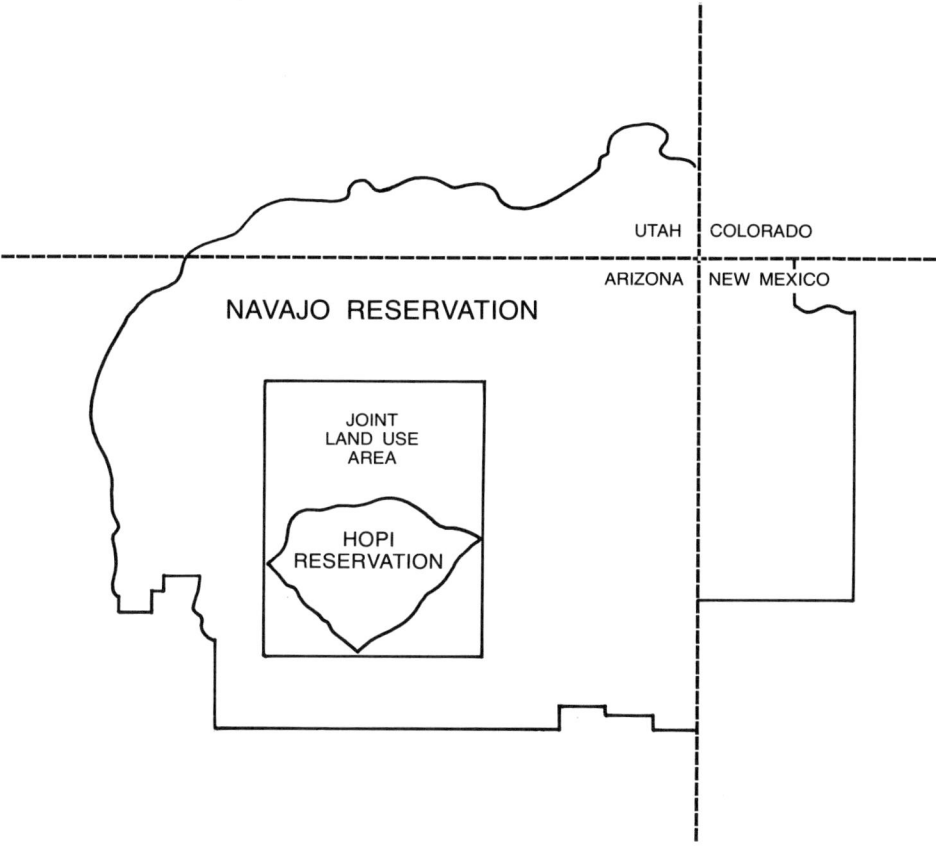

Fig. 12.1. The Navajo Reservation.

50,000 (39%) live in New Mexico and the rest, 4,400 (3%), in Utah. The population per square mile of about 2.6 makes the reservation the least populated of any significant political entity in the United States.

The Navajo human resources, skills and capabilities are part of the essential factor of production upon which Navajo development depends. In general Navajo workers have proved their ability and capability wherever they have found work, but many lack skills and a considerable number have little or no knowledge of the English language.

The median age of the Navajo is approximately 16 years of age, with 57% of the population 18 years and under. About 2,000 young men and women enter the Navajo labor force each year. The significance of this fact is the relatively low ratio of adults to children and the effect of this on family incomes. As consequences of the proportion, there are too few Navajos in the labor pool who are available for productive work and higher investment in schooling is required.

A 1969 survey of the Navajo labor force indicated that 32,350 persons were employed with approximately equal proportions of men and women of whom 23% were engaged in traditional pursuits such as sheepherding, rug-weaving and silversmithing. This study also indicated that 24% of the labor force were classified as unskilled, 8.8% as skilled, and 3.7% as semiskilled. Another study in 1970 estimated the Navajo labor force at 40,000 men and women. Of these, only 15,000 were employed with any regularity. While sources may vary somewhat in their statistics on the employment status, the conclusions are, that as bleak as the national unemployment rate may seem, the Navajo unemployment rate of 63% is "scandalous" (Boyle, 1973). Table 12.1 compares selected economic and social indicator and population characteristics of the Navajo with the rest of the United States.

TABLE 12.1
Selected Socio-economic Characteristics*

	Navajo	All United States
Median Age	17	30
Average Family Size	6	4
Annual Population Growth Rate (%)	3	1
Labor Force as % of Population	31	40
Rate of Unemployment (%)	60	6
Per Capita Income	$759†	$3,700†

*Adapted from MacDonald, 1972.
†Boyle, 1973

According to census data collected in 1969, of 5,734 male Navajos who were 25 to 34 years old, only 1,466 had completed high school and only 467 had done some college work.

The Navajo Tribe's Ten Year Plan summarized the overall educational status and educational facilities on the reservation as follows:

1. About one half of Navajo adults over 25 years of age are illiterate in English—neither read nor write; and one third of Navajo adults do not speak English.
2. Pre-school education is recognized to be essential for Indian children. Headstart Programs are funded uncertainly and reach only a portion of pre-school Navajo children.

3. There is no skill center and there are only minimally equipped technical, paraprofessional and skilled trades training programs for Navajo
Indians.
4. Training in business management and aid to businessmen is essential
if the Navajo are to develop their economy.
5. In terms of agricultural training almost no provision has been made by
any government agency or land-grant university.
6. Because of isolation, health, language and cultural factors, Navajo
children have especially urgent needs for special education programs.

Not only is the level of formal education of the labor force abnormally
low but also the occupational skills are low. The Navajo occupation distribution shows a disproportionate number of workers employed in service and
clerical jobs while relatively few are employed as professionals and managers. This maldistribution becomes clear in a comparison of the Navajo labor
force on the reservation and the New Mexico labor force, as shown in Table
12.2.

TABLE 12.2

**Percent Distribution of Reported Occupations of Employed Persons,
1970: Navajos on the Reservation and New Mexico***

Occupation	Navajos on Reservation (percent)	New Mexico (percent)
Professional, Technical & Kindred	10	19
Managers and Administrators	4	10
Sales	2	6
Clerical	18	16
Craftsmen	13	13
Operatives	15	8
Transport Workers	4	4
Laborers	9	4
Farmers	1	2
Farm Laborers	2	2
Service	20	13
Private Household	2	2

*Adapted from Boyle, 1973

NAVAJO TRIBAL GOVERNMENT

Up to the time when the United States took over the Southwest, there was no unified government under one Navajo leader. Each band (clan members) or extended family unit had leaders called Naat'Aanii ("Speech Makers"). They were chosen by the decision of the group or clan and were removed by the same process.

In fact, it was not until 1923 that the Navajo tribe had its first formal government, composed of six delegates (handpicked by the BIA), created in part so that oil companies would have some legitimate representatives of the Navajos through whom they could lease reservation lands on which oil had been discovered.

Reorganized in 1938 and since that period, the Navajo government has expanded its delegation and broadened its powers. The council has passed extensive codes outlining electoral processes, and powers and duties of its legislative, executive, judicial, and administrative branches of government (U. S. Commission on Civil Rights, 1973). The governmental system and its functions are patterned after the United States government, with the exception of a constitution which has never been established (Taylor, 1972).

The executive branch of government is vested in the chairman and vice chairman and their administrative staff. The people select and elect their two executive officials every four years through an electoral process similar to that everywhere in the United States. These two individuals are elected outside of the Tribal Council; they must be at least 35 years old and members of the tribe, and they must not be in trouble with the law in any way (Navajo Tribal Council, 1970a).

The executive heads, the Chairman and Vice Chairman, head up a sophisticated executive administration which includes: Office of the Auditor-General, Navajo Tribal Legal Office, Office of the Prosecutor, Office of Program Development, Office of Administration, Office of the Comptroller, Office of Business Management, and Office of Operations. These offices are further subdivided to include all other services required by the tribe.

The Chairman and Vice Chairman have the authority with the approval of the Council to select groups of five to eleven members from the council composition to serve on the following committees: Budget and Finance (10 members), Central Loan (5 members), Codification of Laws (5 members), Judiciary (5 members), Labor and Manpower (5 members), Police (5 members), Resources (11 members), Economic Development (5 members), Utility (7 members), and Transportation and Roads (5 members). These committees are part of the executive cabinet and the members carry out specific assignments, make recommendations, and work closely with the tribal administration as a back-up support for the chairman and vice chairman.

An advisory committee selected from the Tribal Council membership by the chairman with the approval of the legislative branch works as a right arm to the executive officials. This committee oversees all tribal business and financial affairs, reviews all land leases and purchases, approves contracts worth over $500 made by the chairman. In addition, it makes rules and regulations for commerce and trade, advises the tribal enterprises, and advises the agricultural, educational, health and welfare, highway, mining, public parks and monuments, wildlife, and numerous other activities.

The judicial system is administrated by the Tribal Court under the guidelines of the Code of Federal Regulations, Title 25 Chapter 11. The system consists of a Tribal Court and a Court of Appeals. The chairman appoints the seven judges, who preside at Tribal Courts, with the approval of the Council, for a two year probationary period. After this period, the Chairman may nominate the probationary judge to be a permanent judge, with "the advice and consent" of the Council. Permanent judges can serve until they are seventy years of age.

The Court of Appeals consists of a Chief Justice of the Navajo Tribe and two tribal court judges who are appointed to hear particular cases as requested by the Chief Justice. The Court of Appeals has jurisdiction over all Tribal Court final judgements except those criminal cases where the defendant is sentenced to 15 days imprisonment or labor and/or fined less than $26, in which case there is no appeal. The courts have the authority to make their own rules of pleading, practice, and procedure outside of those matters that are handled by the federal agency (Navajo Tribal Council, 1970b).

The tribal courts and non-Indian courts off reservation have been challenged numerous times in recent years by individuals represented by DNA, a legal body started by the poverty program. Most of these charges against the tribe and others have been in relation to a violation of individual Navajo civil rights. The role of DNA has brought about an awareness of individual rights and how such rights can be protected.

The legislative powers are derived from the 74 Tribal Council members who meet four times a year at Window Rock, Arizona, to formulate new rules, regulations, and policies, and to enact new laws to govern the reservation. The council can be called into a special session by either the chairman or the BIA area director. The council also approves personnel appointments, budgets, contracts, and all other matters that affect tribal government.

The councilman, or delegate to the Tribal Council, is elected every four years from his or her respective chapter. A representative must be thirty years of age or more and be a member of the Navajo Tribe. There are 101 chapters on the reservation and since there are only 74 Council members in the central government some of the legislators represent two or more chapters.

The Navajo Tribe has an annual budget of $16 million but usually overspends this amount every year to pay for all the services provided by the tribe. Oil leases and royalty revenues have been the major source to pay the cost of tribal government and administration, law and order, established and growing tribal enterprises, chapter projects, and welfare costs, such as the clothing for school children and welfare assistance. The government is having a tough time trying to balance the budget each year because of the rising prices and new growth demands.

The implementation of the tribal programs and services to the people are channeled through a governmental organization patterned and parallel with the BIA, as shown in Fig. 12.2. The two organizations assist each other by coordinating and implementing the services available from the two agencies, although there are many duplications in both organizations. The BIA programs include all or certain phases of: education, employment assistance,

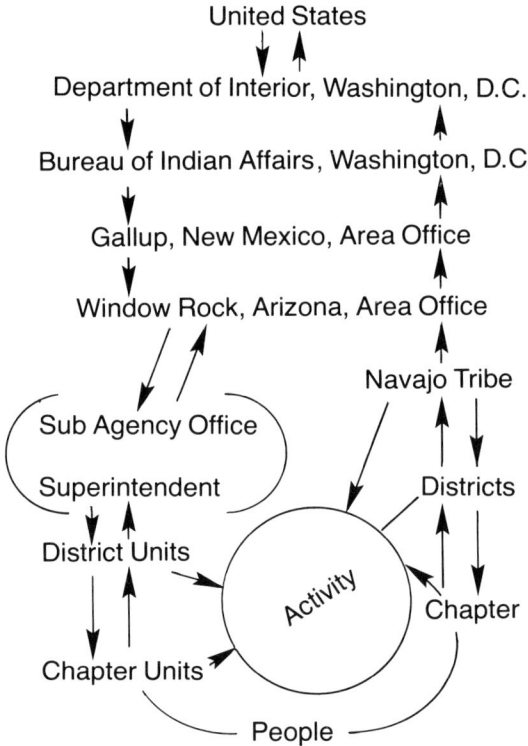

Fig. 12.2. The organization through which tribal programs and services are channeled to the local Navajo.

training programs, credit, housing, road construction, land and range management and other programs in connection with colleges and universities, states and federal agencies.

The administration of the BIA program filters down to the chapter level through the following organizations (see Fig. 12.2):

1. The BIA Office in the Department of the Interior, in Washington, is where the total Indian budget is consolidated and prepared for annual funding. Once the budget is passed by the lawmakers, the funds are separated and these separate funds start their route down toward the local level.
2. The area office administers the allocated budget for the tribe. The Gallup, New Mexico, Area Office handles the funds allocated for Navajo use and expeditures.
3. The Area Director's Office, at Window Rock, Arizona, provides the technical staff to oversee the BIA operations on the Navajo Reservation by properly allocating the funds and administrating the reservation programs to meet the needs of the Navajo people.
4. The funds available for the reservation are divided among five main subagencies, divisions of the main agency, located in different parts of the reservation. At each subagency, the BIA officials include a superintendent and his technical staff to provide services to the local people within that subagency boundary.
5. The subagencies are in turn subdivided into districts, which are the next level of administration. At one time, the district had a full staff to provide the total services provided by the BIA to the local people. Today there are few or none of these offices at the district level. If there are any staff members at the district level, they handle all of the local cases. Such staff members are directly responsible to the superintendent.
6. From the district level, the BIA staff members work with the people in the chapters through the chapter officials. The chapters are subdivisions of a district. The chapter then is the smallest governmental unit.

The operation of the tribal government system filters from the Tribal Council to the chapter in the same fashion as the BIA pattern just outlined. The main agency is at Window Rock where all decisions are made about the annual budget and the social and economic services that will be provided to the people. Rules and regulations are also directed from this office. The approach to tribal government includes tribal officials appointed or elected at the district and chapter levels to work with the tribe and coordinate all efforts with the BIA staff at all levels.

The tribal government administration channels its services to the local people through the following units (see Fig. 12.2):

1. The main subagencies include Tuba City, Chinle, and Fort Defiance, which are all located in Arizona, and Shiprock and Crownpoint, which are in New Mexico. Tribal representatives are located at these sites to provide assistance.
2. The districts and chapters are usually served by the subagency tribal staff.

At the chapter level, the local governmental organization elects a president, vice president and secretary every four years. Each chapter has a chapter house, which the officers use to hold meetings, parties, and other local activities. These houses are furnished by the tribe and have an annual operating budget from which the officials and operating expenses are paid.

The chapter grew out of the 1930 era when extension work was introduced to control and manage smaller units of grazing areas. Each chapter area was assigned a definite number of sheep based on the carrying capacity of the range area. These sheep numbers were designated as sheep units assigned to individuals in the chapter through permits. The theory of the system was to manage each chapter through proper livestock numbers in relation to range conditions.

Gradually the chapter grew into a governmental unit with definite boundaries within which tribal programs and services can be administered. Today the chapter officers hold periodic meetings to discuss tribal programs, domestic problems and local grazing conditions with the local people, and to make regulations and rules regarding local problems and conditions, provide information about local conditions and needs, and make recommendations to the tribal council about budgets, education, employment and other services.

Every four years the chapter people elect a representative to the 74-member grazing committee, which is responsible for the surveys, initiation, development, and passing of regulations and rules regarding the management of grazing, livestock numbers, collecting fees (off reservation only), and land use for total reservation. The central grazing committee is subdivided and grouped by district level for local administration. The district grazing committee consists of members elected from the chapters within the district, and acts as a committee to settle land, water and range disputes, and to work with livestock within the districts and the chapters. In the Crownpoint subagency (off reservation), the grazing committee is referred to as the Land Board Committee. Their duties are similar to the district grazing committee in all respects except that they also help manage land in coordination with state, federal, Indian allotment and private lands off the reservation.

There are six irrigation areas on the reservation and each project has a land board elected every four years to work with the Navajo Tribal Staff to operate the irrigation facilities and to assign land to individuals who would like to farm. Most of the irrigated areas are subdivided into 10- and 20-acre blocks for assignment. The members of the board are considered the managers and they are elected every four years from the chapter in which the irrigation project exists.

Two other organizations that have been created and are functioning at the district and agency level are the district council (tribal organization) and the agency council (BIA organization) organized primarily to coordinate the district and agency programs and services.

The present chapter system is very similar to the late nineteenth century non-Indian range town council in structure and function. In general, the governmental development stage is also at about the same level. Other characteristics of the chapter system include: (1) the chapter leadership and staff are usually older Navajos who have vested interest in land, livestock, and politics; (2) one vote in the chapter, in contrast to a majority vote in the council, can kill or stop any proposal; (3) the young people are generally not involved because of traditional views and lack of interest. Young people today are still considered "wet behind the ears" by their elders, although this cultural trait is breaking down. If the vote in tribal elections is given to the Navajo youth through a change in the voting age from the present 21 years of age to a lesser age, the youth can control the voting, if they so desire; (4) many individuals have served in the chapter over and over for so many years that they are part of the fixture. These are usually individuals whom the chapter people really support or to whom a dominating group in the chapter gives full support; (5) the chapter is very dependent on the Tribal Council for its support and survival; (6) the range rules and regulations associated with the chapter conflict with town growth and development. This is being recognized so that there is a slow moving away from this conflict of interest by the range people; (7) lack of proper management continues to plague chapters; (8) chapter "10-day projects" for which money is allocated by the tribe usually are not effectively used, or allocated funds are so small that starting an effective economic project is difficult; (9) the chapter continues to keep the people dispersed by its site location; and (10) the chapter boundaries are difficult to move across because new residents are usually considered outsiders for a long period.

Another organization, that uses the BIA and tribal governmental pattern by paralleling and working closely with these two entities, is the Public Health Service which administrates free services in health, medical education, clinics, promotes better health standards in the school, homes, and industrial facilities, and provides technical assistance to the BIA and tribe.

The Public Health Service has done a tremendous service since it took over the BIA health facilities in saving lives and promoting longer life expectancy for the people.

The Navajo, like most reservation Indians in America, have a unique legal entity. While they are citizens of the United States (granted in 1924), the people are not generally subject to the jurisdiction of state governments. This, however, is not totally true because the 1968 Civil Rights Act (25 USC. 1321-1326 of Title IV) indicates that state control of Indian judicial matters can be assumed after approval by a majority vote of Indians within the affected areas. This matter, however, cannot be approved by the Tribal Council but must be approved by local people only. The legal status then is one of partial sovereignty. Furthermore the federal government, through the Secretary of the Interior, Commissioner of Indian Affairs and the Bureau of Indian Affairs, retains control (by federal laws) over the judicial affairs of the Navajos.

Thus, while the tribe may be free from the cloak of state regulation, as long as the people do not vote for state control, nevertheless, broad federal laws and administrative codes keep it under the watchful eye of Washington.

THE NAVAJO ECONOMIC SYSTEM

Economically, the Navajo nation operates in a heavily oriented traditional and command economic system steadily working toward a market type of economy. The traditional strength lies in the forces of cultural values dictating or directing either the BIA or tribal regulations. An example of one traditional value over tribal regulation is the ineffectiveness of the tribal code to enforce and regulate the grazing controls. Furthermore, the chapter, district, agency, and tribal decisions are influenced greatly by traditional values and views.

The government sector, primarily the Bureau of Indian Affairs, the Public Health Service and the Navajo Tribe, are the major employers on the reservation (Harmon, O'Donnell and Henninger, 1969). Data compiled in 1969 showed that 66% of those Navajos employed on the reservation work for government agencies (Boyle, 1973). The area's economy is therefore very dependent on federal funding and guidelines.

Tremendous effort has been made in the past and additional effort is still being exerted to develop tribal industries or to get large industrial firms to provide jobs for the available labor force. In spite of these efforts, there are few such industries on the reservation. Two of the larger firms are Fairchild Semi-Conductor and General Dynamics, both manufacturers of electronic devices, who employ large numbers of Navajos. Other enterprises such as Grace-Davidson Chemical Division, the Navajo Block Company, Navajo Optics, United Electric Company and Eastern Navajo Prefab Homes Company are other industries that contribute to the Navajo economy.

Initially these industries were recruited on the basis that they would be free from tribal taxation and government funds would be made available to the company to train their needed manpower. The absence of services and facilities on the reservation, such as housing, small businesses, private property rights, schools, roads, recreation facilities, churches, makes the tax breaks and government benefits inadequate to offset the lack of a fully-serviced town.

Manufacturing based on Navajo resources and mining, is the second major employer, followed by agriculture, trade, and services. The weakest links in the economy are the trade and service sectors, and these are so weak that the Navajos do most of their trading and buying in off-reservation cities such as Farmington and Gallup, New Mexico; Blanding, Utah; Holbrook and Page, Arizona. The economic weakness is primarily due to the limited numbers of small businesses, particularly retailing services. There are only 171 retail establishments on the reservation as compared to two or three times as many of these establishments off reservation in counties adjacent to the reservation according to Gilbreath (1973).

The limited number of small businesses and other services also forces Navajo and non-Navajo workers on the reservation to spend almost all their income off reservation. As long as this continues, only limited primary, secondary or tertiary effects of such dollars can help upgrade the economy. Having no tax base to generate revenue, the Navajo economy suffers and little can be done to provide these needed services and increase the numbers of small businesses.

The predominant market economic system that exists around the Navajo reservation provides the residents of New Mexico, Utah, and Arizona with the basic economic tools for development and growth. These same tools are also used to check reservation development. Business interests at the national and local levels are not very favorable to seeing banks, car dealerships and other types of businesses develop on the reservation. This is partly the fault of the Navajo Tribe and the BIA who are not planning effectively on a long range to overcome the political economy that keeps the lid down on all sound economic reservation development.

The factors of production are non-existent because they are all essentially tied up by the neocolonial reservation context described by Jorgensen (1972). The nature of the neocolonial model has been aptly summarized by Blauner (1969) as follows:

Colonialism traditionally refers to the establishment of domination over a geographically external political unit, most often
inhabited by people of a different race and culture, where this
domination is political and economic, and the colony exists sub-
ordinated to and dependent upon the mother country. Typically
the colonizers exploit the land, the raw materials, the labor, and
other resources of the colonized nation; in addition, a formal
recognition is given to the difference in power, autonomy, and
political status, and various agencies are set up to maintain this
subordination.

Land resources are tied up by the federal trust status. Capital resources are
also very limited due to circulation of the poverty trap and chronic disequilib-
rium shown in Fig. 12.3. The tribe has not acquired the technical knowledge
nor does it have the institutions through which savings, loans and investments
can be made. The bureaucrated and dominant society is unable or is not
willing to provide capital resources for reservation development.

Saving is also difficult because almost all incomes generated are used up
in subsistence living (see Fig. 12.3). Management tools were lost during the
livestock reduction period in the 1930s. Large numbers of managers were
phased out by the livestock reduction process. The large operator's livestock
sizes were reduced to similar livestock sizes of marginal operators (Kelly,
1968). This not only forced large operators out of business but also made
them lose their status in the Navajo culture.

Of all the factors of production, labor is the most plentiful resource
available on the reservation (Boyle, 1973). The Navajo Tribe and BIA have
spent millions and millions of dollars on training and helping trainees get
jobs. However, such efforts are not even keeping up with the population
growth. Not all trainees get jobs for which they have been trained nor
are there other types of jobs available at which they can work. The quality
rather than the quantity is also becoming an important factor. There are jobs
available but because of the lack of proper training in depth these jobs are
not filled.

There is presently no overall manpower program to coordinate training
to upgrade the labor supply to meet the projected job demands. Because of the
absence of such a program all of the training institutions and funding agencies
react only to crises instead of to a systematic long-range program. The labor
resource and its characteristics are also not integrated with the other factors
of production.

A typical community on the reservation exists around a BIA compound
(with government quarters and administrative buildings and/or a school) and
a trading post. Hogans and houses are usually scattered around these with a
poor design for future expansion or growth. The housing available in such a

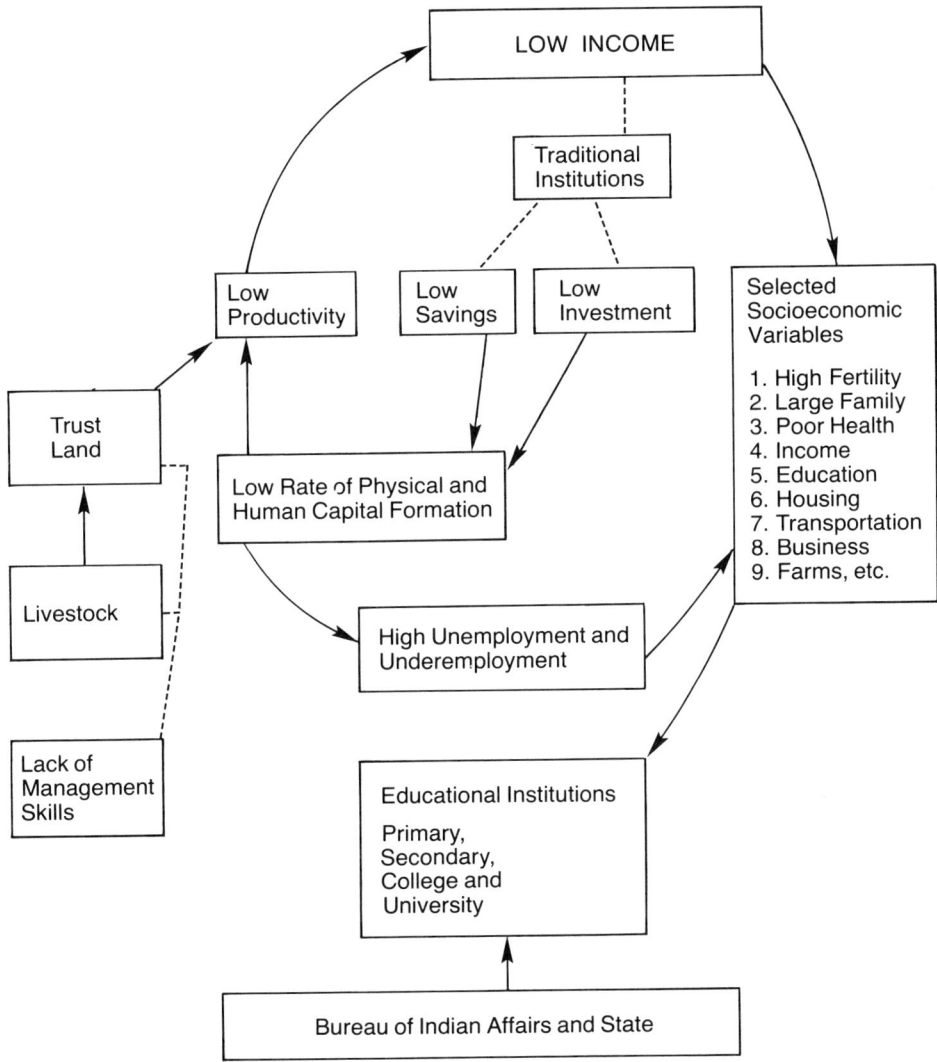

Fig. 12.3. Poverty trap on the Navajo Reservation.

community is either personal or low rent tribal housing or BIA housing. The latter two types are usually not privately owned. These government compounds are usually fenced to keep livestock out. Such an arrangement breaks the ties between the compound and the Navajo community.

The scattered homes across the Navajo reservation have few of the human comforts such as running water, electricity, heaters, gas or electric stoves, and telephones. Recent data show that 61% of Navajo homes are without electricity and 80% are without water and sewer service (U.S. Commission on Civil Rights, 1973). Navajo Tribal Utility Authority has the authority to provide some of these services but is unable to keep up with the growth and economic expansion.

Few major paved highways provide access onto and off the reservation. The lateral roads are mainly dirt or graded roads that become impossible to travel during rainy seasons. Roads generally have been built only to link to various government facilities such as schools, hospitals, and government offices. Only 1,370 miles of roads are paved; this is little more than one-third of the ratio of paved roads to square miles in the rural areas of the states surrounding the Navajo reservation. Other modes of transportation are even more scarce. For example, no major rail facility has shipping depots on the reservation and air transportation is provided only by smaller charter services. These conditions seriously hinder economic development.

The tribe has created the following enterprises: Navajo Aviation Authority, Navajo Broadcasting Authority, Navajo Arts and Crafts Authority, Navajo Engineering Authority, Navajo Health Authority and Navajo Housing Authority. In theory, these tribal enterprises in the future will provide revenues to support tribal social services and economic development. Whether they will do the job is yet to be seen.

The Navajo economic condition is undergoing changes. On tribal lands are sizable deposits of oil, gas, uranium, coal, and other minerals. These are being exploited—coal, particularly, by large corporations. However, only limited numbers of Navajo workers are employed and limited royalty income for the tribe is realized. The cash income is largely absorbed by the expense of tribal government leaving little if anything for economic development.

In general, with few exceptions, the Navajo economy is similar to other underdeveloped countries in just trying to keep up with growth and trying to curtail their worsening natural resources, particularly the land resources.

FUTURE ECONOMIC DEVELOPMENT

In 1962, Congress authorized the Navajo Indian Irrigation Project (NIIP), but funding has been grossly slow and inadequate. The Ten-Year Plan estimates that $150 million is needed in funding through 1982. Its completion would bring 110,000 acres into irrigated agricultural production and provide outstanding recreational opportunities (see Fig. 12.4). The project is expected to begin producing crops in 1976, when the canals under construction will be completed to the first 10,000-acre block of land. Increments of 10,000-acre blocks will be put under cultivation regularly until the total 110,000 acres are phased in by 1987 or thereabouts (Reno and Billy,

Courtesy of U.S. Bureau of Reclamation.

Fig. 12.4. Navajo Unit, Colorado River Storage Project. Navajo Reservoir, a favorite recreational area of the southwest, is pictured in this panorama. A part of Navajo Dam can be seen on the left.

1973). Adequate water resources of 508,000 acre-feet per year have been set aside for this project under the NIIP.

The Navajo Tribal Council has assigned to the Navajo Agricultural Products Industry (NAPI) the responsibility for development and operations to be gradually taken over from the Bureau of Indian Affairs and the Bureau of Reclamation. Navajo Agricultural Products Industry is an enterprise independent of tribal government.

To fully realize the Irrigation Project's potential for maximizing income and Navajo employment, an effective planning machinery is required (Franklin and Resnik, 1973). The other most important missing link is the capital investment needed to start and maintain an effective agricultural enterprise. Since the Navajos have little savings to invest in this venture, new credit forms and a substantial amount of seed money and agricultural technology will have to be obtained to overcome this deficit.

During the years the irrigation project will be developing, from four to seven coal-gasification plants will be built adjacent to project lands. This is the same area for which the tribe has entered into a contract with the Exxon Corporation as a partner to develop uranium mining. All of these developments cast new and favorable light on Navajo economy.

There is some opposition inside and outside the tribe to the strip mining operations, which would supply these gasification plants. Critics say mined lands can never be reclaimed and therefore the tribe is not receiving adequate compensation for them or for the coal mined and the water used. "Replying to this charge, the Peabody Coal Company, who operates a large strip mine in the west part of the reservation, says the royalty rate paid to the tribe is 'very high,' amounting to nearly 10% of the price of the coal'' (Redhouse, 1973).

A local chapter in the uranium mining area has recently voted against the tribal council partnership contract with Exxon Corporation (Farmington Daily Times, 1974b). This type of opposition will force a show down between the Tribal Council and this chapter if the pressures continue to mount.

SOCIAL CONDITIONS

To a degree the people have been able to maintain their social system because of the tradition-command economic system and the fact that the command sector has interfered little or has been lax in the area of social conditions. However, today, the social organization is starting to break down very fast due, in part, to the phasing in of new programs in health, welfare, employment programs, training, education, tribal enterprise, industry, Indian and non-Indian churches, increases in income, credit unions and banks, schools (BIA, church, and public), and other programs that require social adjustment and changes.

The automobile is at the top of the list in causing a breakdown of the Navajo social system (Billy, 1972; Pitts, 1974). Other machinery or items that are also causing similar changes include the radio, television, movies, clothing, music and games. The positive societal effects of the automobile include the following: (1) reduction of time in travel and in doing chores around the home; (2) expanded access to educational, recreational, and other social activities that were hard to attend due to distance; (3) a tool to earn additional income; and (4) a status symbol for work incentive. On the negative side of the ledger, the auto has the following effects: (1) it has impoverished Navajos by a stream of continuous payments; (2) it has changed moral standards; (3) it has caused friction and disruption of family ties; (4) it has released the horses from bondage to roam at will and add to the over-utilization of the range; (5) it has created a series of unplanned roads crisscrossing the reservation and introduced at least two or more roads to each hogan; (6) it has polluted the landscape around hogans, especially if the owners failed to fix the car or if the auto is stripped of its parts for use elsewhere; (7) it has poisoned the air; (8) it has started new social habits; for example, when a car starts in any direction, men and women carrying their shoes get in and go with the automobile regardless of where it is going; (9) it has created a profitable business for off-reservation car dealers and garage owners; (10) it has killed many people and livestock on the reservation; highways are becoming dangerous because Navajos under the influence of alcoholic beverages are driving and causing many accidents; (11) costly fences have been put up to keep sheep, cattle and horses off the main highways; and (12) it has affected the economy tremendously; money that could be used for saving or investments is constantly being utilized to maintain the cars, a factor accentuated by the dirt and rough roads (Pitts, 1974). The free movement of young and adults with a car has left livestock unattended.

The Navajo language has been maintained in spite of the strenuous efforts to phase it out, especially in the schools of the past. The perpetuation of the language is the rule today. The BIA schools on the reservation are in part administrated by Navajo adults who do not speak English. These Navajos are asking the BIA schools to teach Navajo language to those children who want to learn it. The language has been standardized by language scholars and a series of written materials has been developed. Certain schools, such as Rough Rock Demonstration at Rough Rock, Arizona; Borrego Pass at Borrego Pass, New Mexico; Rock Point, at Rock Point, Arizona, are pioneering in preparing educational materials, text, and other materials for young and adults speaking Navajo. Navajo Community College, the first reservation college, and Ganado College, two reservation junior colleges, along with other colleges teach the Navajo language as part of their curriculum (Pitts, 1974). The Navajo language is still a vital tool to help the youth and adult Navajos overcome their fear of economic development; particularly the gasification plants and other future developments.

The extended family is starting to show the stress and strain of high cost. It was common courtesy to provide visitors or relatives with food, even if they were just passing by. However, it is not uncommon for family members in the 1970s to wait until their visitors are gone to serve food. It is also not uncommon for extended family members to refuse to take part in ceremonials and other social activities. Navajos are even refusing to make car payments for extended family members. Landownership, probate land and sheep permits, and other property are no longer shared with extended family members as in the past.

The family clan and its proper marriage relationship are no longer rigidly followed according to earlier social customs. Many close clan members, boys and girls, are getting married in churches, cities and elsewhere without proper regard for the earlier social customs. Many educated Navajos who have grown up away from the reservation do not even know any of their close relatives, few even know the dos and don'ts of the older customs. The traditional rules that were once useful, as a social guide, are no longer used nor are they effective.

The sites of many, many sacred ceremonials are becoming the centers of drinking brawls, sexual activities, fights, and stealing. Even the loss of the proper sequence of prayers and steps in treatments is commonplace. Many activities such as the squaw dances and the rodeos occur all summer long.

These continuous events lead to fatigue and health problems for teenagers and adults who try to keep up with them. For those that hold and sponsor these events, the costs in dollars and social cost are continuing to rise.

Alcohol is becoming a very crucial social problem that is affecting young and adult Navajos alike. When such activities are concurrent with the automobile, the situation becomes dangerous not only for the driver but also for other drivers using the same road between border towns and the reservation. In general, many Navajo alcohol users try to consume all the beverages that they can before entering the reservation because it is still against tribal law to have alcoholic drinks on the reservation. In certain reservation areas, the bootlegging businesses are very prosperous and are causing serious problems. Many well-trained Navajos have serious alcoholic problems and if they get a job, it is difficult to keep them employed. The problem is serious enough to warrant additional alcoholic rehabilitation centers because there are presently too few centers to serve the total reservation area.

The whole school environment to which the Navajo students are exposed tends to lead toward a generation gap between students and their parents. The availability of human comforts such as running water, food and shelter, leads students to acquire a taste for these modern conveniences and when they go home to a different environment it affects their ability to perform at home. The knowledge gap is in part caused by the student performance at home. Another knowledge gap is caused by the student's ability to read, talk, and write the English language. This sometimes makes the parents feel they are too "dumb" in comparison with their children. An experience gap may also form between students and parents. This is due to the parents' experiences of the past, the child's experience of the present, and future projection with respect to the concepts of sheepherding, money, careers, marriage, customs, wages, credit, and others affecting daily living. The free facilities, clothing, food, education, and medical treatments also form a difficult type of philosophical thinking and doing things.

The trading posts, still the primary centers of activities, are changing their policies and practices such as credit, loaning money, taking pawn, buying wool and sheep pelts, buying lambs, and even the type of services they are now willing to perform. The young wage earners, who have not used the old credit system or have no knowledge of it or have been denied credit, have condemned the traders for taking advantage of the poverty and ignorance of the Navajos (Deseret News, 1973). In spite of the critics, many Navajos still prefer to do business with the traders, and the trading post is still the center for social gathering and has affected the social lives of new Navajo couples who cannot expect the same services that the trader gave to his customers in the past.

The boundary lines associated with agency, district, chapter and customary land-use area have made the Navajos more aware of the concepts of ownership of land, water, minerals and even tribal funds given for ten-day projects. A man who marries outside his chapter and goes to live in another chapter may be told to go back to his own chapter to vote, obtain free commodities, or secure a ten-day project job. This is an example of the changing concept of land use.

There are numerous young and adult Navajos who are very critical of the governmental, economic, political, and social services available on the reservation. One very important item is the extent of the authority of the chapter versus the Navajo Tribal Council's authority (Farmington Daily Times, 1974d). Can the Navajo Tribe decide what to do with resources located within a chapter boundary without proper consultation and without a vote by the chapter people? This is a very important question that is going to have to be answered soon.

Another social relationship that has been established over the years is the dependency of the people upon the Bureau of Indian Affairs. This dependency is based on either going hungry or getting something to eat. The old tools that the BIA gave back to the people, after the treaty of 1868, created a dependency on the federal government. The giving away of commodities, free hospital treatments, free distribution of hay and grain for livestock, free tribal clothing paid by the tribe, free scholarships, free education, and a whole series of free goods are the basic elements that have contributed to this social dependency.

Another dependency is the relationship between the chapter and the tribal government in which money is allocated to each chapter annually for chapter use and for projects. In general, most project funds are spent on creating a few jobs in each chapter. The councilmen are continuously trying to get project money from the central government for their own areas. The chapters are also always assisting tribal departments to get larger budget requests than the proposed budget by voting for higher budgets in behalf of the departments. Such votes sometimes influence council members to reconsider the resolution.

All the social changes that are occurring certainly have not been listed in this section. The ones that have been listed are examples used to point out the trends away from the old traditional concepts.

FORCES LEADING TO THE SOCIO-ECONOMIC CONDITIONS DESCRIBED ABOVE

The most important cultural difference between the Navajos and the settlers was the concept of landownership. The view of the Europeans was that land should be defined and located on the earth from a central point. Laws established to provide protection for the owners was a secondary corollary. On the basis of these two cultural factors, lands were claimed in the names of kings and nations. Ownerships were even attached to Indians. Indians in North America were referred to as Spanish Indians, French Indians, English Indians and finally American Indians depending on who was in power. All of these concepts were foreign to the concepts of the Navajo Tribe as well as to other Indian tribes.

The Anglo landownership concept is removed one step in scale from the concept that "the earth is the Lord's" taught during the dark ages (Farmington Daily Times, 1974b). This concept is a number of steps in scale from the Navajo cultural concept of land.

Another major cultural difference was of national leadership. The early settlers assumed that the Navajo nation had one chief. Instead the Navajo nation was divided into clan groups with their own leaders. On many occasions, the settlers attacked the wrong parties because of this one leader concept.

"The idler should not eat the bread of the laborer" was another concept used as part of the policy formulation in the use of non- and renewable resources. Because the Navajos were not working the land, they should move aside for developers to use their resources. The settlers assumed that all Navajos were lazy and that they should not stand in the way of economic development. The Navajos on the other hand felt that the land would furnish them food without transforming its environment.

Another cultural force that was introduced by the European among the Navajos was the concept of the superiority of the European God. Due to this concept many thousands of natives were Christianized and killed before they fell back into their old habits. Particularly is this true among the Pueblo Indians. Even today, there is still keen competition between the various churches to gain and hold native interest. A secondary corollary to this is the white supremacy notion that was exhibited toward the "Navajo savage." Such an attitude still exists in many quarters across this nation. These two concepts enforced the notion that civilization was brought to the savages to save their souls and to change their habits.

Probably one of the most important forces exerted on the Navajo people is the formal schoolroom education with textbooks, structured courses, and grades as tools to assimilate them into a go-getting American culture. Before the Treaty of 1868, the Navajos only used the informative type of teaching about household, farming, livestock, manners, sex education, morals, and so forth. In the Treaty of 1868, the following provision was made in regard to education:

> In order to insure the civilization of the Indians entering this treaty, the necessity of education is admitted, especially of such of them as may be settled on said agricultural parts of this reserva-

tion, and they therefore pledged themselves to compel their children, male and female, between the ages of six and sixteen years, to attend school; and it is hereby made the duty of the agent for said Indians to see that this stipulation is strictly complied with; and the United States agrees that, for every thirty children between said ages who can be induced or compelled to attend school, a house shall be provided, and a teacher competent to teach the elementary branches of an English education shall be furnished, who will reside among said Indians, and faithfully discharge his or her duties as teacher.

This program, however, was not readily accepted by the people. Teachers were sent and schoolhouses built by the government but children were reluctant to attend, so the government sent police literally to round up school-age students. Their parents often hid them. Those that were caught were sent away to boarding schools where they were often beaten and even shackled when they attempted to run away, when they failed to do assigned chores or violated school rules.

The U. S. Government saw nothing wrong with this policy; after all, the treaty did say that one teacher would be provided for every thirty Navajo children who were sent to the boarding schools. But to the Indian parents the idea of having their children taken and kept from them for weeks or months at a time was unthinkable, and they became even more resistant to education.

The situation improved somewhat with the establishment of day schools in the 1930s. Parent and child both preferred this system to the boarding schools because at least the child was home at night and the parents could keep track of his/her welfare and health on a day-to-day basis. Even with modern day improvements in facilities and more enlightened discipline, the Navajos do not like the regimented dormitory life, which is alien to the Navajo concept of freedom.

Many more Navajos are convinced that illiteracy is a bar to economic advancement and mobility. Because of this, the Navajos are beginning to attend school in greater numbers in every phase of training and education.

A very important political force is the concept of land utilization. The settlers moved through two periods of land utilization during which important policies in regard to land use were formulated. The first includes the mining of soil and water, ore, oil, or minerals from the earth without regard for conservation. After all there appeared to be a large supply available for the taking. The second period included the conservation era that started as a result of the dust bowl and other heavy land misuse. The experiences from the first period started a series of conservation laws that were enforced across the United States as well as in Indian territories. Although these laws have been very effective as the mother of federal and states' policies to plan better land and water utilization, the private property concept still makes it difficult to administer all of these laws effectively to safeguard the nation's natural resources.

The advanced technologies that the settlers brought with them were a tremendous asset in combining land and other resources to increase production. These technological tools were utilized to put together skillfully the factors of production to enable the economic system to function effectively to accumulate wealth. Savings, investment and taxation were also part of this development in national growth. Research was added to these forces to create better and more efficient economic tools.

Numerous studies about the Navajos are accumulating dust in educational institutions across the United States. Very few such studies have been used by large corporations to formulate well-thought-out plans to really help the Navajo economy. Usually what happens in most cases is that each corporation says to the Tribal Council, "We will put your labor force to work if you will let us lease your land." The agreements are usually made and then the company decides or finds out that there are no qualified Navajos that can be employed or that they have to belong to a union to work for the corporation. This example can also be applied to the other social and educational programs that exist across the reservation. These studies are full of directive suggestions to aid the Navajos but yet few of these materials are ever applied in action.

A very important cultural trait of the intervening, non-Navajo society is reflected in the Foreign Aid policies of the nation. The U. S. military bombs nations, and then after everything is "cleared" away, we open our bank accounts to rebuild the economy for those survivors who were once enemies of this country. Maybe it is too late to note that the Navajos were also beaten by the Americans, but the Navajo people received only old tools that they already had before. If the Navajos get anything with no strings attached, it is usually free clothing, food and gifts on Thanksgiving, Christmas or during an emergency, where "Project Navajo" comes to the rescue (Farmington Daily Times, 1974d). These gifts are appreciated to be sure; but better economic tools would be more useful to break dependency on government and state handouts.

The United States declared its jurisdiction over Indian affairs immediately upon adopting the new nation's constitution. The states ceded to the federal government the power to regulate commerce with Indian tribes. The federal administrative agency to deal with Indian affairs was originally in the War Department and later (in 1849) transferred from this agency to the Department of the Interior. The federal government's relations with the American Indians generally were dominated by efforts to demolish the original tribal structure and ties of the Indians and to assimilate them into Anglo society. The legislative acts that are benchmarks in these efforts include: (1) the Indian Reorganization Act of 1934; (2) the Dawes Act or the Allotment Act of 1887; (3) the Indian Termination Policy; and (4) the Indian Policy of self-determination with termination (Billy and Reno, 1973).

Under the Indian Allotment Act of 1887, Indian tribes principally in the Great Plains, Pacific Coast, and the Great Lakes states, surrendered their reservations to be broken into allotments to be parceled to individual Indians. Many of the Indians sold their allotments to whomever could give them the right price. As a result of these transactions, the Indians lost a good share of their landholdings and checkerboarding of Indian land became a reality. Surplus unallotted lands were opened to non-Indian occupancy. The Indians lost 90 million acres out of the 140 million acres which were theirs immediately prior to the Act. Their assimilation into the Anglo society was, however, a failure. Their land was assimilated, they were not. However, the Navajo nation was little affected by the Allotment Act of 1887. The Indian Reorganization Act of 1934 was a vehicle through which the BIA was commissioned to try to stop allotting Indian lands, consolidate the remaining fragmented reservation lands and get additional reservation lands back. This law also aimed to help the Indians get on their feet by encouraging group progress through tribal organizations.

The 1934 act was accepted by more than two-thirds of all the Indians in the United States. The Navajos, on the other hand, rejected it due mainly to the reluctance of the people to reduce their sheep and goats. To do so meant a cultural transformation as well as an economic change. This rejection also meant a loss for the Tribal Council in that the council would continue without a constitution and by-laws for self-government for years to come. Other factors such as the way the livestock reduction was handled, and the invasion by non-Indians under the Taylor Grazing Act led to the defeat of the 1934 act. However, in spite of the defeat of this bill, the sheep reduction program progressed over Navajo objection and the Indian Reorganization Act was applied with little modification to the Navajo reservation.

By House Concurrent Resolution, No. 108, adopted August 1, 1953 by both the House and Senate, Indian termination was advanced in the 1950s to solve the Indian problem and terminate federal responsibilities for Indian affairs (U. S. House of Representatives, 1953). Two main tribes—the Klamath and Menominee—and several minor tribes were terminated. The Paiutes in Utah lost their reservation when the government terminated their federally recognized status (Moes, 1972). When the smoke had cleared, the Klamath Indians had sold most of their rights and the result was depressing. On the other hand the Menominee organized themselves as a county to provide services, but lack of capital and almost no tax base forced this group to be restored recently to status as an Indian Tribe (The Eagle's Eye, 1974).

The Navajos have not been affected by the termination policy except for the fact that such a program cannot be arranged quickly. In fact, in 1972 the Bureau of Indian Affairs through its Navajo Superintendent, Anthony Lincoln, challenged the Navajo Tribe to take over the total BIA operation. Quite reasonably the challenge was not accepted by the tribe.

The Nixon policy of self-determination without termination has not been able to get off the ground due mainly to the constant changing of the Indian commissioners. Whether this policy will reach the Navajos and other Indians remains a question.

All of these factors have had a tremendous impact on the Navajos and other Indians in the Colorado Basin. Some were affected more by certain factors than others. During the two centuries since America became a nation, there has been an unceasing struggle against a complete assimilation of Indians by the non-Indians. This certainly has been true with regard to the values, culture, languages and, of course, with Indian lands and resources. Furthermore, some of these values from the past have affected the Navajos in a way that they are no longer independent, but depend on assistance from the U. S. government in order to survive.

NAVAJO LANDS

Before the arrival of the Spanish and Americans into the Southwest, land use rights in Navajo country were governed by traditional beliefs and practices. Navajo land tenure during this period was based upon a simple concept of usufructuary rights in land and water.

Before the signing of the Treaty of 1868, the Navajos were free to move unrestricted in search of food for their stock. These moving patterns did not follow any regular design. During this period heavy growth of vegetation, horse-belly height, was a common sight on Navajoland (Kelly, 1968).

After the Navajo Treaty of 1868, the United States laws and Anglo property rights precepts were imposed on and interposed within the traditional Navajo social framework. However, the Navajos did not really understand the new system imposed by the Bureau of Indian Affairs. The bureau itself was not sure of the system either and it did not have any effective program to follow until some half century later, when a Navajo Council was organized on July 7, 1923. This handpicked Navajo Council continued in operation until 1938, when the Navajo Tribal Council was officially organized and formalized.

The Navajo Tribal Council had difficulties in self-government from the start because of a number of factors:

1. The Tribal Council was an outgrowth of Bureau of Indian Affairs plans.

2. The Council did not truly represent all of the people.
3. Livestock reduction was one of the first orders of business for the Tribal Council.
4. The tribe was not equipped or ready to assume such a great task as livestock reduction. They are still suffering because of this unfortunate event.
5. The Tribal Council was not able to fully provide proper communication to its people.

Because the tribe was unable to recover fully from the livestock reduction and from their rejection of the 1934 Indian Reorganization Act, which would have given them greater power, the measures they adopted to control overgrazing were ineffective. Furthermore, it was not until 1958 that a regular grazing handbook was adopted (Navajo Tribal Council, 1970c).

Navajo land has been tremendously overgrazed, and if control systems are not implemented soon, this overutilized land will blow or wash away. The Navajo land crisis is due to overutilization, droughts, and in part, to: (1) the effects of Navajo tradition; (2) the illusion of land abundance; and (3) the trust and executive order status of the Navajo reservation (Billy, 1972).

The Navajo traditional concept of land use has been expressed accurately by Kelly (1968) as follows:

> Water, timber, and patches of salt-bush are considered communal property, open to all who need them. Farm or grazing land, once it has been appropriated by members of a given family, becomes their special preserve so long as they continue to use it. This is not the same as private ownership. Anthropologists call this kind of ownership "inherited-use ownership," which means simply that so long as a given plot of land is used by a particular family, it "belongs" to them and the right to use may be passed on from one generation to the next.

No right-minded Navajo, for instance, would intentionally trespass on land which he knew was being used by another. This type of land use was effective in the past, when the Navajos were few in number and additional land could always be utilized while the exploited land was recovering from overuse. The Navajos were able to conserve and use desert vegetation successfully by this method.

The traditional system of land use was fine as long as the population was low compared to the available land acreage, but as soon as the population increased, a land crisis arose. Aggravating the problem was increasing land pollution from the livestock. All of these factors led to the present population, land and pollution crisis described by Billy (1972).

The Navajos, over the years, have responded to the population and land crises, by adapting the traditional system into a system referred to as the "customary land use" area. This system differs slightly from the older model only through a mutual agreement with extended family members or other neighbors as what area belongs to whom as far as dominant usage is concerned. In the olden days no permission was ever needed to move around.

Fig. 12.5 shows the placement of a customary use area on one location near the Navajo Indian Irrigation Project area (Land Administration Department, 1974). This figure illustrates how the outline of customary uses relates to available water in one location. The land users have to use the central location jointly. Back away from the water hole the customary area widens

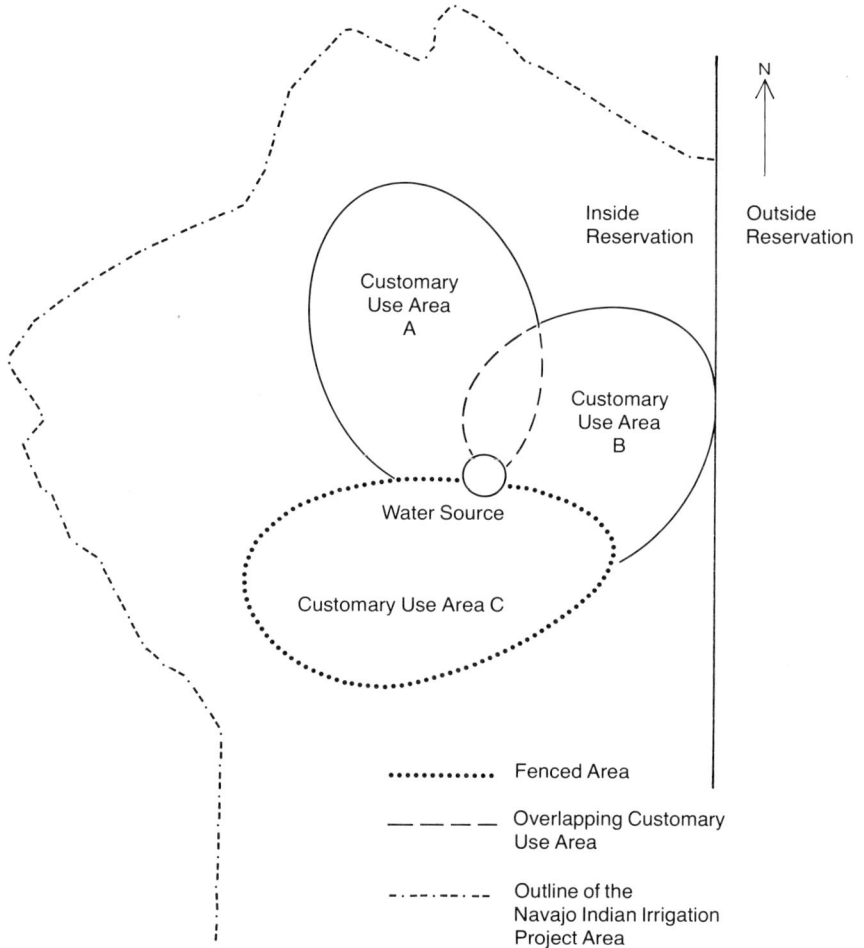

Fig. 12.5. Three customary use areas.

and common usage tends to diminish. The common usage area around the water source is overutilized. These areas can become smaller use areas if the original user subdivides his area between two sons or two daughters. The boundaries of many of these agreements are not definite because no survey instruments have ever been used to draw the lines.

Because of tradition and the lax manner in which the Tribal Council and the BIA enforce land use policy, the regulations regarding grazing and customary use areas are still being formulated.

The basis for control started with individual owners in 1937. All owners were contacted and their livestock numbers were recorded by the BIA. A survey of their range use area was made to determine the proper carrying capacity. Using the carrying capacity, the livestock numbers were reduced to a number that the range could actually carry year around without hurting the use area. On this basis the owners were issued sheep permits tied to a traditional use area.

The loose traditional use areas became permanent restricted areas (customary use areas) when the land crisis became more pronounced. The customary use area concept was stabilized further as a result of the passage of the grazing law in 1958.

In the 1970s, customary use area has the following characteristics: (1) the sheep permits are correlated with a customary use area; (2) these rights are fixed and can only be changed at the BIA agency headquarters; (3) sheep permits can be sold without the consent of other land users living close to the customary use area; (4) the buyer is buying not only the sheep permits, but also the use area; (5) individuals cannot have two customary areas in two districts except when there is a prior customary summer use area on the mountain in another district; and (6) if a man buys sheep permits and wants to move the permits to another district, he must submit his intention to the district in which he wants to settle. The district grazing committee will first review his application and if they are favorable to his proposal, they will help him establish a location for his permits. If the decision cannot be reached at this level, the case is then reviewed by the Tribal Land Administration on behalf of the Resource Committee of the tribe. If this committee cannot decide on the application, the case is laid before Tribal Courts for final judgment.

Land problems are becoming very common in almost every district. The uncertain boundaries causing overlapping land use often lead to land disputes. Other factors that lead to similar results include: (1) needed right-of-ways for housing, businesses, schools, and industrial sites; (2) feed and water storage due to long prolonged drought conditions; (3) irrigation development; (4) fencing; (5) right-of-ways for roads and dams; and (6) right-of-ways for power lines (Arizona Republic, 1974).

The customary land use concept has maintained a dispersed settlement pattern across the reservation. These distances increase the social costs associated with such dispersed settlements.

Developing towns such as Shiprock, New Mexico, cannot always obtain industrial sites for business development due to the fact that many customary land users absolutely refuse to release some of their land for this purpose. Many other proposed economic developments on the reservation are likewise at the mercy of range law. The solution, then, is one of convincing the older Navajos, who have control of the land, animal permits, and leadership of the chapters, to release some of their controls so that a community with a different land use pattern can be started for business sites, other developments, and for younger educated Navajos who have no land. Most often, however, many older Navajos are in favor of development as long as it does not affect their land use area or livestock numbers. It is difficult to change these attitudes.

Another type of traditional view held by the older Navajos is that land should never be in contention. They feel that only the Creator has the right to own fixed areas or to modify the land. When people start to fight about land, then this will surely lead to greater contention and so it is better to leave things as they are. "God will solve the problem in time for us anyway," is their solution. It is difficult to overcome such an argument with logic (Farmington Daily Times, 1974b).

Because of the tremendous increase in population and the overutilization of range resources coupled with drought conditions, the native plants are not able to produce viable and strong seeds to establish seedlings that can compete with tumbleweeds. Weeds have taken over large acreages everywhere on the Navajo reservation. The processes of overutilization and its control through fencing ran against traditional ideas and views so any conservation control is almost doomed before it starts. Even though the procedures for livestock control have been in existence for a number of years, the enforcement of land tenure controls has been a sticky problem for both the tribe and BIA because of traditional values and views.

The illusion of land abundance, the second factor, is very similar to the effect of traditional usage. Where natural resources seem relatively abundant in relationship to population, there is a temptation for any society, Anglo or Indian, to rapidly exploit such available resources to produce immediate income or to utilize the resources to produce secondary products. A good example of this was the harvesting of four-wing saltbush for firewood, a very common practice among Navajo families around Leupp, Arizona, in the 1930s and 1940s. This practice denuded extensive areas of this natural resource. The exploitation was so complete in certain areas that it is now necessary to haul juniper trees for firewood from the Flagstaff, Arizona, area or from any place where there is a wood supply. Such a solution, the harvesting of bush for cooking and the shortage of firewood for heat, has not only reduced the natural resource (vegetation); but also has created a shortage of food for the livestock (Billy, 1972).

Joe Toledo, a lifetime resident of Torreon, New Mexico, is one of the older Navajo livestock owners who still remembers the good old days. He made the following statement about the past: "When I was old enough to see how things really are, it was beautiful here. There were cornfields by the washes behind the houses and over by the school and everything that I remember was green. There were trees up and down the hills, too. Today, they're all gone."

The poor development of proper communication among Navajos has continued to reinforce the land abundance concept. Educational programs for adults have not been effective in overcoming this persistent concept. The reservation isolation must share part of this blame. Furthermore, the fact that so much emphasis has been placed on the operational procedures and policies of the tribal government and its proper functions rather than effective enforcement and control has also caused a neglect of this basic concept and its continuation. The Navajo people have been left in the dark so long that they are just now slowly becoming aware that land abundance no longer exists and they are finding the adjustment to present conditions very difficult.

The third factor, that of the Navajo trust and allotted land dilemma, stems from the early treaties with the federal government. Under the treaty between the Navajo and the United States the natural resources are held in trust by the federal government until the tribe is terminated.

The concept of trust land has harbored the philosophy that every Navajo is by right a joint owner in land, water, minerals, vegetation, oil, fuel, and tribal monies associated with the Navajo Reservation. In reality, these assumptions are false because the natural resources do not belong to the tribe or the people. The people are only users of the surface grazing rights. The land and natural resources are held in trust and administered under the federal government. The tribal money is held by the Tribal Council to be used for the total Navajo nation's welfare, although individuals can get loans.

Included in the trust land concept is still another land use question associated with the larger portion of the Navajo Reservation. There are the "executive order reservations," which are later additions to the original treaty land (3,300,000 acres). This type of reservation differs considerably from a treaty reservation.

A treaty reservation is the result of a bilateral agreement between Indians and the government. An executive order reservation is a unilateral creation by the President of the United States, which Congress has the power neither to approve nor to reject. In essence, this means that certain lands were simply withdrawn from the public domain for Indian use. Although custom usually accorded the Navajos title in this case, the legal title to land is still open to serious doubt.

Trust land on the reservation is not owned by individual Navajos nor does it belong to the Navajo Tribe, but it is federal property held in trust for the tribe. Individual Navajos exercise control over use rights to the land. The tribe cannot sell land or use it as collateral. Reservation land can only be leased from the tribe with the approval of the Bureau of Indian Affairs. Fencing for range management is difficult because of the common-land use rights that overlap each other with limited land description.

In addition to the problems surrounding the legal status of trust lands, are problems associated with the allotment of Indian land, principally in the eastern portion of the Navajo Reservation. These problems are described by Browing Pipestem (1969) in two categories as follows: (1) checkerboarding, where reservation lands become interspersed with individual parcels of land alienated from Indian ownership; and (2) heirship, where ownership of individual allotments becomes fragmented after passing intestate to the heirs of the original allottee.

Allotment lands can be used for collateral, but in too many cases, they are owned by large numbers of heirs. Before an heir can work or use this land, he must get approval from all the other owners. Obviously this seriously restricts the acquiring of capital, especially if the other owners do not agree to such use.

The Navajo land tenure system is changing, although slowly in comparison with the mounting population pressures and the urgent need for economic development. The processes of change have reached a stage where lack of a clear statement of what exists and of what is happening in land tenure is itself a constraint on rational decision making and on economic development.

The Navajo population has multiplied tenfold in the last century; the area of the Navajo Reservation is a little more than three times its size in 1858. Land use is still traditional, for the most part—grazing small herds of stock and subsistence agriculture. Urbanization is proceeding, but there is no legal basis for community planning of the urbanizing communities. The lack of available land for businesses and industries is often a constraint on economic development. Applications for establishing a business must first gain local chapter approval and then move through a series of approvals often taking two years.

Navajo religious belief still evokes Mother Earth and Father Sky, but strip mining operations are digging up one of the sacred mountains and other operations threaten to evict Navajo families from their traditional homes. Once evicted there is no place for evicted families to go—no Navajoland that is habitable is unoccupied.

PROSPECT FOR THE FUTURE

The future prospects look rather bleak for the Navajo Tribe in relationship to the larger dominant society in the Colorado River Basin. This prospect exists primarily because the old tools that the Navajos have are traditional and inadequate compared to the modern tools that the dominant society has to decide the fate of Navajos and other Indians in every aspect of life in the Colorado River Basin.

The future of the Navajos is dependent on two very important considerations: (1) Navajos will have to accept newer tools at an accelerated pace, culling out old tools and maintaining only the valuable tools, with which to enlarge on desirable cultural values that provide for man's happiness, and with which to adapt to the dominant cultural values; and (2) the dominant society must be willing to give Navajos newer tools without adding bureaucratic strings to limit the use of these tools. The present generation in the Colorado River Basin as well as in the United States must be willing to cull out certain undesirable old tools also. These two approaches mean that there has to be trade-off of cultural, political, social, and economic values on both sides between the Navajos and the dominant society to meet together and plan for the future welfare in the Colorado River Basin.

Some of the important elements that must be taken into consideration in Indian culture and land trade-off are the differences between the Indians' and dominant society's socio-economic system. For example, the Navajo model has the following characteristics:

1. The Navajo economy is a skewed command-traditional economic system including a limited market system.
2. The Navajo on his reservation may deal from time to time with four governments (Taylor, 1972): (a) his tribal government; (b) a nearby local community organized under state law; (c) state government; and (d) the federal government.
3. Navajo rights to land and other resources and Navajo claims on the federal government are spelled out in a treaty.
4. The Navajo Tribe is organized as a distinct entity with a trust and Indian allotment land base.
5. The Navajo Reservation includes a large bloc of land in two states and a smaller bloc in a third state.
6. The tribe has its own government patterned after the United States government, but without a very definite and well-defined constitution.
7. The tribal government is using tribal funds to pay for a large portion

of its governmental operations and societal welfare.Tribal funds are being depleted every year due to high cost and lower returns to the tribal treasury.

8. Navajo cultural identity is still a persistent factor in Indian affairs.
9. The tribal government is a partial sovereign nation.
10. The Navajo people enjoy significantly fewer advantages than the average citizens of the United States.
11. The local government on Navajo land is based on range law and the development stage in government affairs and development is at about the late nineteenth-century stage.
12. The availability of the factors of production on the Navajo Reservation are essentially non-existent because of the neocolonial policies and bureaucratic structure within the federal, BIA, tribe, local and state communities.
13. Even if the factors of production were available on Navajoland the following constraints would still exist: (a) private property does not exist on federal trust land; (b) the near absence of saving and investment institutions; (c) dependency on federal funding to a large degree; (d) very few developers such as managers and professionals are available or are now being trained; (e) the highly dispersed communities that exist on Navajoland; (f) older people's control of the livestock, land, and even the politics; (g) the relationship between the tribal and chapter government is not clear and definite.
14. The Navajos are starting to develop small businesses, tools like taxation and tribal enterprises to move ahead.
15. The Navajo nation does not have a comprehensive manpower program to channel the Navajo manpower supply through the training institutions to meet the manpower demands.
16. The outside market system controls the Navajo economy to a large degree.

The future as far as Indian culture and lands are concerned is dependent on how well and how fast the Indians in the Colorado River Basin move away from the old socio-economic model toward the dominant society's socio-economic model and on what socio-economic advantages the dominant society is willing to give up to help improve the Indians' socio-economic conditions. The solution, perhaps, can only be possible when the Indians and the dominant society are on an equal footing in many areas. Whatever the answers and the trade-offs are going to be will depend on a considerable give and take between the Indians and their neighbors. In the final analysis, the give might be greater from the dominant society due to their control of technology and most of the socio-economic tools that are needed by the Indians to move up the scale in resource development even to meet their present societal needs.

The give-and-take concept can be illustrated very nicely by the energy crisis. The crisis has recently shown just how closely the Navajos are tied to the outside market system. The fuel shortage in BIA schools and at Navajo Community College, and the gas and fuel shortages at trading posts were occurring all across the reservation even before the states in the Southwest felt their first energy crunch. The dependency of the Navajos on the outside world exists for nearly everything that is required to continue their present type of subsistence existence (The Farmington Daily Times, 1974c). For the Navajo this means that if their cultural values are going to survive, they must develop their resources in order to make decisions to have their societal needs met. Otherwise the outside market will continue to use Navajo resources to make capital gain and leave the Navajo economy nothing to build with in the future. This is particularly true with respect to the large corporations and the border town businesses. The important question is are they willing to assist the Navajos in developing their resources or will they continue in the practices of accumulating only profits at the expense of human pollution of a society trying to pull itself up.

Another phase of Navajo survival in the Colorado River Basin is also dependent on which of the following alternatives the Navajo choose to implement for future survival:

1. Stay with the same socio-economic model and hope for the best
2. Slowly and surely phase into a termination policy, which means going from the federal government to state government jurisdiction
3. Use the Indian Reorganization Act of 1934 to establish a tribal government with a constitution and take over the BIA operations and function as other Indians have done
4. Follow the Johnson and Nixon philosophy of transferring the control and responsibility from the federal government to Indian communities rather than to state government

The first alternative is certainly no choice for the tribe for it has had this type of management for over 100 years with no real prospect for the future. It would be a tragic mistake for any tribe to feel comfortable and to continue to rely on the old BIA institution to lead to better socio-economic conditions and self-determination.

The second alternative appears feasible and many Navajos would be in favor of this choice. New Mexico, Arizona, and Utah have all been involved with the Navajo Tribe in many services such as the public schools, agricultural extension services, social services, maintenance of roads, soil and moisture conservation, scholarships, and some phases of health programs. Navajos are presently very actively engaged in Arizona and New Mexico politics. On the other hand, many Navajos are not anxious to move under state authority. Past dealings with states on taxation and trust lands have not always been on friendly terms (Farmington Daily Times, 1974e). The records of Indians going under state jurisdiction have not been too successful either (Moes, 1972). If this is the choice, the Navajo Tribe and the states will both have to give up certain laws and assume greater responsibilities in many areas to bring about the desired conditions.

The third alternative has a poor record (Moes, 1972), although the Zuni Tribe in New Mexico appears to be doing all right. If the Navajos took this choice, their biggest problems would be the limited number and availability of managers and the fact that the BIA will continue to operate behind the scene. A similar constitution under which the Zuni Indians operate would mean less freedom for the Navajos. One Navajo chairman, Raymond Nakai, tried unsuccessfully to sell such a constitution, which he had hoped would provide better guidelines for the executive, judicial and legislative branches of tribal government.

The fourth alternative has two possibilities. Either strive for the development of a commonwealth or a new state under the federal government. Peter MacDonald, the Tribal Chairman in 1974, advocated the reservation becoming a state. According to Carl Todacheene, a prominent Tribal Councilman, the Navajos cannot afford to stop short of either a state or commonwealth. Tax legislation passed by the council was a major step in this direction (Todacheene, 1974).

No matter which alternative the Navajo Tribe centers around in the future, the most important question is, "Does the Tribe have an economic vehicle to work with?" It does and it is probably the most important economic vehicle that is available. This is the Navajo Indian Irrigation Project (NIIP)—the only project that is large enough to affect the social and political structure and influence the release of the needed economic tools to effectively turn the tide in favor of the Navajo nation.

The primary, secondary, and tertiary economic effects of the agricultural sector coupled with the cumulative effects of the gasification plants, coal and uranium mining, and all the associated processing services and transportation, and other support industries will diversify the Navajo economic base.

While the development of the irrigation project is in the hands of the Navajo Agricultural Products Industry (NAPI), the size and complexity associated with the development require better tools and greater vision than are presently available either at the tribal enterprise management level or at the tribal level in planning and development. Based on these assessments Reno and Billy (1973) recommended the following: (1) an organizational structure which would integrate within NAPI all irrigation project planning; and (2) an economical development office on site to integrate and coordinate all development.

The give and take associated with this very important development becomes the most critical element to the establishment of the total projects.

The "give" on the Navajo side includes the following:

1. Is the Navajo Tribe willing to make this the number one project on the reservation? If so they need to give full support to the project in financial aid and political pressures.
2. Are the Navajo individuals willing to relocate, to participate in the farming, invest in the farming, and receive training to work in the enterprise?
3. Are the Navajo willing to set aside land for a community with private property, provide a different governmental organization than the chapter, provide a tax base for the community, provide the institution for capital formation, provide the services required for any community, using Indian traditional design for the city, and so forth?

The "give" associated with the state includes the following:

1. Will the state and counties willingly give support to help develop the roads that will be required for the area?
2. Is the state willing to let the tribe use this as a vehicle to establish a new community which will take business and other services away from the cities of Farmington, Aztec, and Bloomfield, New Mexico?
3. Will the bankers and other businessmen agree to establish Indian banks and other businesses either in off-reservation towns or on the project lands?

The "give" from the federal government is:

1. Is the federal government willing to provide the proper funding to do a good job in implementing the project?
2. Will the federal government support a full blanket funding for NIIP instead of having to depend on scattered funds to get the project going?
3. Will the BIA release the development and operating capital to the Navajo Tribe?

This last portion of the "give" is the American people themselves. Is the dominant society willing to let the Indians gain self-determination without termination at the expense of their own socio-economic interest?

All these are questions that require the "give" from the tribe, state, and the federal government. No attempt has been made to try to relate all the factors that are going to be involved, particularly on the "take" side.

Literature Cited

Aberle, D. F. 1969. A plan for Navajo economic development. Vancouver, Canada: University of British Columbia.

Arizona Republic. 1974. T G & E out of Navajo coercion suit. 12 March 1974, A-5: Phoenix, Arizona.

Billy, B. 1972. Population, pollution, and land use among the Navajos. Four Corners Regional Commission Reports (NCC). 9 March 1972.

Billy, B. and P. Reno. 1973. Etats-Unis: ceques nous, Indians voulons. Preuves 4ᵉ trimestre No 16, pp. 37–50.

Blauner, R. 1969. Internal colonialism and ghetto revolts. *Social Problems* 16:395.

Boyle, G. J. 1973. Economic status of the Navajo on the reservation, 1969. Unpublished manuscript.

Brown, Vlassis and Bain. 1973. First and second quarterly report of 1973 of the General Counsel for the Navajo Tribe. The Tribal Council, Window Rock, Arizona.

Crawford, A. Berry, and Peterson, Dean F., eds. 1974. *Environmental Management in the Colorado Basin.* Logan: Utah State University Press.

Deseret News. 1973. White trader vs. "these young novices." 24 August 1973. Salt Lake City, Utah.

Eagle's Eye, The. 1974. Menominee restored tribal government. Provo, Utah: Brigham Young University, 1 (3).

Farmington Daily Times. 1974a. Tribe, Exxon O.K. uranium search. 21 January 1974. Farmington, New Mexico.

——. 1974b. Chapter opposing uranium proposal. 21 January 1974. Farmington, New Mexico.

——. 1974c. Indians join in Washington for more reservation fuel. 22 January 1974. Farmington, New Mexico.

——. 1974d. Sanostee area residents opposed to uranium deal. 29 January 1974. Farmington, New Mexico.

——. 1974e. County-split revolt feared. 19 March 1974. Farmington, New Mexico.

Franklin, R. S. and Resnik, S. 1973. *The Political Economy of Racism.* p. 85. New York: Holt, Rinehart and Winston, Inc.

Gilbreath, K. 1973. *Red Capitalism, an Analysis of Navajo Economy.* p. 457. Norman: University of Oklahoma Press.

Harmon, O'Donnell, and Henninger, Associates. 1969. *Program Design Study for the Navajo Tribe.*

Jorgensen, J. G. 1972. *The Sun Dance Religion.* Chicago: University of Chicago Press.

Kelly, L. C. 1968. *The Navajo Indians and Federal Policy.* Tucson: University of Arizona Press.

Land Administration Department. 1974. Personal communication with Roy Scrivner, the Director of Land Administration. 1 February 1974 in Farmington, New Mexico.

MacDonald, P. 1972. *The Navajo Ten-Year Plan, Part 1.* Navajo Tribe Copyright. Window Rock, Arizona.

Moes, G. J. 1972. Utes learn to exploit nation's hunger for outdoors. In *Boulder Daily Camera.* 2 April 1972. Boulder, Colorado.

Navajo Agency. 1961. *The Navajo Yearbook.* Compiled, with articles by Robert W. Young, Assistant to the General Superintendent. Albuquerque, New Mexico: Bureau of Indian Affairs.

Navajo Forest Products Industries. 1973. *Fourteenth Annual Report of Progress.* Navajo Tribe, Navajo, New Mexico.

Navajo Parks and Recreation Department. 1973. The Navajos. Window Rock, Arizona.

Navajo Tribal Council. 1970a. *Navajo Tribal Code 1969 Edition titles 1-16.* Navajo Tribe Copyright. Window Rock, Arizona.

____. 1970b. *Navajo Tribal Code 1969 Edition titles 7-16.* Navajo Tribe Copyright. Window Rock, Arizona.

____. 1970c. *Navajo Tribal Code 1969 Edition titles 17-23.* Navajo Tribe Copyright. Window Rock, Arizona.

Pipestem, B. 1969. Indian heirship land. Unpublished paper. 25 July 1969.

Pitts, M. 1974. Pickup's durability fits Navajos' needs. *Farmington Daily Times,* 25 February 1974.

Redhouse, J. 1973. Are Navajo coal royalties too low. *The Navajo Times,* 4 January 1973. A-r

Reno, P., and Billy, B. 1973. Navajo Indian economic planning: the Navajo Indian Irrigation Project. *New Mexico Business.* November, pp. 3–12.

Taylor, T. W. 1972. *The States and Their Indian Citizens.* p. 307, Washington, D. C.: Government Printing Office.

Todacheene, C. 1974. Personal interview at Farmington, New Mexico.

U. S. Commission on Civil Rights. 1973. Demographic and socio-economic characteristics of the Navajo. Washington, D.C.: Government Printing Office.

U. S. House of Representatives. 1953. 83rd Congress, 1st Session. *House Concurrent Resolution No. 108.* Washington, D. C.

Weatherford, G. D. 1974. Basin-wide planning and the problem of multiple jurisdictions in the Colorado River Basin. In *Environmental Management in the Colorado River Basin,* eds. A. Berry Crawford and Dean F. Peterson. Logan: Utah State University Press.

13

Conservation and Preservation of Aesthetic Values: A Matter of Choice

RUSSELL GUM

Conservation and preservation of aesthetic values is a matter of choice in the Colorado River Basin. As man's technology becomes more powerful and its effects more far reaching, man's impact upon the aesthetic aspects of the environment become obvious. These impacts are related to the choices made, directly or indirectly, by man: choices made by many different individuals and groups in many different forums. A simple inventory of the major forums where choices are being made would include: economic processes, political activities, legal decisions and administrative regulations. A similar inventory of the choices made which have significant impact on the aesthetics of the Colorado River Basin would include: water resource development, establishment of national parks, development of energy resources, and urban developments. However, a historical account of the aesthetic choices in the Colorado River Basin is not the focus of this chapter.

The basic purpose here is to put the problem of making aesthetic choices into perspective. Specifically, I will attempt to relate current planning methodologies to the problems of including aesthetic values in resource planning and to speculate about possible improvements in the planning process.

AESTHETIC VALUES

If one were to ask almost any citizen to list those features of the Colorado River Basin which were aesthetically good and, conversely, those features which were aesthetically bad, a list including many of the following would likely result: open spaces, the scenic heritage of the national parks along the Colorado River, open pit mines, urban sprawl, power plant plumes, power lines, air pollution and the other ugly condiments of economic development and progress. If one were to repeat the process for a large number of individuals, different items might appear on the lists. If one were to carry the process one step further and try to develop a consensus among those sampled one would discover that aesthetic values are in large part an individual phenomenon and that it is a difficult task to define the aesthetic characteristics of the basin. To find agreement on the relative importance of these aesthetic values compared with one another and with other elements with human value is even more of a hopeless task.

The important fact is that while aesthetic value is a nebulous, complex, and constantly changing concept, it is becoming increasingly important in the resource planning and decision making process. With increased importance, if resource planners and decision makers are to grapple effectively with aesthetic values the terms must be identified, described, measured in some manner and, at some point in the decision making process, compared with competing values.

As a starting point, the following definition of aesthetic values is offered.

> The "good" people ascribe to a situation, object or environment based on their sensual perception of it and not upon the object's use.

Consistent with this definition are concepts such as: beauty, naturalness, serenity and the other common descriptors of aesthetic values. Also consistent with the definition is the possibility of one ascribing "good" in an aesthetic sense to an object without directly perceiving the object but rather by valuing, in a vicarious manner, the aesthetics of the object. Three important concepts are inherent in the above definition.

First, aesthetics is a "good" to be valued. Current planning methodologies recognize that non-monetary values should be considered as inputs to the decision process. In addition to the role of non-monetary considerations in planning, much of the environmental legislation has been in response to demands that non-monetary values are important and must be considered.

Second, the "good" comes from people ascribing value and is not an intrinsic characteristic of an object. The aesthetic concepts of beauty, naturalness, and serenity do not exist independent of people but within people's heads. Aesthetic concepts are perceived by individuals and one's judgment of the aesthetic attributes of an object are based upon one's individual perception and evaluation process.

Third, aesthetic values can be enjoyed with the consumptive use of only the individual's time. In the case of vicarious benefits of an aesthetically pleasing object or environment, direct contact between the individual and the object is not even necessary.

Keeping in mind the above definition of aesthetic value and the basic problem of rationally including aesthetic values in the resource planning and decision making process, I would like to present a brief description of the evolution of the planning process, to discuss proposed planning methods, and to conclude with a discussion of future directions in planning considering aesthetic values which might be worthy of further investigation.

PLANNING—DECISION MAKING PROCESS

If planning can be defined as providing information to decision makers, planning processes can be described by the types of information considered, the ways the information is processed, and the way in which the information is presented to the decision makers. To illustrate these three aspects of the planning process and their application to planning problems involving aesthetic values, portions of three currently proposed planning methodologies are reviewed: the Water Resources Council's Principles and Standards for Planning Water and Related Land Resources (U. S. Water Resources Council, 1973); the House of Representatives' Land Use Planning Bill—Comprehensive Planning Process (Udall, et al., 1973); and the Technical Committee of the Water Resources Centers of the Thirteen Western States, TECHCOM (Technical Committee, 1974).

Water Resources Council—"Establishment of Principles and Standards for Planning." To describe the basic thrust and the specific relation to aesthetic values of the principles and standards of the Water Resources Council, selected portions of the report will be quoted. "Principles and Standards" established a multiple objective planning methodology with two basic objectives:

A. To enhance national economic development by increasing the value of the Nation's output of goods and services and improving economic efficiency.

B. To enhance the quality of the environment by the management, conservation, preservation, creation, restoration, or improvement of the quality of certain natural and cultural resources and ecological systems (p. 6).

In defining environmental quality, specific recognition of non-material human values such as aesthetics was made.

Responsive to the varied spiritual, psychological, recreational, and material needs, the environmental objective reflects man's abiding concern with the quality of the natural physical-biological system in which all life is sustained (p. 33).

Justification for concern with the non-material aspects of human existence and recognition of the implied trade-offs was given as:

Explicit recognition should be given to the desirability of diverting a portion of the Nation's resources from production of more conventional market-oriented goods and services in order to accomplish environmental objectives. As incomes and living levels increase, society appears less willing to accept environmental deterioration in exchange for additional goods and services in the market places (p.33).

The specific guidance for inclusion of aesthetic values can be summarized by the following quotes:

There are many effects which cannot or should not be expressed in monetary values. This is true of many contributions to environmental quality objectives.

When effects cannot or should not be expressed in monetary terms, they will be set forth, insofar as is reasonably possible, in appropriate quantitative and qualitative . . . measures.

If . . . amenable to quantitative measurement they should be described as fully as possible in appropriate qualitative terms (p.37).

It is not presently possible to anticipate or identify, much less measure, all environmental effects or change. Nor are there in existence evaluation standards that permit full and direct quantitative comparisons and ranking of the conditions of identifiable environmental effects that might be expected to result from a plan. Consequently, reasoned judgments by multidisciplinary teams will be required in many situations. When this is necessary, a frank expression of the state of knowledge and the limitations thereof, as well as the limitations of the analysis in each instance, is essential (p. 61).

After stating the above policy about how aesthetic effects are to be considered, the specific types of aesthetic values are listed and information in the form of "a descriptive-qualitative interpretation, including an evaluation of the effects of a plan . . ." is required for each. Also required are whatever quantitative data relative to the aesthetic area which might be relevant.

Land Use Planning Bill. Within the House version of the Land Use Planning Bill a comprehensive planning process is specified whose major difference from the process proposed by the Water Resources Council is a stronger statement requiring public input to the planning and decision-making process. The bill requires:

> Substantial and meaningful public involvement on a continuing basis and the continued participation by the appropriate officials or representatives of local governments in all significant aspects of the planning process (pp. 7-8).

The addition of a public input requirement is significant to a planning process where aesthetic values are concerned since most questions involving aesthetics are not technical but are questions of values. For questions regarding aesthetic values, the public can indicate preferences since there are no technical measures or standards of aesthetic values.

The two planning methodologies, as represented in the Water Resources Council's "Principles and Standards" and the Land Use Planning Bill, illustrate the state of the art with respect to the types of information considered and the ways in which information is processed. They reflect an evolution of the planning process from one where physical facts were direct inputs into the decision process to one where physical facts were put into the decision process in the form of economic facts supposedly representing human values. Physical facts, and human values in the form of economic facts, social facts, and political facts, combine to form a set of planning facts which are put into the decision process as diagrammed:

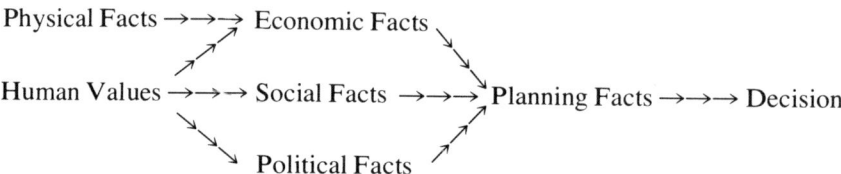

Controversy, of course, surrounds the evolution of planning toward an expression of human values in other than economic terms. Some economists have argued that multiple objective planning is not necessary as all benefits and costs theoretically can be expressed in dollar terms as the appropriate measure of human values. In fact, procedures have been developed and utilized to measure the dollar value of recreation for cases for which real markets do not exist and these same procedures have been applied to environmental quality problems such as air pollution. I do not want to argue that economics cannot be useful in the context of aesthetic values but rather, from the practical and historical view, that economists have often overestimated the real applicability of economics to environmental problems. As John Kenneth Galbraith (1973) states:

> Neo-classical economics . . . did nothing to prepare people for the explosion of concern over the environment—something that might have been expected from a good and competent science. So economists would be wise to be restrained in recommending remedies that grow out of these ideas (p. 288).

A second aspect of the evolution of planning involves the way information is processed. As evidenced by the "Principles and Standards of the Water Resources Council," the planning system has a bias toward information quantified in economic terms. Owing to requirements that objectives other than economic ones be considered, other types of information processing are necessary. Where information in other than economic terms is required the bias is towards quantified information rather than qualitative information. However, because methodologies for quantifying many types of information have not been developed, tested, and/or accepted, qualitative information is a major part of planning in the mid-seventies. This is particularly true for much information on environmental quality and is perhaps most evident in the case of aesthetic values.

In fact, a serious controversy exists concerning the use of quantitative versus qualitative measures in the description and analysis of environmental problems. Opinions range from horror that numbers could possibly be considered as measures of aesthetic values to those who wish every value measured in like terms (usually dollars) so that comparisons among competing resource uses can easily be made. Again from a practical standpoint the use or non-use of quantification should be judged in terms of how well quantification improves the planning-decision making process. While many academics have preferences for quantified versus qualitative planning systems, it is the results of the planning process not the methods used which ultimately must be judged. The following statement made in a context far removed from planning (Rothschild, 1974, p. 46) would seem to apply.

> Nor do I deny that math in itself is a delight in its precision, having its own esthetics based on the precise use of abstract values. I just get a bit upset at its misuse, when all we want is to create a wonderful picture.

The role of public input in defining values or in determining trade-offs regarding aesthetic values is also open to controversy. However, at least in the area of aesthetic values, everyone is an expert in the field (i.e., everyone has his own perception of aesthetics) and the argument that an elite group of experts knows best is difficult, if not ridiculous.

Recognizing the existence of these basic controversies, one may look at a planning process for including aesthetic values from a practical point of

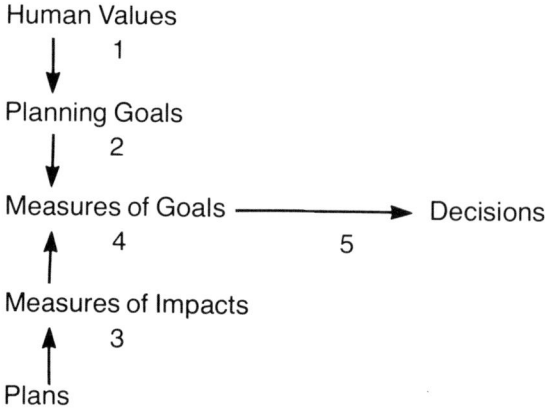

Fig. 13.1. Information flows in the planning system.

view. A reasonable requirement would be that the planning process meet the following criteria:

1. Aesthetic values must be considered as well as other values.
2. The trade-offs implied by alternative plans must be obvious, both in physical terms and in value terms.
3. The amount of information presented must be practical, i.e., the planning process must produce results usable and relevant to decision makers.

While the planning process outlined above, or for that matter any planning process involving multiple objectives, quantitative as well as qualitative aesthetic values, certain practical problems remain. The most important problem involves the planning communication process—specifically how information moves through the planning decision making process. Considering the information flows of such a process as in Fig. 13.1, the following specific problems of information processing are relevant to including aesthetic values in the process:

1. In what form should aesthetic goals be expressed to portray human values satisfactorily?
2. In what manner should these goals be measured and/or compared to other goals?
3. What aspects of aesthetic values will be affected by proposed plans and how can the effects of the plans be measured and compared?

4. How can measures of effects be related to goals?
5. How can the total set of information be effectively and efficiently presented to decision makers?

A system which has proposed answers to the above questions is TECHCOM (The Technical Committee, 1974).

To illustrate the operational paradigm of TECHCOM and its application to planning problems involving aesthetic values, the following hypothetical example of a planning process utilizing TECHCOM is presented. Consider the proposed development of a water quality control project. The first step (step 1, Fig. 13.1) would be to develop a list of human goals that might be affected by the proposed plans. The form of the goal list is that of a goal hierarchy with abstract general goals at the top disaggregated into more specific goals until a level of disaggregation is reached where a lower level goal can be perceived directly. For example, a general goal of environmental aesthetics might have a water aesthetics sub-goal which is further disaggregated into the perceivable components of odor, clarity, and floating objects. The second step (step 2, Fig. 13.1) is to assess the relative importance of the various components of the goals. This is accomplished by empirically developing preference functions that relate the levels of the goals to a person's preference for a situation. For example, a preference function could be defined relating the levels of water odor, clarity, and floating objects to satisfaction in terms of water quality. Additional preference functions could be defined relating the importance of water quality to other goals which, taken as a set, composed the environmental aesthetics goal. This step can be accomplished either by public input in the form of surveys or by a group of experts developing such a function. However, since technical matters are not involved in determining the relative importance among perceived goals, public input is feasible. In addition, if public input is obtained, the differences among members of the public can be considered as information to the decision maker about potential conflicts or equity considerations.

The next step (step 3, Fig. 13.1) is to propose a list of social indicators, aspects of the physical or social environment which can be measured and are expected to be impacted by the proposed plan. The list of social indicators is then related to the list of human goals and measures of the impact of changes in the social indicators on the human goals are made (step 4, Fig. 13.1). The social indicator to goal information transfer is perhaps the most difficult of any process as it involves both technical facts and perceptual facts. Therefore, both technical skill and an understanding of perceptual processes and human values or an experiment designed to estimate such, is needed for this step.

At this point the planner can make a rough estimate of the impact on social indicators of the alternative plans and, utilizing the trade-off information, determine which changes are likely to have the greatest impact upon the goals. From this preliminary information an efficient data collection or estimation program can be developed to provide a final set of impacts to the decision maker (step 3, Fig. 13.1).

The final step in the TECHCOM process includes the way information is presented to the decision maker (step 5, Fig. 13.1). The presentation of planning information is based upon a computerized information system. The design of the system allows the decision maker instant access and choice of both the type of information and the level of detail of the information. Information contained in the system would include an increasing level of detail for each alternative plan considered.

1. Impacts on general goals by proposed plan
2. Impacts on components of the general goals by proposed plan
3. Impacts on social indicators by proposed plan

In addition the levels of the goals and sub-goals for each plan can be reported using alternative sets of human values; for example, value sets representing conservationists, businessmen, planners, and so forth. From this aspect of the system the decision maker can gain insight into conflicts among interest groups.

Of particular value are the efficiency and range of information available to the decision maker. By offering the decision maker such a system it is likely that more alternatives will be considered as well as more aspects of human value, both important improvements in the decision making system and both especially relevant to decisions involving aesthetic values.

CONCLUSIONS

Aesthetic values are matters of choice which are becoming increasingly important as our culture seems to be evolving from one where material values were dominant to one where the quality of life is being defined in terms of material as well as non-material dimensions.

Planners and decision makers, realizing the importance of identifying and assessing the importance of trade-offs among diverse human goals, have begun to develop planning methodologies to provide information on the impacts of plans upon a wide range of human values.

Of critical importance to the aesthetic future of the Colorado River Basin is the way planning efforts provide information to decision makers on aesthetic values. If planners can identify and describe aesthetic values in terms both useful to the decision maker's task of making trade-offs, and valid in terms of reflecting human concern about aesthetic values, choices involving aesthetic values will be much better ones.

Literature Cited

Galbraith, John Kenneth. 1973. *Economics and the Public Purpose.* Boston: Houghton-Mifflin.

Rothschild, Norman. 1974. Offbeat. *Popular Photography,* March.

Technical Committee of the Water Resource Centers of the Thirteen Western States. 1974. Water resources planning, social goals and indicators: methodological development and empirical test. Final report to the Office of Water Research and Technology, U.S. Department of the Interior, Utah Water Research Laboratory. Logan: Utah State University.

Udall, M., et al. 1973. Land Use Planning Act of 1973. H. R. 10294, House of Representatives, 93rd Congress, 1st Session.

U. S. Water Resources Council. 1973. Water and related land resources, establishment of principles and standards for planning. *Federal Register,* 38 (174), Part III. Washington, D.C.

14

Carrying Capacity and Planning

A. BERRY CRAWFORD and A. BRUCE BISHOP

For its size, the Colorado is probably the most utilized, controlled, and fought over river in the world. As Russell Freeman of the Environmental Protection Agency put it, "People expect the Colorado to be their fountain and cesspool. They put demands on it beyond its capabilities"(Findley, 1973, p. 568). The phrase "beyond its capabilities" points to carrying capacity of the basin's resources as a key factor in choices about the region's future growth and development, and recognizes that the quantity, productivity, and assimilative capability of resources limit or constrain the region's ability to support the many competing demands of man's social and economic activities.

Considering the need for better mechanisms to understand and analyze the impact of growth and resource development, the carrying capacity theme serves both to stimulate and to direct discussion toward more consistent bases for planning and decision making. With this purpose in mind, this chapter explores the concept of "human-oriented carrying capacity," describes a possible framework for its use as a regional planning tool, and discusses some of the considerations involved in applying a carrying capacity-based planning process to the Colorado River Basin and its subregions.

ANALYSIS OF THE CARRYING CAPACITY CONCEPT

Developed initially as a concept for describing the growth and dynamics of species populations, "carrying capacity" was defined as a limit on the number of a species that could be maintained within an ecosystem or habitat.

[292]

Given an environment in which a certain population must obtain its life-requirements, the term was interpreted as the maximum population that the environment is capable of supporting over a specified time horizon. Thus, the use of carrying capacity as a tool for the management of biological resources focused on the temporal dimension of the problem and the idea of maximizing production. As a management objective, the carrying capacity concept is usually cast in terms of "maximum sustained yield," i.e., achieving a maximum sustained level of production of some desired good from a given resource system—the number of cattle that can be marketed from a given range, the number of board feet that can be harvested from a given forest, or the number of user-days that can be provided by a given recreational facility—subject to constraints designed to guarantee that production can occur without impairing or depleting the resource base.

Generally, then, "carrying capacity" has meant "population carrying capacity." The appropriate question was: "At what population level does a given community exceed the capacity of the environment to support it?" Applied to biological systems, such a question has proved to be tractable and useful. Applied to human beings and human environments, however, the question must be approached from a different perspective. It makes little sense to attempt to establish an upper limit for a human population since man is capable, to a degree far exceeding all other species, of changing both himself and his social and natural environment. Since man is able to adjust to as well as change the social and physical conditions under which he lives, any view that regards these conditions as inflexible constraints on an upper population limit is too simplistic.

Yet, it is obvious that growing populations are placing increasing demands on the capacity of life support systems and that breakdowns in these systems are occurring with increasing frequency. Shortages in raw materials, energy resources, and essential goods and services; stoppages, delays, and congestion in transport and other distributive systems; pollution and the diminished assimilative and productive capacity of air, water, and land resources—these are among the many symptoms of a breakdown in the carrying capacity of human and natural systems.

Accordingly, the concept of carrying capacity needs to be broadened to include the interactions that occur between human and natural systems, and the concept has to be articulated in a regional planning process oriented toward the management of change. Carrying capacity depends on a host of factors and therefore has to be viewed in a systems context. At a high level of abstraction, Fig. 14.1 exhibits some of the key relationships that would be involved in an analysis of the carrying capacity of a man-dominated regional environment. In this context, there seem to be four general areas in which carrying capacity has to be measured and analyzed. These general areas are:

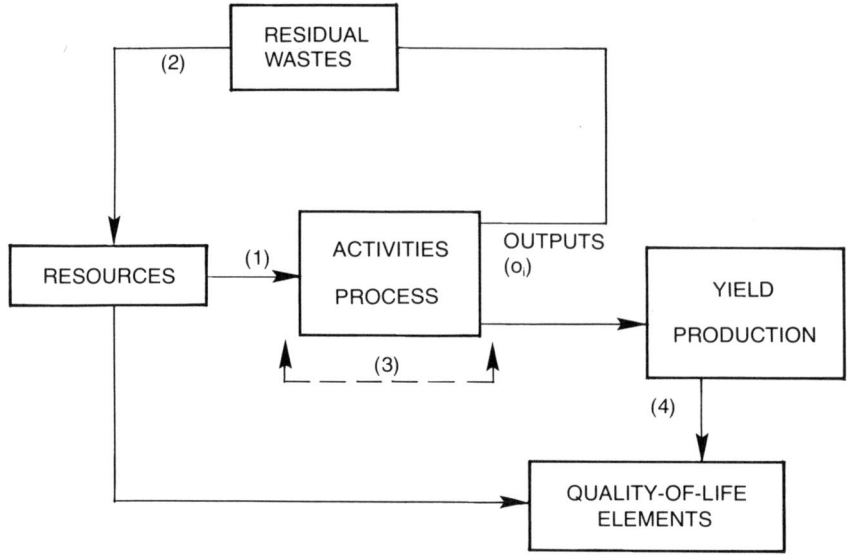

Fig. 14.1. Overview of carrying capacity relationships.

1. Resource-production relations: the capacity of available resources to sustain rates of resource use in production
2. Resource-residuals relations: the capacity of the environmental media to assimilate wastes and residuals from production and consumption at acceptable quality levels
3. Infrastructure-congestion relations: the capacity of infrastructure resources (distribution and delivery systems) to handle the flow of goods and services and resources used in production.
4. Production-societal relations: the capacity of both resources and production outputs to provide acceptable quality-of-life levels

These four dimensions of human-oriented carrying capacity provide a convenient categorization of many particular capacity relations. For example, numerous processes and economic activities are involved in the production of goods and services, and hence many particular carrying capacity relations constitute or contribute to the first general carrying capacity relation. The same can be said of the other three. The specific capacities of a region to produce the conditions of quality life are obviously dependent on the adequacy of the region's resource base and the efficiency of infrastructure or distributive resources.

In analyzing the concept of a human-oriented carrying capacity, several key points should be noted:

A normative concept. The concept of human-oriented carrying capacity has an essential normative component in the sense that it refers to the capacity of a man-organized and man-manipulated environment to serve human ends in the long as well as the short run. If a region's productive and infrastructure resources (including imports) fail to provide the conditions that constitute acceptable quality of life, or if waste loads cannot be assimilated by the media (excluding exports) at acceptable environmental quality levels, carrying capacity is exceeded. Such breakdowns (shortfalls and excesses) are thus identified by comparing actual or achieved conditions to a norm or standard.

Limits and constraints. Carrying capacity is a boundary-oriented concept involving the ideas of limits and constraints. In analyzing carrying capacity relations, one should attempt to identify conditions that limit or constrain the capacity in question. Scarcity of water for irrigation limits farming in arid regions, for example, and hence constrains the capacity of agriculture in these regions to supply agricultural products. Lack of water operates (along with other limiting factors) to place an upper limit on agricultural production. The desired level of agricultural production is established by the demand for agricultural products. If agricultural production, limited by the availability of water, is lower than the desired production level, a carrying capacity breakdown occurs. Here, as in most man-controlled systems, however, limits are subject to change and modification. Means exist for increasing water supply and for decreasing demand; the carrying capacity relation in question is elastic.

Measurement and indices. To be effective as a planning tool and for decision- and policy-making purposes, indices for measuring carrying capacity have to be developed. Ideally, these indices should range from the general to the particular and should describe not only the four general carrying capacity relations noted above, but also the particular carrying capacity relations which these general relations aggregate and summarize. By scaling the indices on a common basis with capacity standards, a means will exist for identifying shortfalls and excesses and for making adjustment and trade-offs.

Tools for analysis. No single model would be capable of analyzing the kinds and levels of relations that have been discussed and of generating appropriate indices. Rather than relying on a single carrying capacity model, various kinds of analytic techniques and models that are already operational and tested should be modified and synthesized. This observation is prompted in part by the need to provide analysis at various scales of spatial and temporal resolution. An overview of the analytical relationships and levels of spatial and temporal resolution that would be involved in determining carrying capacity changes is discussed in the next section.

Trade-offs and controls. The ability to make adjustments and trade-offs presupposes not only that the decision maker understands how carrying capacity relations are changing in relation to both baseline conditions and capacity standards, but also what controls would be effective in correcting capacity dislocations and how these controls can be implemented. Technological innovations and improvements in natural resources extraction, production, waste treatment, resource substitution, and waste recycling; cost (price) changes in relation to the above activities as well as to exports, imports, goods and services, and polluting activities; organizational, design, and efficiency improvements in distribution and delivery systems; coercion and education in relation to consumer demands, conservation, pollution, procreation, and migration—these are some of the factors that decision makers must consider in effecting the adjustments and trade-offs that might be necessary to insure stable and safe carrying capacity relations.

The observation that "human-oriented carrying capacity" essentially means "capacity of human resources to serve human ends" does not imply that these ends (desires, needs, aspirations) are static and unchanging. Indeed, from the standpoint of a carrying capacity-based management perspective, controls that are effective in scaling down these ends will sometimes be recognized as the controls most needed.

Regional Descriptions. Use of the term "region" in regional carrying capacity also requires comment. There are, of course, many ways of delineating a region. A region may be defined in hydrological terms as a river basin, in meteorological terms as an airshed, or in political terms as a governmental jurisdiction. It might also be defined as a socio-cultural unit (e.g., Sioux country) or as an urban-suburban complex (e. g., greater Phoenix). It makes little sense to speak of an "ideal" planning region, although certain regional delineations are more suitable for certain planning purposes than others. In Fig. 14.2, for example, region A, designated as an area of planning jurisdiction, consists of two large river basins, B_1 and B_2, with subbasins, B_{2_1} and B_{2_2} within B_2. A also contains three counties, C_1, C_2, and C_3, two of which are in river basin B_1, and an airshed, D, that extends the length of the two counties in B_1. Recognizing that other region-types might be defined for planning purposes as falling within A, two points have to be made: The first is that human-oriented carrying capacity analyses do not presuppose any particular regional typology (provided, of course, that people live in the region-type in question under some appreciable degree of social organization). The second, more significant, point is that if A is the area of planning jurisdiction and carrying capacity analyses are to be undertaken at various scales of spatial resolution, it is essential that common units of analysis be employed so that information at the macro-level of analysis can be employed at the micro-level of analysis, and vice versa.

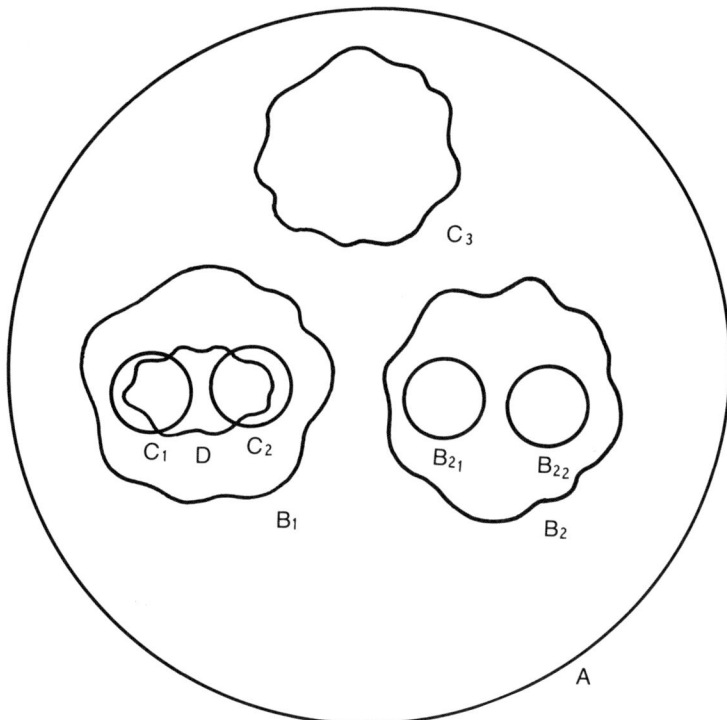

Fig. 14.2. Relationships among region types where A = a planning jurisdiction, B_1 and B_2 = river basins, B_{2_1} and B_{2_2} = river sub-basins, C_1, C_2 and C_3 = counties, and D = an airshed.

Closed and open systems. Import/export relations must be accounted for since no region operates as a closed system. Fig. 14.3 expands upon the simple closed system of Fig. 14.1 to indicate some of the areas in which imports and exports occur. If wastes are exported out of the region, the waste loads to be assimilated by the intraregional environmental media are obviously diminished. Similarly, the importation of goods and services serves to relax the danger of shortfalls in intraregional production. Any analysis that excludes import/export relations and attempts to define a region as entirely closed is bound to be arbitrary and unrealistic. Within the system, salvage and recycling of resources may also occur as a means of maintaining the resource base. Technology and price levels will be important determinants of recycling rates within a region.

CARRYING CAPACITY-BASED PLANNING

When carrying capacity breakdowns occur—e. g., when the demand for essential goods and services exceeds a region's productive capability and

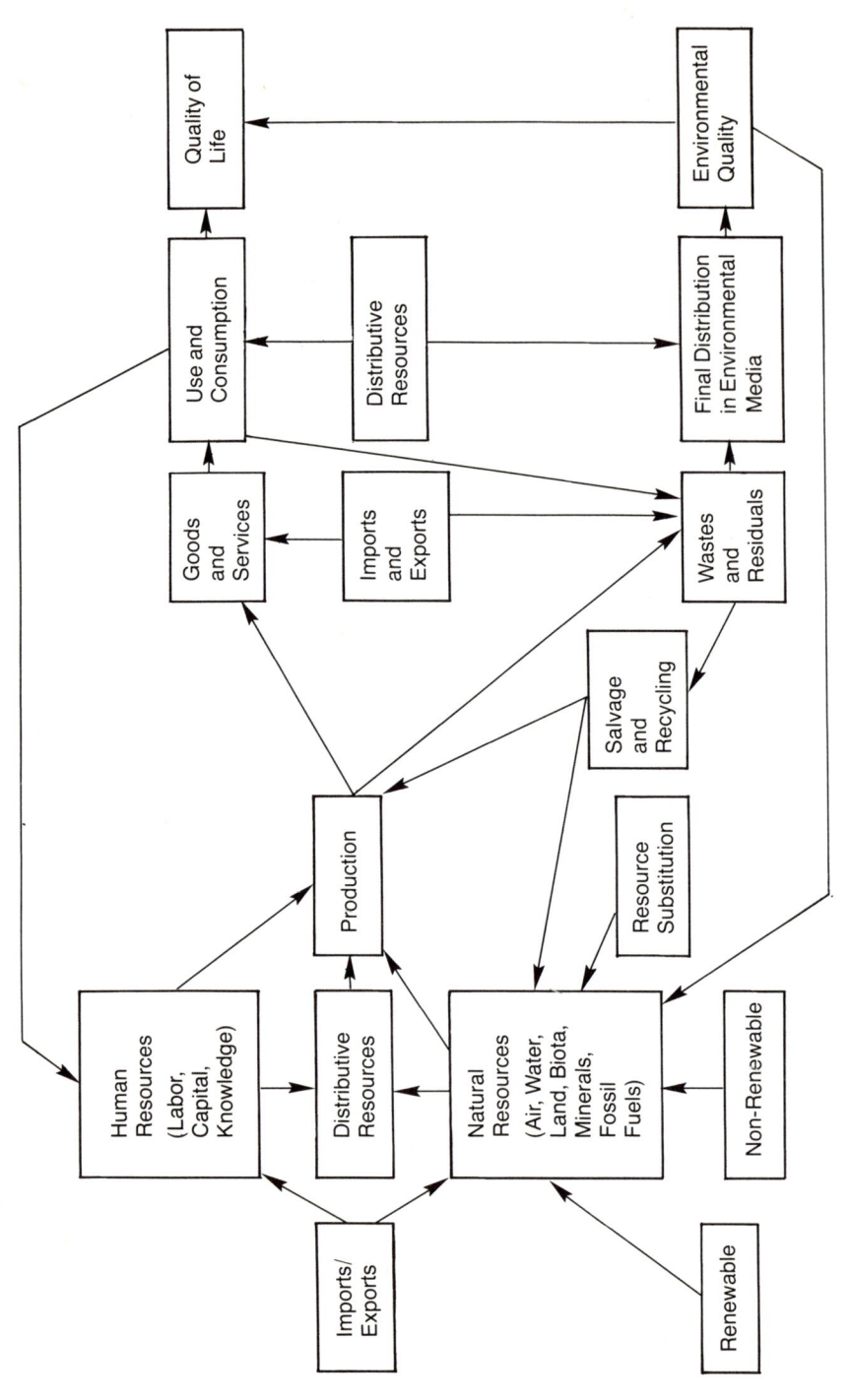

Fig. 14.3. Components of carrying capacity in an open regional system.

quality-of-life expectations are not satisfied; or when congestion and inefficiency in a region's infrastructure and distributive resources results in frequent and recurring blockages, stops, and delays; or when the atmosphere and
terrestrial and aquatic environments cannot assimilate additional wastes at
safe and acceptable quality levels—society as a whole suffers. Carrying
capacity-oriented planning seeks to guard against such breakdowns and the
not uncommon phenomenon of shortfalls in one resource area precipitating
shortfalls in other areas. Such planning would aim at maintaining a resilient
social order: a system capable of withstanding the forces of change and
providing in perpetuity the conditions of a healthy human and natural
environment.

A carrying capacity-based planning process is conceived as a means for
making the adjustments and trade-offs that would be necessary to direct
change in safe and socially useful directions. The conceptual basis for such a
process is suggested in Fig. 14.4. The carrying capacity concept described
earlier recognizes that in order to improve the "quality of life" relative to
both natural and human environments, the pattern and level of production and
consumption activities must be compatible with the capabilities of the natural
environment, as well as with social preferences. Recognizing that society is a
composite of a wide range of values and expectations, clearly conflicts (carrying capacity differentials) will exist between changes associated with manipulation of regional systems and the desired future conditions of the natural and
social environment (carrying capacity bases). To reconcile the attitudes and
expectations for the human environment and the quality and stability of the
natural environment, decision makers must analyze trade-offs, develop plans,
and implement controls that will equitably balance the two.

In operational terms, such planning would involve (1) identifying the
forces of change—present and future, regional and extra-regional—that will
probably result in significant regional changes; (2) analyzing how these driving forces will affect the quality and quantity of the region's material, productive, infrastructure and environmental resources and how such resource
changes will affect the basic carrying capacity relations identified previously;
(3) formulating alternative policies and programs for maintaining and
safeguarding these essential areas of carrying capacity; (4) analyzing and
evaluating the trade-offs that would be involved among the alternatives as a
basis for selecting the most promising course of action; and (5) designing an
implementation strategy. Keeping in mind the task of considering how such a
planning approach might be applied to the Colorado River Basin and its
subregions, a brief discussion of each of the above planning steps follows.

Identification of Driving Forces

The planner has the ongoing responsibility of identifying the events and
decisions that are having or might have serious and extensive impacts in the

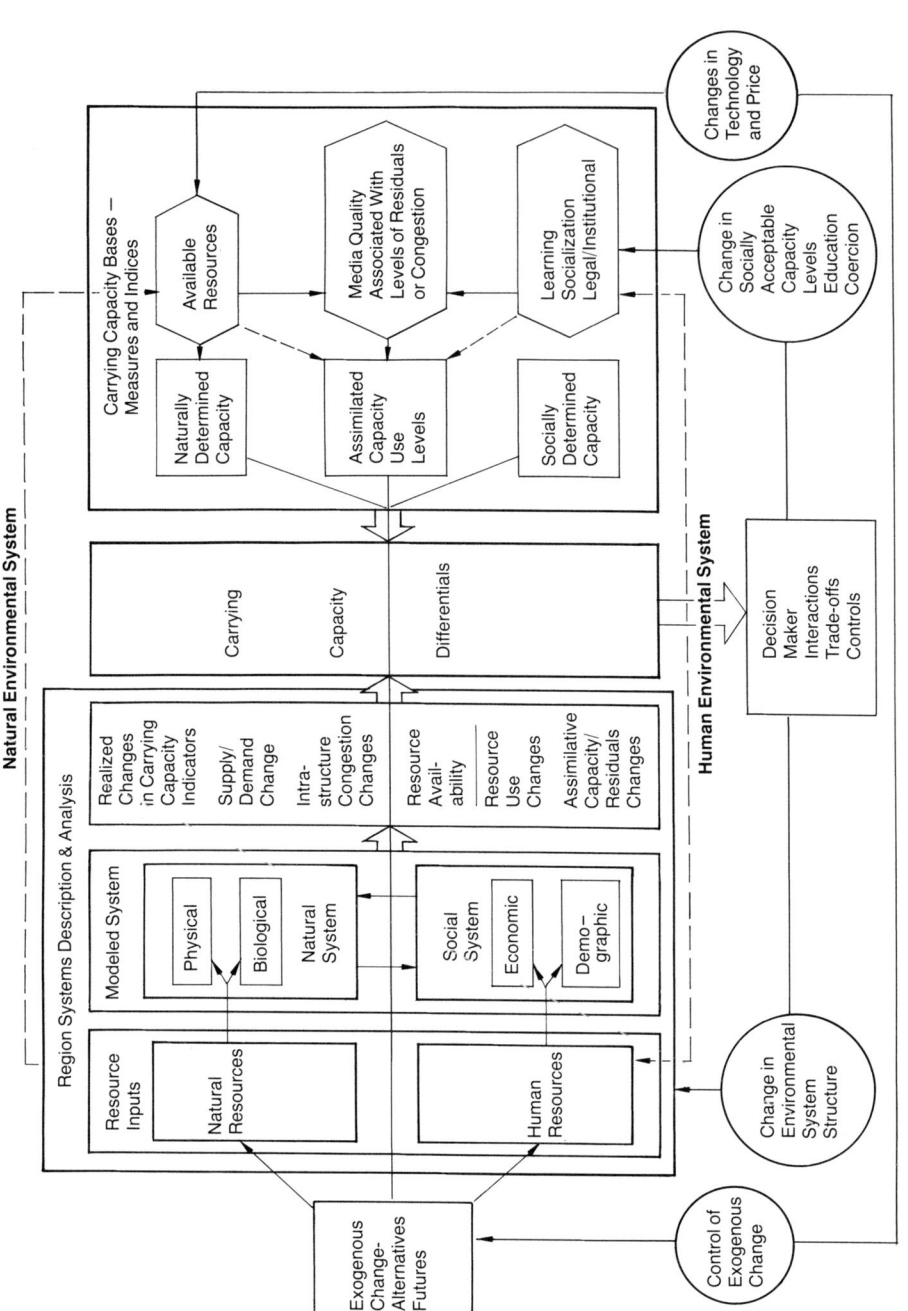

Fig. 14.4. The carrying capacity planning process.

area of his planning jurisdiction. These events and decisions might occur within the region, or they might occur outside the region and operate as exogenous influences. They might be considered individually, or, more realistically, they might be considered in various combinations. To characterize the various modes and levels in which driving forces might operate in the region, it is useful to develop descriptions of alternative futures. One future might describe a maximal growth situation: a situation in which one or more forces of change will result in intensive changes in extensive areas of the region. Another scenario might describe a minimal growth situation. In any case, a number of scenarios might be considered based on various assumptions about changes in driving forces and trends in the future. As time elapses and new and unforeseen driving forces enter the picture, updated scenarios will have to be constructed. Planning based on an ''alternative futures'' approach places an emphasis on identifying the effects of possible changes (as opposed to producing a forecast) and is aptly characterized as ''adaptive'' and ''contingency'' planning, focusing on the planning process rather than a ''plan'' as a static product.

Analysis of the Effects of Driving Forces on Carrying Capacity

Once the forces of change are characterized in alternative future scenarios, the planner has the task of estimating how the region's resources would be affected and how, in turn, changes would occur in the levels of goods, services, and amenities. Fig. 14.5 provides a summary overview of the ways in which such an analysis might be organized. The diagram suggests four primary levels of analysis, each representing a greater or lesser degree of spatial and temporal resolution. Beginning at a broad regional level, regional economic and demographic models would describe production, consumption, and resource exchange in terms of both requisite production inputs (natural resources, raw materials, infrastructure resources) and production outputs (goods and services, wastes and residuals). These four classes of resource state changes are then seen as inputs to other models and analytic techniques representing higher degrees of spatial and temporal resolution, including materials transformation models, distributive models, and impact models. The outputs from these linked models would be cast as changes in capacity indices vis-à-vis the four general areas of carrying capacity relations, viz., supply and demand, resource use and resource availability, infrastructure and congestion, and waste loadings and assimilative capacity.

The kind of analysis envisioned is described by the following four questions which are germane to a carrying capacity-based planning process: (1) How would x (a driving force in an alternative future) affect a region's capacity to produce acceptable levels of goods, services, and amenities? (2) How would x affect the capacity of the region's environmental media to

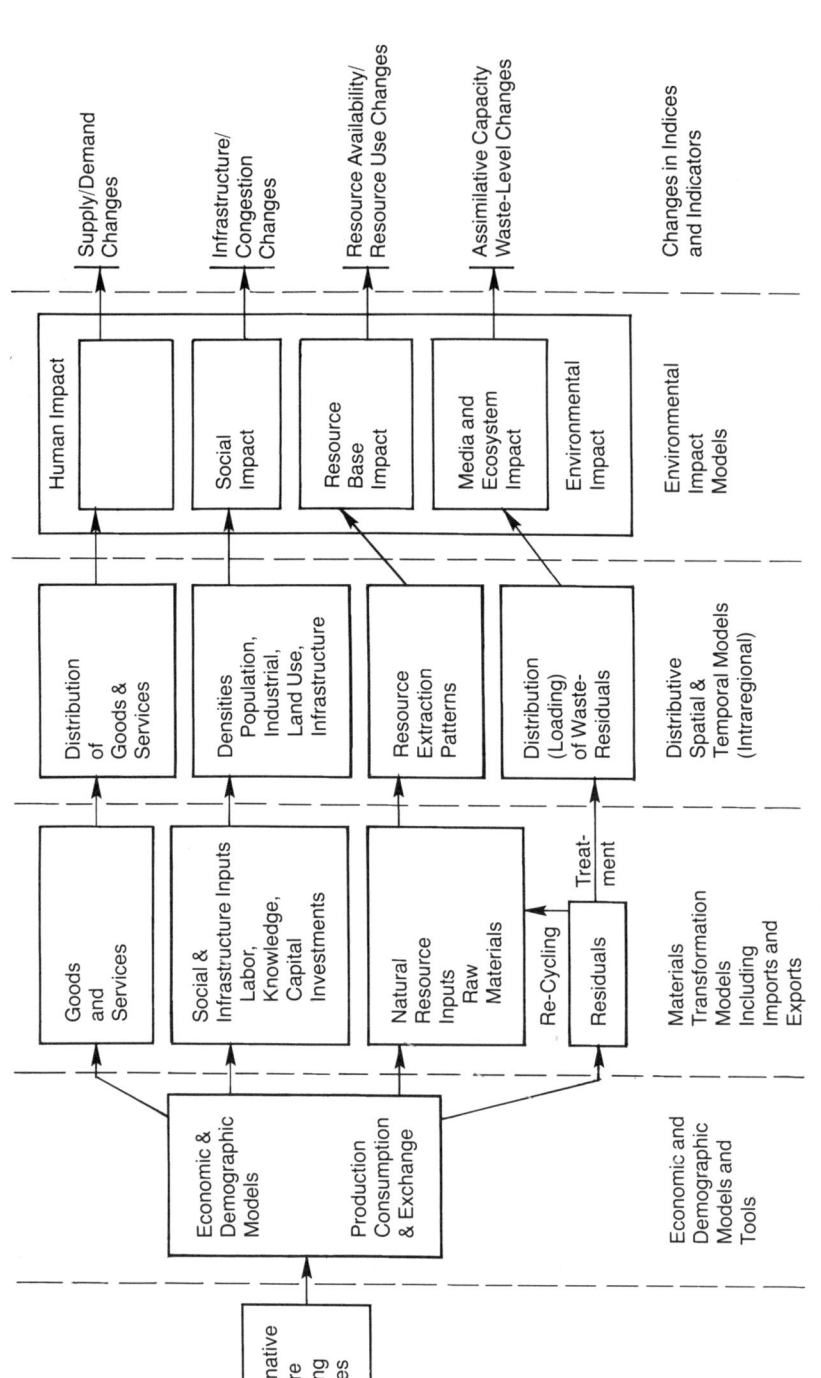

Fig. 14.5. Relation of modeling and analytical techniques for carrying capacity.

maintain acceptable quality levels? (3) How would x affect the capacity of the region's natural and human resource base to maintain requisite levels of production? (4) How would x affect the capacity of the region's infrastructure resources to distribute production inputs (raw materials, etc.) and outputs (goods and services, wastes and residuals) efficiently, i.e., in ways that maintain or do not diminish each of these other capacities? The interdependencies in analyzing carrying capacity are also illustrated since question (1) leads to question (3), and questions (1), (2), and (3) all lead to question (4). Indices are required to provide summary answers to these questions—answers that would be comprehensible and useful to policy and decision makers. Although a state-of-the-art discussion of index construction cannot be undertaken here, a few of the promising efforts to answer these complex and general questions should be mentioned briefly.

The Council on Environmental Quality (CEQ) has sought to develop indices in the area of the natural environment that will facilitate answers to question (2) above. Under CEQ contracts, several air quality indices have already been developed and other indices are being developed for other areas of environmental concern, notably water pollution, pesticides, toxic substances, land use and wildlife. For each of these areas, CEQ has identified what it assumes are critical indicators. If satisfactory indices can be developed for the above areas of environmental concern, intermediate level indices could presumably be constructed and aggregated into an overall environmental quality index. Relying in part on the lead of CEQ, the County of San Diego plans to develop an environmental quality index and has made substantial progress in developing requisite subindices. These efforts offer hope for the kind of carrying capacity analysis suggested here. With satisfactory techniques for calculating environmental indices and (minimal) standards with which they can be compared, a basis will exist for assessing the effects of alternative futures on one important carrying capacity relation.

Similar efforts have been directed toward the development of indices relating to the social environment. The Technical Committee's effort in 1971 to develop the TECHCOM is a good example of one such project. In this study, an interdisciplinary task force identified nine general social goals, disaggregated these goals into several tiers of subgoals, and identified social indicators under each lowest-level subgoal. This hierarchical array is intended to approximate what is meant by the phrase "social well being." To illustrate, the tentative disaggregation of one of the general goals (the economic opportunity goal) is provided below:

Given (1) values to represent the changes that would result in the social indicator set from the implementation of actions described in an alternative future scenario, and (2) indexing functions for relating changes in social indicators to changes in the goal set, TECHCOM is being developed as a methodology for comprehensively analyzing the trade-offs that the alternative future would involve. By comparing these computed indices (at various levels of abstraction) with acceptable standards, an effective means would exist for answering the question: How would x affect a region's capacity to maintain acceptable quality of life?

A. Present standard of living
 1. Median per capita income
 2. Prices of goods and services
  —Cost of Living Index*
  —Consumer Price Index
 3. Quality of goods and services
  —Repair costs per capita as a percent of purchase price
  —Cars recalled annually as percent of total
 4. Selection of goods and services
  —Percent change in the number of new patents issued
  —Retail employees per capita
  —Retail per capita sales receipts
 5. Leisure time
  —Average weekly working hours
  —Per capita receipts of amusement and recreation service establishments
  —Per capita attendance at state parks
  —Per capita sales of hunting and fishing licenses
 6. Stability
  —Percent growth rate of per capita income
  —Inflation rate
  —Unemployment rate
  —Business failures as a percent of the total number of businesses
B. Future standard of living
 1. Employment potential
  —Employment growth rate (percent)
  —Unemployment rate (percent)
  —Net migration as a percent of total population
  —Median education level (years)
  —Median income growth rate (percent)
 2. Savings and investment potential
  —Economic growth rate (percent)
  —Population growth rate (percent)
 3. Retirement potential
  —Social insurance contributions per capita
  —Private insurance contributions per capita

C. Equality of economic opportunity
   —Gini coefficient for income distribution by income class
   —Median education for ethnic groups
   —Employment rate for ethnic groups
   —Ratio of female unemployment rate to male unemployment
     rate

*The items prefixed by dashes (—) are social indicators.

Quality-of-life parameters and standards can be defined in a simpler conceptual model than the TECHCOM example. For purposes of analyzing carrying capacity, it might be preferable to work with social minimums, i.e., the concept of minimally acceptable quality of life.

An approach for analyzing the effects of alternative futures on the carrying capacity of a region has been suggested in outline form. Only two of the four dimensions of carrying capacity have been examined in any detail, and this discussion has centered on the output side of production. How economic, demographic, materials transformation, and distributive models might be utilized to analyze the other two carrying capacity relations and to output the kinds of information required for calculating environmental and social indices has only been placed within a conceptual framework, the development and testing of which are of course subject to further research.

Formulation and Implementation of Strategies
for Managing Change

Given an understanding of how regional carrying capacity would be affected under alternative futures, the planner's first consideration in formulating strategies for managing change should be that of insuring that serious carrying capacity failures or breakdowns do not occur. This will involve guarding against resource shortages and depletions; wasteful and extravagant consumption; the unwarranted siting and distribution of population, commercial, and industrial centers; congestion and inefficiency in transportation and other distribution and delivery systems; pollution levels that virtually eliminate the capacity of environmental media to safely assimilate additional pollutants; and the lack or inability of institutional means to augment, mobilize, and redirect resources on short notice and to prevent sporadic shortfalls from reaching crisis proportions and causing a rash of shortfalls elsewhere in the region. The provision of such safeguards is, of course, a complex and difficult matter. Besides the institutional inertia that blocks the implementation of even the most well-conceived plans, the planner must know how the kinds of carrying capacity safeguards just enumerated could be provided through the technological, economic, educational, and legal controls that might be brought to bear. Besides knowing how conditions in the regional system might be changed through the operation of driving forces, the planner must know how these forces can be managed (channeled, counteracted, forestalled, damped, or whatever) in appropriate ways through the application of available controls. This requires an understanding of complex cause and effect relationships and a modeling capability for analyzing these relationships at various scales of temporal and spatial resolution. The blocks in Fig.14.4 represent areas in a regional system in which various combinations of controls operate.

CARRYING CAPACITY-BASED PLANNING ISSUES IN THE COLORADO RIVER BASIN

Using the planning approach outlined above, we will make some observations and comments about carrying capacity in the Colorado River Basin. Although most of the discussion will deal with the Colorado River Basin as a whole (it recognizes that the urbanized areas of Denver, Salt Lake City, and southern California will also have significant impact), reference will be made as well to the upper and lower basin, the urban areas of Phoenix, Tucson, and Las Vegas, and potential growth centers adjacent to oil shale developments.

Seven important areas in which the region is likely to experience significant development or decline are urbanization, Four Corners coal, Green River Formation oil shale, recreation, environmental quality, agriculture, and water use. Whether each of these areas experiences development or decline, it is worth noting that many of the forces of change will likely be initiated from outside the region. Perhaps this is not surprising if one considers that nearly half of the land in the basin is publicly owned, many of the region's basic industries are export-based, and the region is a vast storehouse of energy resources that the nation and the world as a whole could draw upon to alleviate acute energy shortages. In this context, a carrying capacity-based planning approach seems to be entirely appropriate. Even if change in the basin stems largely from exogenous forces, future contingencies can be anticipated and dealt with in a management frame that seeks to prevent carrying capacity failures and the hardships that attend such failure.

To illustrate the kinds of carrying capacity issues and problems that may call for resolution in planning for the future of the basin, the relationships of a few very gross driving forces, environmental constraints, and regional outputs were examined in a simple simulation. Consistent with the planning approach described in the last section, three alternative futures were constructed around the following eight variables: urbanization, Four Corners coal development, oil shale, recreation, environmental quality, agriculture, water development, and gross regional product. A modified KSIM simulation technique was used to generate thirty-year time plots for the eight variables for three alternative futures (see Figs. 14.6, 14.7, and 14.8).

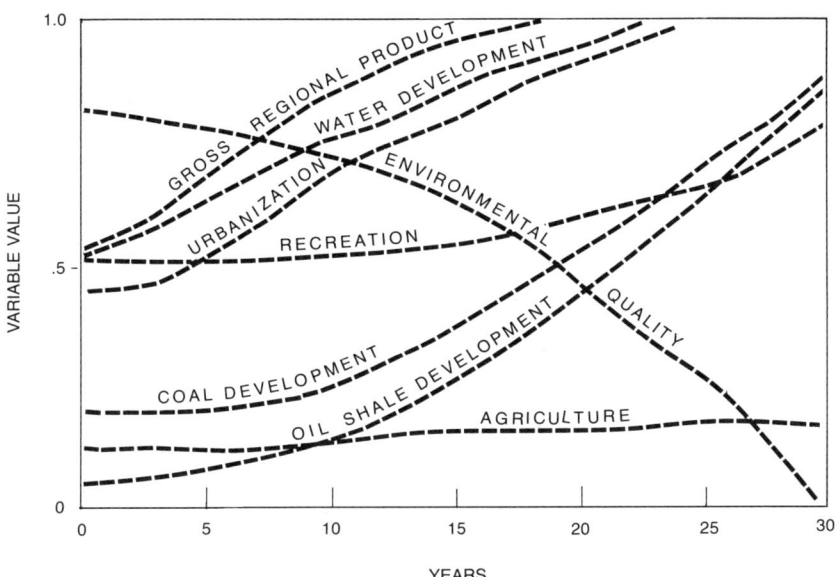

Fig. 14.6. TREND FUTURE: the result of continuing
the relationships that are assumed to exist currently.

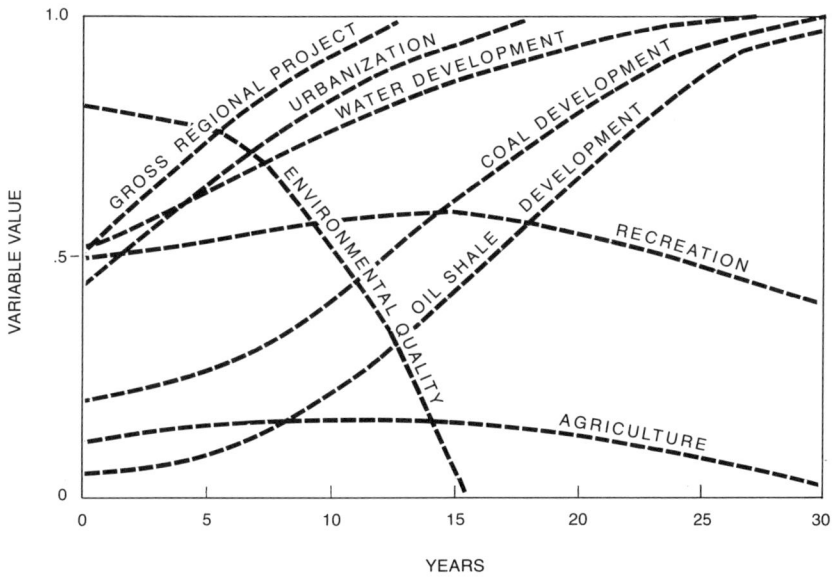

Fig. 14.7. ENERGY DEVELOPMENT FUTURE: the result of relationships
assumed to exist under a strong energy development policy orientation.

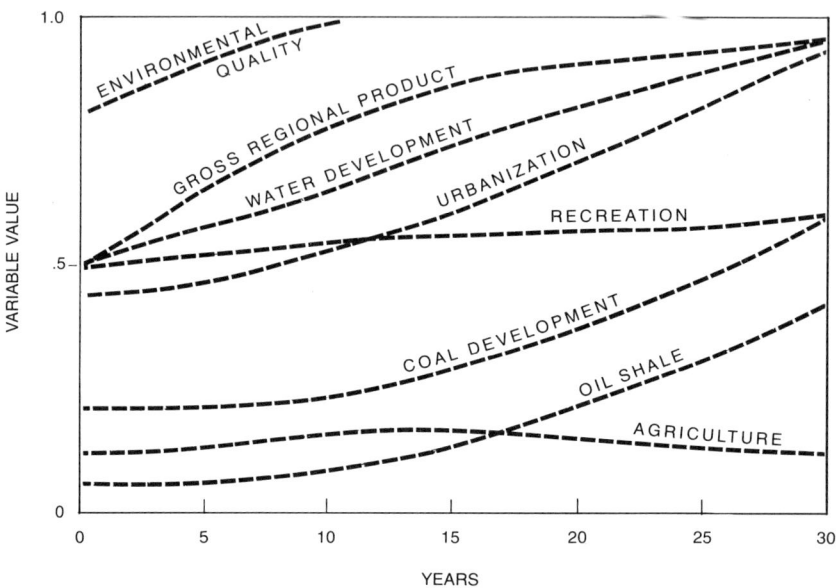

Fig. 14.8. ENVIRONMENTAL PROTECTION FUTURE:
the result of relationships assumed to exist under
a strong environmental protection policy orientation.

Briefly, the KSIM technique (Kane, 1973) involves (1) scaling the variables of interest to initial conditions (i.e., conditions estimated to prevail at the time of simulation); (2) developing a cross-impact matrix to describe direct causal relations among the variables; and (3) using KSIM mathematics to convert the interaction matrix into time plots for each variable. A different cross-impact matrix was developed to characterize the interactions that would likely occur under three different policy orientations. The first sought to describe the "policy climate" of the recent past (*Trend*), the second that of a strong energy development policy orientation (*Energy Development*), and the third that of a strong environmental protection policy orientation (*Environmental Protection*).

Table 14.1 contains estimations of initial conditions of the selected variables. These values were obtained by (1) assigning the value of 1.0 to an estimated maximum potential for each variable over a thirty-year time frame and (2) scaling the present level of each variable with respect to the maximum.

A seven-point scale was used to describe the direction and magnitude of direct causal relations in the cross-impact matrices. Our estimations of causal relations under the three policy orientations are contained in Table 14.2. These were +3 = strong positive impact, +2 = moderate positive impact, +1 = slight positive impact, 0 = null or indeterminate impact, −3 = strong negative impact, -2 = moderate negative impact, and -1 = slight negative impact.

These alternative futures are, of course, based on rough estimates of initial conditions and variable interactions, and are only meant to stimulate and provide a general basis for discussion and future investigation of carrying capacity considerations under various policy orientations. Used in this vein, the three alternative futures pose the question, "What will happen if...?" and the remaining discussion is exemplary of the types of analysis needed to identify areas of potential carrying failure and point out some of the trade-offs and issues that would be involved in preventive action.

TABLE 14.1

Assignment of Initial or Baseline Values to the State Variables

	Variables	Initial Conditions
A	Urbanization in the CRB	0.4
B	Four Corners Coal Development	0.2
C	Oil Shale Development	0.05
D	Recreation in the CRB	0.5
E	Environmental Quality in CRB	0.8
F	Agriculture in CRB	0.1
G	Water Development in CRB	0.5
H	Gross Regional Product	0.5

Three carrying capacity areas are singled out for discussion in relation to the other variables: (1) the capacity of a nongrowing agricultural industry to supply intra- as well as extra-basin demands for agricultural products; (2) the capacity of the basin's environmental media to maintain acceptable quality levels in view of rapid urban and energy resources development; and (3) the capacity of the basin's water resources to support such development.

Agriculture

With reference to the first carrying capacity area, all three alternative future scenarios portray agriculture as experiencing little or no growth over the next thirty years. In both the *Energy Development* and *Environmental Protection* scenarios, in fact, a gradual decline occurs. Assuming that proportionately greater amounts of capital and water will be allocated to urban and energy resources development than to agriculture—increasing amounts of capital and water will be available, according to the

TABLE 14.2

Estimation of Causal Relations Between Each Pair of State Variables Under the *Trend, Energy Development,* and *Environmental Protection* Policy Orientations

	A	B	C	D	E	F	G	H
A	1	2.5	1.5	1.5	−2.5	−1.5	1.5	1.5
	1	2.5	0	1.5	−2.5	−1.5	1.5	1.5
	1	1	0	1.5	−1	−1.5	1	1.5
B	1.5	0	− .5	−1.5	−2.5	−1.5	2	2
	1.5	0	0	−3	−3	−2	2.5	2.5
	1	0	− .5	−1.5	−1.5	−2	2.5	2.5
C	1.5	0	0	−1.5	−2.5	−1.5	2	2.5
	1.5	0	0	−3	−3	−2	2.5	3
	1	0	0	−1.5	−2	−2	2.5	3
D	.5	.5	.5	0	−1	−1	− .5	1.5
	.5	.5	.5	0	−1	−1	− .5	1.5
	.5	0	0	0	− .5	−1	− .5	1.5
E	−1	−2	−2	1.5	.5	.5	− .5	−1
	.5	0	0	2	2	1	.5	0
	0	0	0	2	2	1	.5	0
F	− .5	−1.5	−1.5	0	−1.5	.5	2	1
	−1	−2	−1	1	−1.5	1	2	1
	−1	−2	−1	1	−1	1	2	1
G	1.5	2	2	1	−2.5	1.5	0	2.5
	1.5	−2	2	1	−2.5	1.5	0	2.5
	0	1	1	1	−1	1.5	0	2
H	1.5	1	1	1.5	−1.5	1	1.5	.5
	1.5	1	1	1.5	−1.5	1	1.5	.5
	0	0	0	1.5	2	1	1.5	.5

Trend scenario—the decline of agriculture is not surprising. The problems which this situation poses are: Will stable or decreasing agricultural production create agricultural product shortages within the basin? Will agricultural production levels fall short of the national demand? Many observers answer in the affirmative to both of these questions and are predicting serious shortfalls in the agricultural products industry. For example, in viewing the problem as national and international as well as regional in scope, Thomas points out in Chapter 8 that the Colorado River Basin is one of the great agricultural land reserves in the nation and that the basin's great potential for agricultural production should be viewed as an invaluable national resource. It has been estimated that the total number of irrigated acres in the basin in 1960 was 2,638,164. The total land suitable for irrigation is estimated to be 46,818,600 acres. Other important carrying capacity questions are: Will development in non-agricultural sectors significantly diminish the basin's potential for future agricultural development? Will agriculture in the basin be capable of responding in a timely way to a rapidly increasing demand for food and fiber? Certainly these are questions that should be included in management and resource allocation decisions.

Environmental Quality

Trend forbodes ill for environmental quality in the basin, but not noticeably so until the last five years of the thirty-year time horizon. Large capital investments will be made in projects and programs relating to energy resources and urban development. Not only will this result in proportionally small investments in environmental protection, but also an accelerating development in urban and energy producing activities will generate increasing pollution levels. (Under *Energy Development*, this situation is intensified and environmental quality "bottoms out" in fifteen years.) Although the repercussions of diminishing environmental quality would apparently not be reflected in the other variables in any significant degree during this time frame, there seems to be little doubt that various media quality standards would be exceeded in thirty years. (Under *Energy Development*, violations of standards would occur in fifteen years.)

While it is informative to examine the agriculture production/demand for agriculture products relation at an aggregate regional level, it is obvious that the pollution load/media quality relation has to be analyzed at much higher levels of spatial and temporal resolution. Will increasing urbanization in the counties of Maricopa (Phoenix area), Pima (Tucson area), and Clark (Las Vegas area) create serious air pollution problems during peak travel times and during inversions? Will stack emissions violate air quality standards in the immediate airsheds? Although the problem of salinity in the Colorado must be viewed in a total regional perspective, the question "Does the level of salinity exceed 850 ppm?" has spatial reference to conditions at Lee's Ferry, Hoover Dam, Imperial Dam, and other specific locations in the river system.

Water Supply

The third carrying capacity relation which merits careful consideration is the capacity of the basin's water supply to support urban and energy development. *Trend* suggests that water development will reach or exceed the development potential in approximately twenty-five years. In 1960, approximately 97 percent of consumptive water use in the basin occurred in the agricultural sector. If the basin's urban population increases 253 percent (from 1.7 million in 1960 to 4.3 million), by the year 2000 as some predict, municipal water use should increase more than threefold. The use of water in energy development will be even more dramatic under *Trend* (and certainly under *Energy Development*). A million barrel a day oil shale industry would consume approximately 150,000 acre-feet of water per year and the amount of water used for coal development (steam, scrubbing operations, dust abatement, revegetation, etc.) is not expected to be less. The question which all this poses is whether the limited supply of water is adequate to meet the demand for development. Since the potential for additional water development in the basin will probably diminish to zero within thirty years, and with water use in the lower basin already exceeding entitled amounts, carrying capacity failure vis-à-vis water supply appears almost certain. Augmentation of supply through importation or weather modification would seem to be a key factor in future planning and management.

Other Capacity Areas and Analyses

The foregoing discussion has briefly suggested a few of the possibly critical areas of carrying capacity in the Colorado River Basin in relation to three alternative future scenarios. Other capacity areas could also be examined, of course, such as changes in recreation levels and types with respect to increasing urbanization and declining environmental quality under *Trend* and *Energy Development,* or the capability of regional infrastructure (highways, pipelines, power transmission lines) to support urbanization and expansion of gross regional product.

Such a preliminary survey of possible problems might be used by planners as a guide in undertaking more detailed studies to determine which sets of contingencies described in the scenarios are most likely, in collecting reliable data for examining key carrying capacity relations in greater detail, and in determining which shortfalls might critically affect the conditions of the quality of life in the basin's human and nonhuman environments.

SUMMARY

In developing environmental management strategies for the Colorado River region, planners and decision makers must continually assess the social and environmental implications of alternative proposals. Recognition of capacity limits in regional activity support systems along the lines of the carrying capacity planning process described could provide decision makers with a workable approach to assessing the natural and human viability of proposals.

Indices are being developed as a means of providing a working knowledge of environmental quality, and of charting trends and changes in quality levels. The development of carrying capacity concepts could help to extend the usefulness of these indicators into the realm of making comparative evaluations of environmental quality dimensions in terms of ranges and limits of acceptable levels, and the impact of various regional growth policies.

Regional environmental management which incorporates a carrying capacity planning process could thus be used to examine the character of changes that will occur under different levels of social and economic activity, how changes in the physical environment relate to the social objectives and values for resource development and use, and whether such changes are within acceptable limits of environmental and social carrying capacity.

Literature Cited

Findley, Rowe. 1973. The bittersweet waters of the lower Colorado. *National Geographic* 144(4): 540-569.

Kane, J., Vertinsky, I., Thompson, W. 1973. KSIM: a methodology for interactive resources policy simulation. *Water Resources Research* 9(1): 65-79.

15

Toward a Natural Resource
Management Policy

JACK M. CAMPBELL

Against the backdrop of the nation's emerging long-term concern for land use, water quality, and energy, the time has surely come to explore new possibilities for improved management of natural resources within the Colorado River Basin. The Federation of Rocky Mountain States is a regional organization of the states of Montana, Idaho, Wyoming, Colorado, Utah, and New Mexico. The governors, together with private sector and academic representatives, serve on the board of directors and participate, with others, in the Federation's seven operating councils: Regional Planning, Telecommunications, Human Resources, Transportation, Market Development, Arts and Humanities, and Natural Resources. Although four member states—Wyoming, Colorado, Utah, and New Mexico—have areas lying within the Colorado River Basin and are known as "states of the upper division," the Federation does not address matters directly related to water.

The Federation believes that cooperation should begin at home, between states and between regions. At a time when international cooperation is becoming an increasing imperative of a sane and livable world, a parochial point of view is, clearly, no longer appropriate. Sheer numbers of inhabitants in the United States now demand far greater regional and interregional cooperation than when the nation was candle-lighted two hundred years ago. The complex nature of the nation's problems compels the citizens of the region to look beyond state boundaries without violating state sovereignty and public responsibility. They live in a light-bulbed, flush-toileted, kitchen-sinked, inanimate-energy-activated, mobile, mass civilization which they cannot escape, even should they wish to do so, without inviting social disaster. Rather than losing that civilization, it should be refined, with constant environmental awareness.

As a former public official of a single state of the mountain West and now, as President of the Federation of Rocky Mountain States, I cannot speak for all of the individual states with regard to the future of the Colorado River Basin. Two of the six Federation states are not within the basin. Three basin states are not members of the Federation. As has been stated, there are sharp differences of opinion on a wide range of matters of management and administration between the upper and lower basin states and there are international complications as well. However, I believe that the Colorado River Basin with its vast, unoccupied, mostly arid lands, can become a new national treasure in its own right and not just an exporter of food, fiber and energy resources.

None of the vast new energy potentials within the Colorado River Basin will ever come to full fruition without careful conservation and management of water. Western energy resources and water are like Siamese twins; they are inseparably joined together. Water is needed for the production of energy, from whatever source, and for the people who produce the energy.

RESOURCES OF THE BASIN

Before attempting to address the matter of a management process for the Colorado River Basin, which will suggest some new arrangement, one needs to put into perspective the magnitude and importance of the natural resources within the basin.

Energy Potentials

The energy situation has brought about an intense national interest in the energy resources of the Colorado River Basin. These resources include the region's oil and natural gas (with their derivatives), its coal and oil shale, its potential for solar energy and its potential future geothermal sites. The current energy short-fall stems partly from the lack of an organized and coordinated program of short- and long-term research and development in the 1960s and early 1970s. There has been, in a time of assumed surplus, a failure to recognize the fundamental importance of energy to our civilization, a need for better management arrangements in order to understand the impact of the present decisions upon future consequences, and the development of wasteful practices in the use of depletable resources.

In order to understand the new pressures upon the basin, the impact of future energy fuel requirements on the Colorado River Basin states must be examined with particular attention to water requirements. This must be done with an awareness that land, water, and energy, in relation to the quality of life, cannot be separated.

Oil and gas. The states of Wyoming, Colorado, Utah, New Mexico, and Arizona contain substantial additional reserves of oil and gas, which are either totally undeveloped or can be recovered from existing wells by secondary and tertiary methods. Particularly, if secondary and tertiary methods of production through water flooding are accelerated, water will be required.

Whatever decisions may be made with regard to the acceleration of recovery of oil and gas reserves lying within the Colorado River Basin, the rising costs of both domestic and imported oil and gas quite obviously will generate increasing pressures for the use of coal and oil shale, both of which are found in abundance in the Colorado River Basin.

Coal. The vast low-sulphur coal reserves in Wyoming, Colorado, Utah, New Mexico, and Arizona inevitably will be the object of serious pressure for development to meet the short-term requirements for energy in the United States. Much of this coal can be strip-mined and the environmental/economic trade-offs must be very carefully examined. Mining of this coal and reclamation of land following it, will require some water. New technology, which permits the transportation of coal by means of slurry pipeline could put some added pressure on the water supplies of the Colorado River Basin. Coal development, like oil shale development, will attract more people and, perhaps, industry to the Colorado River Basin with further impact on water supply.

Several coal gasification plants are planned for construction in the Four Corners area of the basin and coal gasification may very well become one of the major uses of the region's coal as its oil and gas reserves decline. This view is reinforced by the fact that energy in the form of gas can be transported more cheaply than in the form of coal and that air pollution control in a coal gasification plant is more easily accomplished than in a coal burning power plant.

Oil shale. Parallel with increased activity in the production of oil and gas in the basin and the acceleration of coal production, there is already rising pressure for development of the vast oil shale reserves in Wyoming, Colorado, and Utah. Very high bids, in January and February, 1974, for 5,000 acre leases of federal lands containing oil shale deposits demonstrated the enormity of this energy fuel potential. The development of these oil shale reserves will require very substantial amounts of water in the production and retort process and will have a major effect upon the economic/environmental balance in this portion of the Colorado River Basin. Perhaps just as significant will be the tremendous influx of people and industries associated with the oil shale development. Here again, there will be increased need for water supplies.

Uranium. Paralleling the oil, gas, coal, and oil shale developments will be increased reliance upon nuclear energy. The Colorado River Basin states of Wyoming, New Mexico, and Utah contain most of the developed uranium resources of the country. If all nuclear power plants in the planning state, under construction, and partially operational become fully operational, the known, economically recoverable, uranium reserves in these states will not be adequate to supply the necessary fuel for the plants. This situation is accelerating the exploration for uranium and increasing its price. Mining and processing uranium also requires the use of water. Inasmuch as a substantial portion of the Pacific coastline appears unsuitable for nuclear power plants for a variety of reasons, there will be pressures to locate inland power plants to supply the metropolitan areas of southern California with adequate electric power; particularly, in the desert areas of southeastern California.

Geothermal. Technological breakthroughs in the development of geothermal energy add another, longer-range dimension to the production of needed energy from the Colorado River Basin area. Some potential sites lie within the boundaries of the basin, particularly in the Imperial Valley and Rio Grande Basin where Colorado River water is being used. Research is being conducted which involves injection of water into holes drilled into hot rock bodies lying near the surface to fracture the rock in order to improve percolation and generate steam which can be withdrawn and utilized for the production of electric power. Subterranean sources of heat, particularly in the southwestern part of the United States, generally have been estimated to have a potential for supplying almost twice the power the United States currently obtains from all hydroelectric sources. But here again, some water may have to be consumed.

Solar. Finally, continuing research must be vigorously pursued for the development of solar energy for the production of electricity to be used within and without the basin, and for the electrolysis of water to produce hydrogen as a fuel, as an industrial chemical for use within the basin, and even for exporting out of the basin. One of the more attractive ideas for making solar energy production practicable during times when the sun is not shining is to use a part of the electrical energy to create hydrogen to be stored and later fed back into the same steam boiler that produced the electricity in the first place, and to produce oxygen as an important industrial chemical by-product.

Rapid acceleration of research and development in the large-scale use of solar energy for production of electricity seems to be at least as feasible as the development of the nuclear fusion process which is encountering substantial technological difficulties. However difficult, it is imperative that both of these processes emerge as soon as possible to provide almost inexhaustible supplies of new energy. Even in the fusion process the Colorado River Basin area will continue to be deeply involved.

MANAGEMENT DIFFICULTIES

Discussion of energy fuels was important for the obvious reason that the nation's determination to become at least "security-sufficient" through the use of its own energy fuels is a major and current driving force that will not be denied. The impact of this force upon the general political, social, and economic conditions of the Colorado River Basin states is of great importance. But one must not lose sight of the fact that unless wise planning and action take place, new shortages of food and fiber and many non-energy fuel minerals and metals which abound in these states will occur. Nor should one forget that the basin is not only the largest national "storehouse" for food, fuel, timber, minerals, and metals, but is also the "playhouse" for those who live there and the many who tour there to enjoy, for a moment, the quality of life that basin residents have found.

In all of this, water is the critical ingredient. Community wars have been fought and men have died because of it or because of lack of it. It is perhaps the region's most highly sensitive political matter and only the very brave among political decision-makers even mention it unless particular circumstances compel them.

Part of this sensitivity is a result of very deep traditional roots for management of the basin. Any institutional changes are difficult, but particularly difficult are those which affect water in the West. The Colorado River Basin has, for many years, been considered in two distinct parts—an upper basin and a lower basin. It has been managed in this fashion, often with confrontation and always with some degree of suspicion between the two areas. Through the years, even this division has been complicated further by obligations to the Republic of Mexico.

Social emphasis upon environmental protection, the use of much of the land in the basin for recreational purposes, and the pressures of population growth have added dimensions heretofore only lightly considered. The impact of the National Environmental Policy Act, the Water Resources Planning Act, the National Water Commission Act, the Federal Water Pollution Control Act, and the developing National Land Use Policy Act, have strong implications for the future management of the Colorado River Basin.

To compound the problem of effective management of the basin, one must recognize that there are in the mid-seventies at least twenty-one federal departments, bureaus, or agencies which have operational responsibilities and authority. There are also at least fifty state agencies among the seven river basin states which have some degree of responsibility or authority. A partial list of these federal and state agencies is attached as Appendix A. With respect to water, it seems obvious that no major structural reorganization of the Colorado River Basin can take place in the immediate future without very serious repercussions to existing organizations or institutions having some measure of management responsibility.

MANAGEMENT ARRANGEMENTS

One is sorely tempted to leap all of the very difficult political and institutional questions and suggest an ideal management system incorporating the latest theories in organizational management; however, this would be neither practical nor useful. All institutions, particularly those dealing in public matters, must accommodate to the real world. The more than 400 years which have elapsed since the discovery of the Colorado River have resulted in institutional roles and in decisions or non-decisions which have had, and continue to have, substantial impact upon what may be accomplishable now and in the future. In this country there is an emerging new set of social values, or at the very least, social values are being assigned new priorities as society changes. Of course, any institution must accommodate to these changes, but accommodation can come about only if the practical implications of institutional change are recognized and understood. At least in the West, there appears to be no institutional mechanism less amenable to change than the management and allocation of the region's limited water supplies. By the same token, the importance of water as an important natural resource in the Colorado River Basin compels accommodation within a reasonable time-frame.

A new phrase descriptive of an approach to natural resource management policy for the Colorado River Basin might be "accelerated evolution." While this may seem to be a contradiction of terms in itself, it is an appropriate description. Compelling developmental pressures on the basin do not permit the luxury of time which may be required if evolution is to take its natural course. On the other hand, to attempt to leap immediately to a substantially changed management arrangement in the basin seems quite impractical.

Four inescapable historic events have a profound bearing upon what can be done to provide new arrangements for improved administration of the Colorado River Basin:

1. At Bishop's Lodge, in Santa Fe, New Mexico, itself a historic site, a small plaque commemorates the signing of the Colorado River Compact on November 24, 1922, by representatives of the upper and lower basin states and federal representative Herbert Hoover. The Compact is discussed more fully in chapter 2.
2. On February 3, 1944 a treaty with Mexico was signed, which was

ratified on November 8, 1945, allocating and providing for annual delivery to Mexico.

3. On October 11, 1948, again in Santa Fe, the states of the upper basin entered into the Upper Colorado River Basin Compact. After setting aside 50,000 acre-feet for use by Arizona in that small part of Arizona within the upper basin, the remainder of the water was allocated by authorizing 51.75% to Colorado, 23.00% to Utah, 11.25% to New Mexico, and 14.00% to Wyoming.

4. The lower basin states of Arizona, Nevada, and California were unable to reach an agreement on allocation of waters of the Colorado River and in 1952 Arizona filed suit in the Supreme Court of the United States, asking the Court to determine that allocation. This suit continued for some 12 years before a final decision was reached allocating the waters of the lower basin in 1964 (*Arizona vs. California*, 367 U.S. 340).

It is clear from these four critical events that the water from the Colorado River Basin is allocated through four different arrangements:

a. Between the upper and lower basin states by the Colorado River Compact of 1922

b. Between the United States and Mexico by an international treaty of February 3, 1944

c. Among the states of the upper basin by an interstate compact dated October 11, 1948

d. Among the states of the lower basin, by a decision of the Supreme Court of the United States on June 3, 1963, followed by a final decree dated March 9, 1964

The Colorado River Storage Project Act (70 Stat. 105) of 1956 and the Colorado River Basin Projects Act of 1968 must also be considered. The Colorado River Basin Projects Act was the result of an unusual degree of seven-state effort marked by confrontation and cooperation. It typified the adversary system at work but demonstrated that united efforts and a willingness to negotiate and compromise can bring constructive results. This experience may have been the forerunner of basin-wide endeavors to solve problems in the area on a regional scale. One of the most significant impacts of these federal acts was the specific granting to the Secretary of the Interior of very substantial powers with regard to the Colorado River Basin. These powers are a warning to the states of the basin.

Salinity Management

For many years managers of the basin had been aware of increasing salinity problems and had made periodic and usually piecemeal efforts to address them (see chapters 1, 2, 6, 11). Finally, in January, 1960, at the first session of a joint federal-state "Conference in the Matter of Pollution of the Interstate Waters of the Colorado River and Its Tributaries" the Colorado River Basin Water Quality Control Project was established. This was a very significant basin-wide step forward. In 1963, based upon recommendations of the conferees, the project began detailed studies of the mineral quality problem in the Colorado River Basin. In April, 1971, the Environmental Protection Agency transmitted its draft report on "The Mineral Quality Problem in the Colorado River Basin" to the conferees and water resource agencies in all of the basin states for review and comment. As a result of these reviews, the agency revised its draft report and submitted a final report dated later in 1971. By a resolution dated February 17, 1972, the seven states expressed their views to the Environmental Protection Agency and to other interested federal and state departments and agencies (see Appendix B). At another meeting in Denver, Colorado, state and federal conferees reached additional conclusions and recommendations and set these out in a document dated April 27, 1972 (see Appendix C). At last, it was clearly demonstrated that, given the right circumstances, the seven states could close ranks and arrive at common positions on a critical issue.

Amendments to the Federal Water Pollution Control Act in 1972 (P.L. 92-500) resulted in efforts by the Environmental Protection Agency to deal with individual states rather than treat the basin as a unit with the resulting possibility of adverse effects. Conditions required that some new arrangement be developed for addressing further the problems of salinity and mineralization of the Colorado River waters. On November 8-9, 1973, representatives of the water quality and water resources interests of the seven Colorado River Basin states convened in Denver, Colorado. Those in attendance designated themselves as the Colorado River Basin Salinity Control Forum and on November 9, 1973, entered into what they called the "Seven Colorado River Basin States Accord," which expressed the consensus of the states with respect to regulations proposed by the Environmental Protection Agency delineating requirements, procedures, and a plan of implementation for salinity control in the Colorado River Basin (see Appendix D).

The November 8-9, 1973 meeting lacked the formality of the meeting in Santa Fe, New Mexico, in 1922, and it is doubtful that a plaque will ever be dedicated in the meeting room. It is, however, quite possible that this first meeting of the Colorado River Basin Salinity Control Forum will be remembered as a significant breakthrough for effective new arrangements for state-federal cooperation on a region-wide problem of the Colorado River Basin while at the same time recognizing the historic respective roles of the individual states. The basin, for the purposes of the forum, was treated as a total social and geographical unit. Some may question the authority of those who attended the meeting to enter into an accord and others may question the content of the accord itself. Whatever the degree of informality, the participants somehow represent their respective states, and the "accord" reflects the position of each and all of the states. Consensus on this matter was achieved because common interests existed. This demonstrates "accelerated evolution" can work and that strong political unity can be achieved.

Salinity in the Colorado River Basin has been the most persistent, and is perhaps the most immediately pressing problem in the basin, but it is certainly not the only one. Equally important matters, such as conservation, augmentation, environmental management, energy requirements, changes of water use, and land use planning, continue to arise. Obviously, these are matters which must be addressed collectively by the seven states of the Colorado River Basin.

Advanced management techniques, such as management by objectives and designed integration of planning into the decision-making process do not seem adoptable as of the mid-1970s. The history of the basin and the complexities of its management requirements compel dealing with the "process" first. This must be done in the light of the "possible." Sophisticated management techniques must follow after the new arrangements become acceptable and useful to the participants.

Proposed Colorado River Basin Forum

With the approval of the governors of the seven states, I suggest the Colorado River Basin Salinity Control Forum should become the "Colorado River Basin Forum." Its members, consisting of not more than three from each of the seven states, should be appointed by each governor. Initially, emphasis should be upon water resource and land use management officials as members, but consideration should be given to appointment of state officials working in functional areas such as physical environment and energy. Appropriate provision must be made for active participation by federal officials in order that the forum shall be a state-federal activity. Arrangements should be provided for input from elements of the public, such as private industry and "public interest" groups. The forum should have no legal power or authority but should attempt to arrive at basin consensus. Implementation should be recommended to, and carried out by, the existing mechanisms in full compliance with the "law of the river" as established by the Colorado River Compact, the Treaty with Mexico, the Upper Colorado River Basin Compact, and the decree in *Arizona vs. California*.

Several steps should be taken immediately after the forum is established:

1. Select officers to serve for one year and arrange staff support and minimal funding on the basis of state contributions
2. Clearly identify its role and establish its procedures, subject to the approval of each of the governors
3. Identify matters in which there is basin-wide commonality of interest, such as water as an element of land use planning and management, energy and physical environmental balance, water conservation and augmentation, impact of energy fuel, agriculture, recreation, and population growth upon requirements and uses for water
4. Establish task groups with seven-state and federal representation to address these and other basin-wide matters and to provide reports and recommendations to the forum for discussion and action
5. Provide a synthesizing and integrating influence for the task group reports to avoid fragmentation
6. Provide a continuous appraisal of longer-range, comprehensive problems and opportunities in the Colorado River Basin with regard to the use and management of resources; and, in this regard, to monitor federal legislation and regulations for the purpose of injecting basin-wide views and positions agreed to by the governors
7. Maintain close and direct communications with the offices of the governors on all matters

For those who prefer to insist rigidly upon the status quo, this approach may conjure up fear of future consequences. For those who want immediate radical overhaul of existing structures, it will seem far too timid.

The alternatives to "accelerated evolution" toward new management arrangements are not pleasant to contemplate. To stand fast, resisting change, would be to invite—almost to insure—increasing federal intrusion into matters historically reserved to the states. To catapult into a totally new management arrangement would accentuate the adversary approach and lead to confusion and turmoil when sound management is imperative.

Appendixes

**Appendix A. Agencies That Have Some Responsibility or Control
Over Management of the Colorado River Basin**

   I. Office of the President
      Office of Management and Budget
      Water Resources Council
  II. State Department
      International Boundary and Water Commission, U.S. and Mexico
 III. Department of the Interior
      Bureau of Reclamation
      Bureau of Land Management
      Bureau of Indian Affairs
      Office of Saline Water
      Bureau of Mines and Mining
      Bureau of National Parks
      Bureau of Sports, Fisheries and Wildlife
      Bureau of Outdoor Recreation
  IV. Department of Agriculture
      Forest Service
      Soil Conservation Service
      Agricultural Research Service
      Rural Electrification Administration
   V. Department of Defense
      Army Corp of Engineers

   VI. Department of Transportation
      Coast Guard
  VII. Health, Education and Welfare and Housing and Urban Development
      Agencies dealing with grants and loans, water supplies, and sewage
      disposal.
 VIII. Environmental Protection Agency
  IX. Council on Environmental Quality

    The following is a résumé of state offices involved in Colorado River
Basin matters:
   I. Colorado
        Colorado Water Conservation Board
        Colorado State Engineer
        Northern Colorado Water Conservancy District
        Southeastern Colorado Water Conservancy District
        Colorado River Water Conservation District
        Southwestern Colorado Water Conservancy District
        Uncompahgre Valley Water Users Association
        Colorado State Department of Health
        Ute Mountain Ute Indian Tribe
        Southern Ute Indian Tribe
        Colorado Water Congress (unofficial)
        Other small, numerous conservancy districts, such as Tri-county,
        Savery-Pot Hook, and Dolores
  II. New Mexico
        State Engineer
        New Mexico Interstate Stream Commission
        New Mexico Water Quality Control Commission
        Navajo Tribal Council
        Jicarilla-Apache Tribe
 III. Utah
        Utah Division of Water Resources
        Utah State Engineer
        Central Utah Water Conservancy District
        Utah Water Users Association (promotion; unofficial)
        Uintah Utes
        Department of Health, Division of Pollution Control
 IV. Arizona
        Arizona Water Engineer
        Arizona Interstate Stream Commission
        Arizona Power Authority
        Central (Arizona) Project Association (unofficial)

Salt River Project
Department of Health—Water Pollution Control Commission
V. Nevada
Nevada State Engineer
Colorado River Commission of Nevada
Southern Nevada Water Conservancy District
Department of Health—Water Pollution Control Commission
VI. California
Colorado River Board of California
Division of Water Resources of California
Colorado River Association (promotion; unofficial)
Coachella Valley County Water Users District
Metropolitan Water District of Southern California
Palo Verde Irrigation District
Imperial Irrigation District
Southern California Edison Company
City of Los Angeles Department of Water and Power
California Water Pollution Water Control Commission
Coachella Irrigation District
VII. Wyoming
State Engineer
Wyoming Department of Economic Planning Development
Wyoming Water Development Association (unofficial)
Department of Health—Pollution Control Commission

Appendix B.   Resolution of the Conferees of the Colorado River Basin States, February 17, 1972

WHEREAS, the Colorado River Basin Water Quality Control Project was established as a result of recommendations made at the first session of a joint Federal-State "Conference in the Matter of Pollution of the Interstate Waters of the Colorado River and Its Tributaries," held in January of 1960 under the authority of Section 10 of the Federal Water Pollution Control Act (33 U.S.C. 466 et seq.); and

WHEREAS, in 1963 based upon recommendations of the conferees, the Project began detailed studies of the mineral quality problem in the Colorado River Basin; and

WHEREAS, the Environmental Protection Agency transmitted in April 1971 its draft report on "The Mineral Quality Problem in the Colorado River Basin" to the conferees and water resource agencies of the Colorado River Basin States for review and comment; and

WHEREAS, all Colorado River Basin States reviewed and commented on the draft report on the mineral quality problem in the Colorado River Basin; and

WHEREAS, the Environmental Protection Agency has revised its draft report and transmitted to the Colorado River Basin States a final report dated 1971; and

WHEREAS, the said report constitutes a necessary step toward the solution of the mineral quality problem of the Colorado River system; and

WHEREAS, the States and Federal agencies have implemented measures to control salinity of the Colorado River; and

WHEREAS, the Bureau of Reclamation is authorized to make, and has feasibility investigations underway, to determine additional measures to reduce the salinity of the waters of the Colorado River under present and future conditions; and

WHEREAS, during 1971 the States of the Colorado River Basin urged committees of Congress to appropriate funds to the Bureau of Reclamation to accelerate feasibility investigations of salinity control projects on the Colorado River; and

WHEREAS, additional funds were appropriated to the Bureau of Reclamation for these feasibility studies; and

WHEREAS, in the interest of comity between the United States and Mexico the State Department has given its support to a basinwide salinity control program:

NOW, THEREFORE, BE IT RESOLVED by the Conferees of California, Arizona, Nevada, New Mexico, Colorado, Utah and Wyoming that:

(1) A salinity policy be adopted for the Colorado River system that would have as its objective the maintenance of salinity concentrations at or below levels presently found in the lower main stem;

(2) In implementing the salinity policy objective for the Colorado River system the salinity problem be treated as a basinwide problem that needs to be solved to maintain Lower Basin water salinity at or below present levels while the Upper Basin continues to develop its compact-apportioned water, recognizing that salinity levels may rise until control measures are made effective;

(3) To guard against any rise in salinity the Congress and the Administration be urged to accelerate the entire salinity control program and, in particular, to augment the F.Y. 1973 budgeted amount of $1,005,000; and

(4) The Bureau of Reclamation have the primary responsibility for

investigation, planning, and implementing the basinwide salinity control program in the Colorado River system;

(5) The Environmental Protection Agency continue its support of the program by (a) consulting with and advising the Bureau of Reclamation, (b) accelerating its ongoing data collection and research efforts, and (c) transferring funds to the Bureau of Reclamation;

(6) The Office of Saline Water contribute to the program by assisting the Bureau of Reclamation as required to appraise the practicability of applying desalting techniques; and

(7) The adoption of numerical criteria be deferred until the potential effectiveness of Colorado River salinity control measures is better known;

BE IT FURTHER RESOLVED that the Environmental Protection Agency be commended for performing the necessary studies and completing the 1971 report on the mineral Quality Problem in the Colorado River Basin; and

BE IT FURTHER RESOLVED that copies of this resolution be transmitted to the Secretary of State, Secretary of the Interior, Administrator of the Environmental Protection Agency, Governors and Members of the Congress of the Colorado River Basin States, the Commissioner of Reclamation, Director of the Office of Saline Water and other interested entities.

Appendix C. Conclusions of the Conferees of the Colorado River Basin States

Denver, Colorado   **April 27, 1972**

The State and Federal Conferees have unanimously reached the following conclusions and recommendations.

I. It is recommended that:

A salinity policy be adopted for the Colorado River system that would have as its objective the maintenance of salinity concentrations at or below levels presently found in the lower main stem. In implementing the salinity policy objective for the Colorado River System, the salinity problem must be treated as a basinwide problem that needs to be solved to maintain Lower Basin water salinity at or below present levels while the Upper Basin continues to develop its compact-apportioned waters.

II. The salinity control program as described by the Department of the Interior in their report entitled "Colorado River Water Quality Improvement Program," dated February 1972, offers the best prospect for implementing the salinity control objective adopted herein. Therefore, it is recommended that:

   1) to minimize salinity increases in the river, a salinity control program, generally as described in the Interior Department report, be implemented on an accelerated basis;

   2) the Bureau of Reclamation have the primary responsibility for investigation, planning and implementing the basinwide salinity control program in the Colorado River system;

   3) to accelerate the salinity control program, the Bureau of Reclamation assign a high priority to LaVerkin Springs, Paradox Valley, and Grand Valley water quality improvement projects with the objective of achieving stabilization of salinity levels on the Lower Colorado River at the earliest possible date. The contemplated impact would be to initiate immediate action so as to achieve, by 1977, the removal of 80,000 tons of salt per year from LaVerkin Springs, 180,000 tons per year from Paradox Valley, and 140,000 tons per year from Grand Valley. This would provide a total reduction of 400,000 tons per year and would result in an estimated subsequent reduction of 33 mg/l at Imperial Dam.

   4) the Office of Saline Water contribute to the program by assisting the Bureau of Reclamation as required to appraise the practicability of applying desalting techniques; and

   5) the Environmental Protection Agency continue its support of the program by consulting with and advising the Bureau of Reclamation and accelerating its ongoing data collection and research efforts.

III. To achieve the salinity policy described herein, the long range program of the Bureau of Reclamation shall be directed toward achieving reduction of salinity concentrations that would otherwise exist at Imperial Dam to the extent of at least 120 mg/l in 1980, 355 mg/l in 1990 and 405 mg/l in the year 2000.

*   *   *

The conferees agree that the Bureau of Reclamation's Program as submitted in its report "Colorado River Water Quality Improvement Program," dated February 1972, should be considered as an open-ended and flexible program. If alternatives not yet identified prove to be more feasible, they should be included as part of the program, and if elements now included prove not to be feasible, they should be dropped. In addition, it should be recognized that there may be other programs which could reduce the river's salinity. Since present levels are greater than desirable, an effort should be made to develop additional programs that will obtain lower salinity levels.

The February 1972 report states that the Bureau of Reclamation Mathematical Simulation Model for the Colorado River system will be used to evaluate the Water Quality Improvement Program. This will be an important tool to evaluate the program's progress. The results of this evaluation along with the general program progress should be reported annually to the conferees and other interested State agencies.

Appendix D. Seven Colorado River Basin States Accord, November 9, 1973

The seven States agree with the essence of the draft proposal submitted by Region 9 of EPA to California which is to be found in the fourth paragraph on page 1 and the paragraph numbered (1) on page 3 of the California draft.

The States agree that salinity criteria for the Colorado River Basin would be useful in the final formulation of a salinity control program such as would be undertaken by enactment of pending Congressional Bills, H.R. 7774, H.R. 7775, and S. 1807, and agree further that the States must cooperate with the Federal government and each other in support of such legislation which would implement the Conclusions and Recommendations published in the proceedings of the Reconvened Seventh Session of the Conference in the Matter of Pollution of the Interstate Waters of the Colorado River and Its Tributaries in the States of California, Colorado, Utah, Arizona, Nevada, New Mexico, and Wyoming, held in Denver, Colorado on April 26–27, 1972, under authority of section 10 of the Federal Water Pollution Control Act (33 U.S.C. 1160), and approved by the Administrator of the Environmental Protection Agency on June 9, 1972.

The States have established a mechanism for interstate cooperation (Colorado River Basin Salinity Control Forum) and for preparation of semiannual reports on the development of numeric criteria and the adoption of such criteria by October 18, 1975.

As was concluded by resolution of the Colorado River Basin States Conferees of the Conference in the Matter of Pollution of the Interstate Waters of the Colorado River and Its Tributaries held in Las Vegas, Nevada and the Reconvened Seventh Session held in Denver, Colorado, implementation of the Colorado River Salinity Control Program generally as described in the report of the Secretary of the Interior entitled "Colorado River Water Quality Improvement Program, February 1972" would carry out the most appropriate plan of implementation for salinity control for the Colorado River system. The appropriate objective of the project is the maintenance of salinity at or below levels found in the lower main stem as of April, 1972, while the Upper Basin States continue to develop their compact apportioned waters.

The seven States concur in the goal of compliance with the adopted criteria by July, 1983, with the understanding that the levels of the criteria and the date of compliance are to be conditioned on the degree of effectuation of the Colorado River Salinity Control Program and other Federal, State, and local programs and the understanding that the criteria will not be used to delay or interfere with any State's development of its compact-apportionment of the waters of the Colorado River.

Index